PROGRESS IN WATER TECHNOLOGY

VOLUME 3

INTERNATIONAL ASSOCIATION ON WATER POLLUTION RESEARCH

President
Dr. G. J. Stander

Vice-Presidents
Professor B. B. Berger and Mr. E. Kuntze

Secretary-Treasurer Mr. P. E. Odendaal

Executive Editor Dr. S. H. Jenkins

CONFERENCE OFFICERS

President   Professor H. I. Shuval, Israel

Vice-Presidents   Dr. L. Coin, France
                     Professor S. E. Ulug, Turkey

Secretary   Mrs. R. Tamar

ISRAEL HOST COMMITTEE

Chairman   Dr. G. Shelef, Ministry of Health
                     Mr. E. Balasha, Consulting Engineer
                     Mr. A. Feinmesser, Ministry of Agriculture
                     Mr. M. Fleisher, Ministry of Health
                     Mr. S. Kishoni, Association of Engineers and Architects in Israel
                     Dr. U. Marinov, National Council for Research and Development
                     Professor H. I. Shuval, Hebrew University of Jerusalem
                     Professor A. Wachs, Technion – Israel Institute of Technology, Haifa

SPONSORING ORGANIZATIONS IN ISRAEL

Association of Engineers and Architects in Israel together with

Hebrew University of Jerusalem
Israel Petroleum Institute
Mekorot Water Company
Ministry of Agriculture – Water Commission
Ministry of Health
Ministry of Interior
National Council for Research and Development
Tahal – Water Planning for Israel
Technion – Israel Institute of Technology – Haifa
Union of Local Authorities

PROGRESS IN WATER TECHNOLOGY

VOLUME 3

WATER QUALITY: MANAGEMENT AND POLLUTION CONTROL PROBLEMS

(Jerusalem Workshop Papers)

Edited by

S. H. JENKINS

PERGAMON PRESS

OXFORD · NEW YORK
TORONTO · SYDNEY · BRAUNSCHWEIG

Pergamon Press Ltd., Headington Hill Hall, Oxford
Pergamon Press Inc., Maxwell House, Fairview Park, Elmsford,
New York 10523
Pergamon of Canada Ltd., 207 Queen's Quay West, Toronto 1
Pergamon Press (Aust.) Pty. Ltd., 19a Boundary Street,
Rushcutters Bay, N.S.W. 2011, Australia
Vieweg & Sohn GmbH, Burgplatz 1, Braunschweig

First edition 1973
Library of Congress Catalog Card No. 73–1162.

Jenkins, S.H.   Water Quality:– ☐
            Management and Pollution Control Problems
(Jerusalem Workshop papers.
New York   Pergamon Press, Inc.
                            1-24-73

Compton Printing Ltd, London and Aylesbury
ISBN   0 08 017006 4

CONTENTS

New Analytical Techniques

Virus Problems in Wastewater Disposal

Problems of Developing Countries and Arid Zones

Low Cost Waste Treatment Systems

Municipal Sewage Treatment: Problems of Design, Operating and Maintenance

Municipal Sewage Treatment: Problems of Small Communities: Packaged Treatment

Industrail Wastewater

Synthetic Detergents Environmental Considerations

Renovation and Reclamation of Domestic and Industrial Waste Waters

Ecological Assessment of Marine Pollution

FOREWORD

AT the Sixth Conference of the International Association on Water Pollution Research held in Jerusalem in June 1972, thirteen Workshop Sessions were organised. Consisting of 26 papers on widely different topics of current interest written by experts at the invitation of the Programme Committee, each Workshop was the responsibility of a Convenor who was selected for his wide experience in a particular field. The high attendance at the Workshop Sessions and the stimulating exchange of views they promoted prove that participants attracted to a conference devoted to a discussion of papers on advancements in water pollution research and development have interests that extend beyond their immediate sphere of activity. It is because the Programme Committee believes that the wide ranging interests of engineers, technologists and administrators who are active in water pollution abatement should be fostered that it has been decided to print the proceedings of the Workshop Sessions as Volume 3 in a new series of books entitled "Progress in Water Technology" and the proceedings of papers presented at the technical sessions as a separate volume entitled "Jerusalem Conference 1972, Advances in Water Pollution Research".

The intention is to extend this idea in the future by issuing a separate publication containing Workshop Proceedings. Since the invitations that have now been received from host countries have already settled the venue of the two yearly international conferences for the next ten years the International Association on Water Pollution Research is virtually creating an international forum for the discussion of major issues on water pollution control. Decision and policy makers at all levels of government no doubt recognise the advantage of participation at regular intervals at meetings where important aspects of water management problems are discussed, just as specialists in various fields may seek to influence world opinion on such matters. *Ad hoc* groups of individuals sharing similar interests are already beginning to regard the two yearly conferences as a convenient focal point for the furtherance of these interests and the International Association will encourage the extension of this kind of activity.

Since Volume 3 of "Progress in Water Technology" consists of the 26 papers and discussions of papers at the Jerusalem Conference in 1972, which are included in the present volume, attention is drawn to the fact that the first two volumes in this new series are:

Volume 1 — Applications of New Concepts of Physical-Chemical Wastewater Treatment (Nashville Conference)
Volume 2 — Phosphorus in Fresh Water and the Marine Environment (London Conference)
Volume 4 will appear during 1973 as "Towards a Unified Concept of Biological Waste Treatment" (Atlanta City Conference)

Subsequent volumes will deal with problems associated with the discharge of industrial wastewaters to public sewers; oxidation ditch processes; sludge disposal problems; the effect of heavy metals in the aqueous environment; nitrogen compounds in water management; and the marine disposal of wastewaters. In some cases the volumes will appear as the proceedings of specialised conferences that are to be organised in different parts of the world. Otherwise the subjects will be covered by books written by specialists under the editorship of well-known authorities, acting under a general editor appointed by the International

Association on Water Pollution Research. Under a long term arrangement Pergamon Press Ltd., will be responsible for the publication of all the books in the series "Progress in Water Technology".

The International Association on Water Pollution Research would like to record its thanks to the Convenors and authors of papers at the Jerusalem Conference, to the Israel Host Committee for the excellent local arrangements they made for staging the Conference, to Mr. J. I. Waddington for indexing the conference proceedings and Councillor D. M. Wardley, J.P., D.L., Chairman, Clyde River Purification Board for allowing Mr. Waddington's services to be placed at the disposal of the International Association on Water Pollution Research, and to Miss I. M. Herrick, Secretary to the Executive Editor.

S. H. JENKINS, *Chairman of the Programme Committee & Executive Editor*

ACKNOWLEDGEMENTS

The Israel Host Committee headed by Dr. G. Shelef worked unsparingly to arrange every detail so that the Conference could proceed in a smooth and efficient manner. Special mention must be made of the superb work done by the Conference manager, Mr G. Rivlin, Director of Kenes-Organisers of Conferences and Special Events Limited together with his devoted assistants Mrs Elana Shapiro and Mrs Katrin Tchechik. Mrs Ruth Tamar, Conference Secretary applied her usual skills in helping to see to it that all things were arranged properly and ran smoothly. Many other people contributed actively to the successful organization of the Conference and it would be impossible to mention them all but their efforts are nonetheless fully appreciated and acknowledged.

WHAT PRICE – WATER POLLUTION CONTROL

ABEL WOLMAN

The Johns Hopkins University, Baltimore, Maryland

It is axiomatic that water pollution, in its popular connotation, is sinful. In the climate of public discussion, in the last ten years, the term "pollution" has taken on an absolute character in place of its relative nature. To the specialist, pollution is related to what, where and how. Since water is never static, its ingredients, even without the impact of man and his works, have always been varied. When we speak, therefore, of pollution, it becomes necessary to quantify, to define, and to assess positive and negative values of corresponding methods of abatement.

To the professional, these tasks are complex. To the layman, they are simple – water should be of pristine purity and its ingredients should be zero. The price or the value of these political dicta necessarily places the members of this group in an uncomfortable situation of choosing between supporting the unwise or struggling to maintain the logical. Temptation is always high to bend with the winds of political doctrine, particularly when we have spent a century trying to spark public interest in our discipline. Suddenly, that interest is dominant, even if at times hysterical. Are we then to temper this burst of enthusiasm to re-make the environment in the image of distilled water? The challenge is real. The risk is even greater if that challenge is not discerned and the decision how to meet it is not equally clear. The indecision, so far characteristic of our stance, may result in our being trampled under the charge that we are ecologically immature, are in the eternal business of destroying nature, or are slow in disclosing or implementing sane correctives.

ON OBLIGATIONS

The debate is old as to whether a good end justifies a bad means. Some have recently contended that falsifying scientific data may be justified, if the ultimate purpose is for the common good. The late Leo Szilard has been quoted as suggesting: "don't lie, if you don't have to." De Toqueville, on the other hand, had long maintained that the intellectual's guidance has been very limited in determining public behavior. In general, that behavior has been simplistic, gross and often contradictory.

In such conflicting dilemmas, where should we sit? In my own mind, our position is clear – it is on the side of honesty, logic and wisdom. Our purpose should be to disclose scientific verity, to develop sound technologic implementation, to provide alternative solutions, to clarify choices, and to list tangible and intangible costs and benefits. All of these obligations need to be met to be most useful to society and its political mentors. The politician-statesman, of course, ultimately determines public policy – sometimes wisely and sometimes contrary to the scientists' myopia. Our task remains, however, one of illumination of issues and choices. This has not always been our universal position. The ecologists' contribution has certainly extended the depth and breadth of our horizons, in spite of their tendencies, on occasion, to over-emphasize the evils of science and technology.

This Conference gives us an opportunity to rehearse the considerations which should guide us in making our best contribution to the health and welfare of society. A few months ago, Dr. H. E. Stokinger, in the United States, presented several guidelines for investigators. Parenthetically, Dr. Stokinger is the chief of the Laboratory of Toxicology and Pathology of the National Institute for Occupational Safety and Health, of the United States Department of Health, Education and Welfare. He dignified his suggestions in a series of commandments,* as follows:

"1. Standards must be based on scientific facts, realistically derived, and not on political feasibility, expedience, emotion of the moment, or unsupported information.

"2. All standards, guides, limits and so on, as well as the criteria on which they are based, must be completely documented.

"3. Avoid the establishment of unnecessarily severe standards. This admonishment runs against the current tide of boiling popular enthusiasm for cleaning up pollution completely. But it is time that popular enthusiasm cool down, to recognize the consequences of establishing goals instead of standards.

"4. Determine realistic levels.

"5. Interpret the 'Delaney clause' with informed scientific judgment. This much maligned clause has become an excuse for oncologists to use inappropriate and unrealistically high levels in testing for carcinogenic potential.

"6. Determine trends, not pro tempore monitoring. The most flagrant violations of this commandment are the recommendations ... regarding mercury ... One moment's reflection would reveal that the concentration of mercury in the oceans has not changed perceptibly since the white men reached these shores, and that men have eaten these fish and lived and died without signs or symptoms of mercury poisoning. This is not to say that local, aqueous mercury or other excess pollutants should not be spotted, and, when possible, controlled, but that the thoughtless and irrational extension of a local finding to global dimensions is inconceivable in persons of sound mind!

"7. Delimit banning. . . . The banned food additives were either unnecessary or could be readily substituted with less harmful substances. Not so the totally banned DDT† and alkyl mercury compounds. First, DDT does not present an 'imminent hazard' to public health, despite misstatements to the contrary; second, its use for controlling the spread of malaria and African trypanosomiasis is unexcelled, and equivalent substitutes are not available at this time."

To these precepts, other cautions should be recorded. It is unfortunate that rarely are the consequences of decision making so set forth as to make clear, not only monetary costs and benefits, but the far more subtle resultants. Remote and intangible consequences are rarely assessed, because these so often lend themselves to exaggeration both for good and evil results. For example, the purification of water is often in a low cost category, while the processes for waste treatment of very high efficiency are more expensive. Choices between them are too often determined by emotion rather than by reasonably familiar quantitive parameters.

Similarly, threats of economic disaster due to abatement measures and costs are perennially debated. Estimates of such debacles vary widely, depending upon who makes them and for what ends. Yet impartial evaluations are increasingly required in order to determine the economic consequences of political action. In this area of activity, our

---

*H. E. Stokinger. How to Achieve a Realistic Evaluation. Science, 174, 662-665 (November 12, 1971). Jour. American Water Works Association, 64, No. 4 (April 1972).

†This ban has so far been lifted by the Environmental Protection Agency. (A.W.)

members have a high responsibility for intelligent input, provided they can maintain some kind of intellectual equilibrium in the modern maelstrom of environmental excitement.

The program this week is illustrative of both the positive and the negative contributions of which we speak. The workshops cover wide ranges of significant topics. The public defender, looking over our shoulders, however, might well find it difficult to discover relevance, to determine priority and to evaluate direct and indirect social costs and benefits. One might argue, of course, that these determinations are not within the purview of our assigned purpose or of the disciplines here engaged. The omissions need to be noted, even if perpetuated, because already rumblings regarding the balancing of equities may be detected in various quarters.

In many countries, officials, responsible for public and private expenditures, complain that the lists of research undertakings are not only great in number, but most often lacking in delineation of relative importance or priority. Hundreds of areas of exploration are noted, often with the tacit assumption that all the inquiries are of equal weight and necessity. It is easy to understand why this is so. Every investigator, if he is worth his salt, has his own set of "articles of faith", whether engineer, biologist, chemist, or economist. There should be room, somewhere in this assembly, to debate the merits of short and long range relevance, tangible and intangible costs and benefits, and good and bad externalities.

One of these contradictions is strongly brought to mind as an immediate consequence of the Stockholm meetings. Simply stated, the basic questions are whether pollution is in fact global in nature, whether waters are worse in quality today than 20 years ago, and whether universal monitoring is necessary or not. Are answers to these salient questions part of our deliberations this week? One might raise reasonable doubts as to each of these significant questions. Answers to all of them are already erroneously embedded in the current political literature.

Global pollution, with some important exceptions, is probably not demonstrable. Universal deterioration of waters over past decades is likewise subject to considerable question. Unfortunately, with the exception of recent studies in the United States* and England,** documentation is rare on this score. The studies need to be widely, geographically extended.

Of even greater importance to this group is to engage in a realistic appraisal of the significance of our activities in relation to the problems and policies of the populations of two-thirds of the world. How far, how fast, and how costly are the research findings applicable to developing countries? Is universal implementation desirable and at what price in those countries hungry for moving upward by agricultural and industrial expansions? Are unadjusted standards and goals, now appropriate to developing countries, applicable to the rest of the world? Again, the agenda are relatively bare on these pressing political issues.

The sheer comprehensiveness of our interests and tasks makes selectivity difficult. It has been said, with some truth, that comprehensiveness of plan stands in the way of quanta of implementation. On a broader scale, in some countries the peculiar situation has developed where action is at least temporarily stalled while waiting for total environmental impact coverage and diagnosis. For developing countries, this restraint is a source of increasing concern.

*M. Gordon Wolman. The Nation's Rivers. Jour. Water Pollution Control Federation, 44: 5 (May 1972), pp. 715-737. Science, 174: 905 (November 26, 1971).

**Report of a River Pollution Survey of England and Wales, 1970, Volume 1. Her Majesty's Stationery Office, 1971. London, England.

Even at the risk of sermonizing, this occasion appeared an appropriate one at which to suggest modest undertakings in post-auditing our activities, and in stock-taking of our future objectives and goals. In these days of suspicion of science and technology — in the recent colorful semantics of "the careless technology" — our precepts may well be examined, without indulging in self-flagellation! In so doing, the highly pertinent conclusions reached in the United Nations Conference at Founex, Switzerland, just one year ago, give valuable guidance.

"The major environmental problems of developing countries are essentially of a different kind. They are predominantly problems that reflect the poverty and very lack of development of their societies. . . . In both the towns and in the countryside, not merely 'the quality of life', but life itself is endangered by poor water, housing, sanitation and nutrition, by sickness and disease and by natural disasters."*

The quest for pure water is centuries old. The definition of purity has gone through continuous up-grading, until, in fact, truly tailor-made water is the product. In this practice, social choice has been a dominant factor. The guidance and implementation of such choices remaind our primary responsibility.

*Development and Environment. United Nations Conference on the Human Environment. Founex, Switzerland. June 4-12, 1971.

OPENING ADDRESS TO THE SIXTH INTERNATIONAL CONFERENCE ON WATER POLLUTION RESEARCH

BINYANEI HAOOMA – JERUSALEM – JUNE 19, 1972

PROFESSOR HILLEL I. SHUVAL
Conference President

CHALLENGES FOR THE FUTURE IN WATER QUALITY MANAGEMENT

Since the first Conference of the International Association on Water Pollution Research a decade ago in London, our organization has truly circled the globe, having held conferences in Tokyo, Munich, Prague, San Francisco and Hawaii, and now completing the cycle here in Jerusalem. In many ways this is symbolic since Jerusalem, one of the world's most ancient cities, has struggled to solve problems of water for some three thousand years. Here one can find the remains of marvellous ancient water engineering works side by side with the most modern developments in water technology. Israel is mainly a semi-arid country, but the vision of plentiful flowing water sources has been echoed from the earliest times. The prophet Isaiah yearned: ". . . for in the wilderness shall water break forth and there shall be streams in the desert . . ." (Isaiah 35, 6).

A few kilometers from this building where we sit today, the water engineers of King Hezekiah some 2,700 years ago, laid out a complex subterranean water supply tunnel dug in solid rock, bringing living water from the Gihon spring into the walled city of Jerusalem. The water supply tunnel functions to this day. The crowning glory of the ancient Israel water engineers of two thousand years ago was the construction of an intricate system of aqueducts, tunnels and siphons bringing water to the city over a distance of seventy kilometers with some two-hundred thousand cubic meters of capacity in storage reservoirs. Following in this ancient tradition of water engineering, modern Israeli engineers have drilled thousands of wells, built reservoirs and have transported water by aqueduct and pipeline hundreds of kilometers from the Jordan River in the north to the parched deserts of the south, fulfilling the vision of Isaiah of bringing streams of water to the desert – to make it bloom.

The problem that we shall be dealing with at this Conference is the protection and maintenance of the quality of our vital water resources. A paradox of modern technological society is that more and more water is required as populations grow and the standard of living increases, resulting in greater and greater withdrawals from ground and surface sources. With the growing urban and industrial use of water, greater amounts of organic and inorganic wastes are spewed back into the water sources so that less and less pure water becomes available at the quality required as a result of the self-destructive process of pollution.

By the time one of Europe's major rivers, flowing from its sources in the mountains reaches the sea, its entire flow may be almost wholly made up of water used once or more times by upstream cities. Six million people draw upon this river as their main source of water supply. How long may we continue at this pace? Will all of the surface waters of the world eventually face this fate?

Another paradox of our modern technological world is that our society "is hooked" on the use of a vast array of agricultural chemicals and fertilizers to ensure the plentiful food supply for our growing population. But some of these essential chemicals are polluting our water resources or leading to unexpected ecological imbalances. The situation has reached almost crisis dimensions in Israel, a country which is presently utilizing almost all available ground water. Nitrate levels have been increasing at the rate of two milligrams per liter per year for the past fifteen years and today some five hundred wells already show nitrate concentrations equal to or above the standard recommendation for drinking water by the World Health Organization. It appears from research results which will be presented at this Conference by an Israeli scientist, that the very high use of inorganic fertilizers rich in nitrogenous chemicals may be a major factor in this type of ground water pollution. Can we find a solution to this problem which will both provide protection of ground water while maintaining the right level of food production essential to our existence?

Not all water pollution is a result of the disposal of urban and industrial wastes into the environment. The buildup of naturally occurring inorganic salts in the aquifer as a result of incautions water resources management could in some cases provide an extremely serious threat of water pollution, particularly in arid zones. Heavy pumping of ground water coupled with intensive irrigation practises could cause this problem. While substantial amounts of water are lost by evaporation in irrigation practise, the dissolved salts are completely returned to the aquifer resulting in a slow buildup of minerals in the ground water. Here in Israel, this increase amounts to a few milligrams per liter per year. The full impact will be felt only in twenty or even fifty years from now, by which time major portions of our ground water may no longer be suitable for agricultural or domestic use. Do our water quality management programs face up to the severe implications of such creeping water pollution, or will we pass this problem on to the next generation when it may be too late to reverse the process?

It would be a disservice to our profession to imply that we are recklessly riding an uncontrolled path to total pollution of our water resources and that nothing is being done to reverse this trend. At this Conference we shall emphasize not only the new and often ingenious technology developed in our research institutes aimed at overcoming the problems of pollution, but the real accomplishments achieved with the aid of these innovations. Many of the papers to be presented here this week will report on real progress.

An outstanding example is the paper of Gameson and his colleagues of the U.K. They report on the dramatic improvements in the Thames River over the past twenty years. In 1950, a reach of some thirty-five kilometers of the river was at times devoid of oxygen and fish life was extinguished, while today after major investments in modern wastewater treatment, the river is entirely aerobic once again and more than fifty species of fish inhabit this rehabilitated waterway. From the other end of the world, Fujiki of Japan will report later this week that Minamota Bay — once heavily polluted with deadly mercury wastes has been cleaned up by the construction of plants to remove mercury from industrial wastewater and by technological changes in other plants which avoid the use of mercury entirely. Fish caught in the bay in 1961 showed mercury concentrations of 23 mg/kg, more than forty times the concentration considered safe for human consumption. By 1970, only 0.2 mg/kg of mercury was found in the fish of that once ill-fated bay. This is well within safe limits.

We all rejoice in these historic accomplishments which have been achieved by the co-operative efforts of research scientists and practical engineers who actually went ahead and did the job.

May this Conference serve as a turning-point in the successful campaign to reduce the pollution of our water environment. Let us learn not only to warn of the dangers of uncontrolled pollution, but to point to the successes that can result from the planned and judicious application of our research efforts. Nothing breeds success more than success itself.

With a growing shortage of water in many parts of the world, we will be seeking new sources of water supplies. Can we afford to throw away a vital resource such as water that has been used only once and is still 99.9% pure water? Engineers talk freely of the utilization of seawater, but it is worth noting that seawater contains thirty times more contaminants than municipal wastewater. Israel is one of the few countries in the world that is already utilizing almost all of its ultimate, natural water resources and must look for maximum conservation of its existing supplies. We are already renovating 25% of our urban wastewater for agricultural and industrial purposes. The challenge of the future will be to develop systems to purify wastewater so that they will be safe for unlimited urban use. This issue will be discussed at some length at this Conference.

This international Conference with representatives from thirty-five countries from all continents, from various political, social and economic systems, is symbolic of the role that science and scientists can play in breaking down the barriers of communication that are sometimes artificially placed in our way. The free exchange of ideas is a sine-qua-non for peaceful co-operation of the type that is necessary to assure the proper management of the quality of the world's water resources. Many of the major rivers of the world pass through three or four countries on their twisting path to the sea. Only the most exacting co-operation among the nations sharing the use of the river can guarantee that the maximum benefit can be derived for all concerned. The seas are shared by all nations and here only full international control can prevail in preventing their degradation. May I add that our own region could benefit immeasurably by peaceful co-operation among the nations of the area in developing the limited water resources and to take the measures necessary to prevent their pollution and preserve their quality. May this Conference provide the forum for the meeting of the minds necessary to achieve such true co-operation among nations.

Our profession is called upon to provide scientific answers to the current and future problems involved in preserving the quality of our water. However, technology is not enough. We must guide our administrators and statesmen as to the optimal application of the new technology capable of preventing the deterioration of the environment. On the one hand, insufficient preventive action today may lead to irreversible damage that cannot be corrected at any price in the future. On the other hand, an irrational use of limited financial resources today to achieve unreasonable goals such as a level of absolutely zero pollution as has been recently proposed by some influential ecologists, may lead to a state of disenchantment or even revolt among the public who have so far enthusiastically supported environmental improvement programs. Will they blindly support expenditures to achieve an exaggerated degree of environmental pollution control at the expense of other worthwhile social needs such as housing and education? Concepts of costs and benefits and maximum social gain must be introduced into our formula for improving the environment.

The task that lies ahead is not an easy one and no single mathematical formula will provide the answer to the complex problems of science and society as far as the management of man's environment is concerned. Here I can only suggest a philosophy which I have found useful in my own career of public service: an eminent British scholar suggested these thoughts as a guideline for the public servant: "Give me the strength and fortitude to change those things which can be changed — the patience and peace of mind not to attempt to change those things that cannot be changed — the good judgement and wisdom to tell one from the other . . ." May this Conference provide us with improved knowledge of how to protect our environment for the betterment of man and the wisdom required to guide ourselves and our leaders on the rational use of this new science and technology so as to achieve the most beneficial social gains for mankind from our efforts.

DECISION MAKING IN
WATER POLLUTION CONTROL PROGRAMS

MARTIN LANG
Commissioner, Department of Water Resources,
Environmental Protection Administration, Municipal Building,
New York, N.Y. 10007, U.S.A.

An inexorable trend is now discernible in the planning, construction, and operation of water pollution control programs. It is the inevitable progressive shift of decision making and management from the local City, to the region, to the State, to the national, and even to some degree, to the international level. This will inevitably entail the abdication of the local sovereignty over water pollution control now exercised by cities in most countries, with all that this implies of a conceivably bitter struggle for fiscal, policy, and above all, political control.

For we engineers, who conceived and implemented such programs for our cities, this will be hard to accept. But if we can momentarily forego our proper pride in what we have done for New York City, London, Tokyo, Paris, Berlin, and Haifa, we must concede the logic of managing drainage areas and river basins as overall entities and, in coastal waters, treating estuarine zones as entire biogeographical provinces. After all, each drop of the receiving water and its marine biota only knows its immediate micro-environment, not geographical or political subdivisions.

As manager of one of the largest municipal programs, in New York City, with a current construction budget of 1.8 billion dollars, I am now at the center of a complex, and at times, abrasive, inter-relation of City, State and Federal agencies. This is becoming increasingly typical throughout the United States, as the fiscal burden, and of course, policy control, moves successively to the State and Federal governments. In turn, this has led to the appearance of a new kind of leadership in this field, that of a politically astute, tough minded lawyer.

New York City led the country in gathering all its environmental departments under an Environmental Protection Administration, now headed by Jerome Kretchmer, a skilled lawyer and former State Legislator. New York State followed by consolidating its environmental agencies in a Department of Environmental Conservation, under Henry Diamond, a lawyer with political experience in the Federal government and on Governor Rockefeller's staff. Then President Nixon set up an Environmental Protection Agency, as a new, separate entity, with a politically experienced lawyer, and former Congressman, William Ruckelshaus, in charge, and reporting directly to the President. It is no coincidence that while all these powerful policy makers have markedly different individual styles, they all share certain attributes. Each quickly grasped the interdependence of air, solid wastes, and water pollution control programs, each objectively appraised the multiplicity of overlapping and conflicting jurisdictions, regulations, and standards, and each vigorously sought to devise legislation or influence legislation to retain or obtain environmental power, while hopefully shifting the fiscal costs to the other government levels.

In short, water pollution control is now a major political issue. This can, and has, led to decisions on type of treatment, standards for effluents, standards for receiving waters, and industrial wastes regulations which, at one extreme, are generated by public emotion,

1

and, at other extreme, by narrow political expediency.

Where does this leave the engineers and scientists, who perhaps deluded themselves in thinking they could conceive and execute programs on the basis of pure scientific rationale, without engaging in the give and take of political and legal disputation?

Regrettably, at this time, they have been left on the sidelines, while pollution control decisions, even down to technical minutiae, have been pre-empted by political and industrial interests, and the myriad of self-appointed 'ecological experts' now springing up everywhere. Few of these know the efforts we made in the past twenty years, when fiscal resources and public support were minimal, to at least bring water pollution control to its present level. Instead, on the one hand we are attacked by industry as ivory tower academicians imposing unrealistic standards, and on the other hand, by self-styled environmentalists, as not being sufficiently sensitive to the true needs of society.

There are some very specific things we can do, in the legal and administrative areas, to reassert our proper role of rational leadership.

We must not retreat to cloistered universities and laboratories. In the crude, but forceful, American political adage, 'If you can't lick 'em, join 'em!'. We must demonstrate the same zeal in analyzing current legislative programs and regulations, that we expend in developing new processes. We must not shrink from using all media of communication, all public forums, not just scientific journals (where we only speak to each other), to advance reasoned programs. I sometimes think I did more to win public support and to convey the scope, cost, and aims of our New York City program when I explained it in lay terms to the public at large on a television program with Mayor Lindsay and Administrator Kretchmer, than in all the papers I have written and presented to engineering forums.

The international fraternity of water pollution experts knows that our ultimate mission is the preservation and enhancement of our waters, *not* the treatment of sewage. Treatment is only one of the means to fulfill this mission. This simple truth has not fully penetrated to those who wrangle about effluent standards versus stream standards, who choose up sides between primary, secondary, and tertiary treatment, who think they will restore our rivers or estuaries by selecting the right number of 80% or 90% or 99% removal, who think they will create a new environment by legislative fiat.

We must get this message across in time to influence the decisions of the 1980s and 1990s. The decisions for the 1970s have largely been made.

In New York City we have created a team of an Administrator, with a General Counsel whose staff keeps continual liaison with Commissioner of Water Resources and his consultant scientists to revise or generate legislation, in interaction with industrial dischargers, in assessing and attempting to influence state or federal legislation, in the complex legal problems of state and federal aid programs. More importantly, we are publicizing the fact that new techniques of instrumentation and mathematical modeling are now available to give us powers to predict the response of a given stream, lake, or estuarine system to all permutations and combinations of effluent discharges, thermal loads, shoreline structures, and outfall configurations. Using such a dynamic model, we can adapt to a changing population, a changing industry, changing admissible effluent concentrations and thermal inputs, and still determine the optimum mix of treatment facilities, degrees of treatment, and outfall placements to achieve desired best use of the receiving waters.

Now this is a far cry from current attempts to rigidly fix and imbed in law some magic numbers of percentage removal to guarantee a pristine marine environment. For one thing, estuarine systems are different from landlocked lakes. Also, our cities are not like a fly frozen in amber, static and immutable, but dynamic and changing in their intakes and discharges. The classic parameters of dissolved oxygen, BOD, and coliform are even now being superseded, because of new threats to our environment from new, refractory organic componds, and organic metallic complexes.

To reintroduce such scientific rationale, we must first master the intricacies of the legislative process and gain at least a working knowledge of the newly developing body of environmental law. To do this, we must ally ourselves with legal and political leaders, and buttress each water pollution control agency with competent counsel. For example, in advancing any program in New York City, I must deal with and reconcile the regulations of over twenty different agencies, including such powerful regulatory agencies as the Federal Environmental Protection Agency, the State Department of Environmental Conservation, the U.S. Corps of Engineers, the Interstate Sanitation Commission. Like any planner and administrator, I experience frustration in coping with multiplicity of masters, many with conflicting views. Clear cut and rational guidelines are necessary. The burden is on us to help generate them.

It is my avowed intent to provoke a frank and full discussion of the status of the legal constraints under which water pollution programs are developed in individual countries. Certain common denominators will emerge, and successful procedures in one country may well stimulate similar actions elsewhere. If new processes are developed in our field of water pollution research, it does not follow, as the day follows night, that they will be employed, if a rigid regulatory structure inhibits the application of new concepts.

It is futile to cite the current situation in the United States. It would be outdated before publication and presentation. The Congress of the United States is now debating a major bill, covering a spectrum of activities from fixing the level of toxic metal discharges to the relative role of the states and the central government in controlling water pollution control. Our EPA counsel must subscribe to numerous legal digests to stay abreast of new court rulings every day on environmental matters. In New York City we have set up an Environmental Control Board, with quasi-judicial powers, to compel pollution abatement more rapidly than through the glacially slow conventional court procedures.

Recently Mayor Lindsay personally convened a joint meeting of many agencies in an effort to obtain concerted action on water pollution abatement, without jurisdictional wrangling. This included the Federal Environmental Protection Agency, the N.Y. State Department of Environmental Conservation, the U.S. Corps of Engineers, the U.S. Coast Guard, the Interstate Sanitation Commission, the local U.S. attorneys of the U.S. Attorney-General's office, the New York City Department of Health, and the N.Y. City Environmental Protection Administration. This list gives some indications of the redundancy of regulatory agencies in our field. The significance of such a meeting to this International Association is that it presages such eventual meetings at the national, and, ultimately, international levels. It appears that it will be at best, some five years before cities and industries in the United States will have some orderly matrix of regulatory guidelines.

A current editorial from the New York Times, entitled 'Of Time and the Rivers', included as Appendix A, dramatically illustrates the divergent attitudes among governmental agencies, each of which professes adherence to a better environment.

We must also address ourselves to the new administrative techniques. Programs of the scope of the Tokyo Metropolitan Government, the coastal region of Sweden, the Delaware Bay in the United States, Sao Paulo in Brazil, call for innovative management to coordinate and control many simultaneous design and construction projects. In the past decades, there was not the present great urgency to promptly complete and operate water pollution control facilities. Now the pace of our activities is remorselessly tracked by the public and press.

The same attention we give to esoteric concepts of the kinetics of the complete mixing process must now be devoted to computerized program control, as we translate our dreams into concrete, steel, and equipment. In may country, the Federal government itself has expressed considerable dissatisfaction with the delays in the conventional procedures of preliminary design, final design, competitive bidding, and construction, by advancing the alternative of 'turnkey' construction. This has involved a large segment of the consultant engineering community in a heated confrontation with Federal agencies.

In truth, the pace has been slow. In U.S. experience, at least five years was required from decision to start-up for a plant in the 100-million gallons per day range. Large petro-chemical plants were created in two years. Now the public, once they have provided the money, expect to see sparkling effluents and a dramatic upgrading of water quality immediately.

Perhaps New York City is an extreme example. A program for expanding and upgrading twelve existing plants, two new giant plants, one auxiliary combined overflow plant, many pumping stations, over thirty miles of intercepter tunnels, together with a massive program of trunk and lateral sewers, had to be designed and committed to contract within four years to be eligible for 60% State aid. Much of the design had to start even while the State and Federal Governments were making up their minds on acceptable degrees of treatment. This entire program comprises about 90 individual projects.

To even track the status, and monitor the progress, of each project, called for sophisticated computerization of information flow. Monthly status reports enable the staff to 'manage by exception'. I stress this, because every country will soon be faced with similar situations, as pollution control facilities become a major factor in heavy capital construction everywhere.

So my sanitary engineering colleagues will have to acquire some expertise with 'critical path control' for major projects, even in the design phase. Such intensive management procedures will yield unexpected dividends, for they will be carried over into the eventual operation, process control, and maintenance.

Our field has long been an interdisciplinary enterprise, comprising as it does chemistry, bacteriology, hydraulics, physics, oceanography, marine biology, civil engineering, structural engineering, electrical engineering, economics, and mathematics.

Now that water pollution control is very much in the midst of controversial worldwide debate, with understandable impatience by the public for swift restoration of water quality, it is mandatory for us to add to our armament legal competence and modern administrative techniques.

APPENDIX 'A' Editorial, The New York Times, Sunday, December 26, 1971 'Of Time and the Rivers'.

The dispute now blowing up between the Administration and Congressional environmentalists over the Federal Water Pollution Control Act Amendments of 1971 seems rooted more in rhetoric than in reality. Stripped of some grossly exaggerated charges about the cost of enforcement, the difference boils down to an academic wrangle over whether or not the proposals will make the country's waterways unnecessarily pure fourteen years from now, to the possible detriment of the national economy. The minuscule progress made so far gives to such objections a rather dreamlike quality.

Senator Edmund S. Muskie, who originated the legislation — which his colleagues approved by a vote of 86 to 0 — feels that enforcing water quality standards has not been an effective approach to the problem. After five years many states still lack approved standards. It is difficult, as a practical matter, to relate the volume of effluent to be permitted in a given case to the water uses involved. Controls vary in effect, and there are disputes over Federal and state standards. The new legislation proposes instead to impose effluent limits, decreasing them as control technology permits until no more pollutants are allowed.

Administration spokesmen have taken the view that the end to be sought is water quality sufficient for the purpose involved and that to limit effluents beyond that is to force prohibitive costs, hurt the environmental cause and even unbalance the economy. Industrial representatives, testifying before the House Public Works Committee, went further. They thought too stringent the requirement that industry use 'the best practicable' cleanup technology by 1976 and the 'best available' technology by 1981.

What makes the clash so needless is that both the Administration and supporters of the Muskie legislation are disposed in fact to move a considerable distance from their starting points. The Muskie forces in Congress have made it clear repeatedly that they are talking about goals, not hard and fast deadlines, and phrases like 'best practicable' in relation to technology certainly suggest that they are not unmindful of the economic factor. By the same token, Russell E. Train, who testified as chairman of the President's Council on Environmental Quality, defended 'whatever costs are necessary' to achieve high water quality and would require 'whatever level of treatment' may be needed to achieve the standards set for a particular body of water. It is even possible, if a Federal District Court decision handed down last week should be upheld, that the entire permit system may have to be drastically overhauled in any case.

In these circumstances, it should not be beyond the grasp of statesmanship to forgo political advantage, reconcile a nonconflict, and get moving seriously on the great cleanup of the nation's waters which all agree is desperately needed.

THE POTENTIAL ROLE OF EFFLUENT CHARGES IN WATER QUALITY MANAGEMENT

JEAN-FRANCOIS SAGLIO
Directeur du Secrétariat permanent pour l'étude des problèmes de l'eau
JACQUES GARANCHER
Charge de mission au Secrétariat permanent pour l'étude des problèmes de l'eau
67 Boulevard Hausmann, Paris 8ᵉ, France

WATER QUALITY MANAGEMENT

To assume water quality management means attendance to the pursuance of a pre-established programme aiming at the maintenance or the improvement of the main features of a natural medium such as a water course, a lake or the sea, in order to allow the practice of a specific activity.

Water quality management does not appear usually as a necessity except when we face a sufficiently serious and wide decay of water quality, when making use of rivers for human needs sets more and more difficult problems to the authority dealing with the water supply of some community, when the repulsive appearance of water makes it often unfit for bathing and swimming and when there is less and less possibility to get a fish alive in more and more long sections of the rivers.

Meanwhile, as many growing nations are now well acquainted with the difficulties that, in the field of water problems, every industrial country is facing, they now pay attention to the fact that it is in their interest to devise a water quality management system that will not suffer from the defects of an existing situation.

On the other hand, some big industrial countries are still in possession of very large areas they have to develop and where there is an opportunity to plan for the best water quality management system. They may succeed in the solution of problems which appear to us specially difficult. For instance, there is no problem about warming up rivers with all the consequences implied, more particularly in the field of ecology, if they utilize for the central heating of a new town the available energy of the effluent of a steam generating station.

As they have to build everything, they can establish industrial combines with a succession of works, each of them utilizing the effluent from the previous one. This practice, connected with the recirculation of water will make all problems owing to water shortage disappear. It will be a very effective fight against water pollution at the lowest cost.

A PROGRAMME FOR WATER QUALITY MANAGEMENT

When drawing up such a programme, we usually have to refer to a system which has been planned according to the researches of a team which include economists and experts belonging to various techniques concerned with water problems. They will often take a course which has been laid down according to a law or to statutory statements.

7

Theoretical considerations to which we have to refer at first are, in this field like in others, must be adjusted to the necessaries of practice.

Theory

If we cite the French system as an example, the way which has been shown by items 3, 4 and 5 of the law of December 16th, 1964 (Law relating to the organization and the repartition of water and the fight against its pollution) appears to be perfectly logical.

(1) At first, a 'National Inventory'[1] will enable one to ascertain the state of pollution of the natural medium and to be acquainted with the now prevailing conditions.

(2) According to this state of affairs, and in order to satisfy or conciliate all the requirements (Water supply, Public health, Agriculture, Industry, Transports, Biological life — particularly piscicultural fauna, Leisure, Aquatic sports, Conservation of sites), statutory statements will determine the physical, chemical and bacteriological criteria which are to be achieved for a certain situation within a required time.

(3) According to those purposes that have been assigned, they will settle the constraints which are to be imposed, in order to satisfy those purposes, on the various activities which are liable to lead to impairment of water quality. That may lead to modification of the present regulations relating to existing polluters.

The determination of the constraints which are to be imposed will be settled according to the following procedure:-

We want a water course to remain, or become fit for a special purpose: to reach a favourable result, we need to obtain that each parameter of quality, expressed in term of concentration, will not exceed the given limit of the regulations in force relating to that utilization.

The sum of the products of these data into the flow of the water course gives the highest amount of pollution which this water course is able to admit; any overstepping of this peak value would make the water unfit for the required use.

It is enough to compare this permissible pollution load with that measured when carrying out the Inventory. That will give us the pollution load which we have to remove from the water course.

It remains to agree on a system of dividing the excess load and computing the share that each polluter has to take in the removal of pollution.

In Practice

In practice, things are not so simple as they appear in theory.

(1) A national inventory will not give a truly objective picture of the pollution, unless a long run of results is available giving the concentration of impurity and the load of all constituents under different conditions or of the flood for every distinctive item.

That is the only way to get a good picture of water quality.

It appears to be difficult to carry out such a task in the framework of the hydrographic basins over the whole of French territory.

(2) To give a list of data defining the highest values they can admit for the concentration of some items, in order to define an objective for the quality of a water course is no longer significant. Of course it is necessary to assign limits, but they have to allow for some possible tolerance, expressed as a probability.

---

1. This National Inventory was obtained throughout the year 1971.

(3) Neither is it an easy task to decide on the abatement of the pollution. The discharge of the peak pollution from various sources is problematical. The consequences that follow from the discharge of these loads on the quality of receiving medium depend on many parameters. We can do nothing but draw statistical correlations, the knowledge of which cannot be obtained except from a very large number of measurements.

Therefore, to make a programme of management quite clear seems to be an intricate problem. It would even be an inextricable problem if, in every concrete case, it was not possible to make use of simplifications that will allow one to compute values ranging about assigned data.

But such a process sets a problem in the juridical field: when a programme relating to water quality management has been established, it will involve the possibility of setting civil or penal liabilities. It would be very difficult for a judge to fine anyone when he has nothing to refer to except values that are not absolute but range within certain limits.

On the other hand, any programme has to take into account the possibility of including future sources of pollution. When allowing each polluter an assigned load of pollution, the responsible authority ought not to permit from the outset a discharger to take up the whole of his permissible maximum load.

It is necessary to keep a definite quantity of potential pollution in reserve. This reserve will be progressively used up as new polluters establish themselves. If the reserve were to be exhausted, the authority would have to take into consideration increasing the strictness of the conditions permitting discharges to take place.

PLANNING FOR REGULATIONS

In consideration of what has been stated, the following matters are implied:-

(1) *Concerning the objective*: one has to set limits for each parameter as a test of quality, but with the possibility of exceeding it. In fact, when an objective of quality is to be assigned to a section of a river, we usually have to consider a synthesis of several objectives that we want to reach. For instance we want to obtain water to meet the requirements of pisciculture, while the use of it would be for water supply and the practice of some aquatic sports.

Each of these requirements can be shown in a graph giving limits assigned to the various items that were selected, in order to characterize the requirement.

Each value we have to assign to each item, according to the general objective we want to reach, will be obtained while observing the result of superimposition of the various graphs.

(2) *Concerning the discharges*: to assign as a constraint a limit to the highest permissible load of pollution. With regard to this point, the French rules concerning discharges take into account whether they come from urban or from industrial sewers, whereas hitherto they were based on standards relative to concentration. The latter method is undoubtedly a very bad one, especially in the matter of industrial discharges. It encourages the authority which is responsible for the discharge to mask the pollution that it brings to the water course because of the admission of clean water in the pipe above the discharge.

This practice of dilution is contrary to the principles of good management in the fight against water pollution. In addition, the practice of dilution leads to wastage of water.

With regard to urban discharges, one does not find deliberate addition of clean water to the sewage. But it is a frequent practice that the sewage from a city is finally collected in a main sewer which is permanently carrying a flow of clean water, the amount of which is sometimes very important. (A stream that has been diverted in its passage through the city.) The load of pollution it brings to the receiving waters could be harmful and incompatible with the objective that has been assigned to the quality of a water course.

The use of individual constraints implies the choice of a method to distribute the total load of pollution permissible between the existing polluters. To do that, they can decide, for instance, to reduce the pollution from every discharge, in proportion to the existing load; or they can decide on the basis of an economical optimization of the situation.

They will decide how to divide the reserve load of pollution for new polluters or in the provision of an extension of activity for the existing ones. They will now consider how much time has to be allowed to reach the objective, and compute, spreading out of payments over a period, the cost of investments and operation.

Questions of finance being properly cleared, the authority which is in charge of the water quality management can begin to work in order to carry out the programme. Yet, the results that are to be expected are still problematical in the state of our knowledge. It will be absolutely necessary to check the evolution of the quality in the receiving waters, along with the development of the programme. Thus, early conclusions can be modified according to the results of these investigations.

But it is only when the programme has been carried out completely that one can have an actual appreciation of the results. If the objective has not been reached, one ought to decide either to modify the objective that had been assigned to the quality of the water or to carry out a complementary programme. Thus it appears necessary to periodically reconsider the objectives that have been assigned.

In such planning for regulations, the objective which has been assigned for the quality of a river is a tool to solve problems and to show the responsible authority the way to administrative action and to planning. This objective ought not to have a juridical value.

If it was not so, the responsible authority would have to respect an objective at all costs, running the risk of unnecessary building plants for treatment whose prices would be prohibitive, considering both investment and operation. It could be lead, too, to adopt harmful measures such as closing down of works. Therefore, it would not be better not to respect an objective to which a juridical value would have been granted. This would ruin the credit of the system of regulations and of the responsible authority.

It is the programme this responsible authority has to attend to (that consists of abatement of nuisances from existing polluters, and in regulating new discharges) which makes it necessary to assign standards to the discharges.

In parallel, we have to know the standards for discharges in the existing state of knowledge and apply them within the framework of a policy concerned with objectives of quality.

STANDARDS FOR THE DISCHARGES

These may concern various items such as: (1) the highest concentrations; (2) the highest load of pollution, either absolutely or with a reference to some distinctive quality; (3) the lowest rate of efficiency for a treatment plant; (4) the necessity to utilize a given process or plant.

The easiest way to reach the objective which has been assigned for the quality is to impose standards consisting of a pollution load as an absolute figure, avoiding constraints that would be impossible to satisfy. They ought to try, too, to impose conditions attainable by using the best available plant.

Domestic Sewage

The pollution load from public sewers is liable to large variations according to the users' habits and to rainfall.

The results we can obtain by treatment are appreciably affected by these variations. Accordingly, the authority in charge of the sewers of a city is not master of the load of pollution discharged into a river at every moment. On account of the very variable dilution of domestic sewage, the best constraint to assign in such a case is to operate a treatment plant with a guarantee of standards concerning its efficiency. The authority responsible for water quality management has to estimate the pollution load and to make sure that it allows the objective assigned to the quality of a water course to be reached. Later it may be necessary to review the standards assigned.

Industrial Waste

We have previously argued about the serious objections to the practice still widely used to assign standards concerning concentrations. Of course, the pollution load from industrial activity depends on the processes used and on the size of the works. On account of the knowledge we have obtained in this field during the last years, it is now possible, in the case of a given industrial establishment utilizing a given operating process, to evaluate, with enough accuracy in the existing state of technology, the load of pollution F_1 which is discharged, corresponding to a distinctive characteristic of the process. (On the basis of one metric ton of manufactured article, for instance.)

It is also possible to determine the minimal treatment which, in any condition of discharge and with any objective of quality assigned to the quality of receiving waters, is to be required from an establishment of this kind. They can then calculate the pollution load F_2 ($F_2 < F_1$) which may be discharged to the receiving waters corresponding to the same process.

The assigned objectives may be particularly severe. The main thing is that the authority making the decision does so in full consciousness of the consequences for a given industrial establishment. Therefore, it is necessary to know what are the most exacting constraints it is possible to assign to it, in the existing state of knowledge, i.e., what is the minimal pollution load for a particular process. This may be termed F_3 ($F_3 < F_2 < F_1$), that the authority can assign the establishment to respect.

A management whose basis would only be to develop industry at all costs in a country might be tempted to assign standards relative to the maximum load between F_3 and F_2, according to the objective of quality of the receiving waters. In addition, it is necessary to reconcile conflicting interests and sometimes to give preference to some of them in the management of a whole territory (for instance to guarantee an exceptional quality to the water, up stream some hydraulic basin, to give special protection for piscicultural activities).

In the system the constraint on each of the various industrial activities ought to be an absolute value for the pollution load accepted in the discharge.

In relation to the processes carried out at the works, authorization will be given to the highest value F_p for the pollution load corresponding to the characteristic of the effluents produced. The management authority ought to make its decision knowing the meaning of F_p. This value does not allow the creation of a particular industry under ordinary conditions unless $F_p > F_3$.

In some river basins where it has been decided to exercise special protection, the authority will be compelled to make $F_p < F_3$. Such a decision will normally exclude the development of certain industries. They could only be established at the cost of installing very exceptional means to prevent pollution. That might even include transferring the pollution load out of the river basin under consideration. In these conditions, the industry will not be permitted unless it provides advantages that outweigh the constraints imposed in the field of water pollution.

This policy of managing the location of industrial premises will not be successful unless other types of industry subjected to less severe constraints are used. Therefore it is important to have a sufficient diversity of objectives to provide for all foreseeable conditions.

As it is a necessity to assign maximum pollution load for each industrial establishment, each increase of production capacity will necessitate a correspondingly greater effort to control pollution.

Thus, the formulation of objectives for river-water quality leads to the imposition on the various discharges and to the continuation of traditional action in the regulating discharges.

The success of such action presupposes that the authorities in charge of management are provided with competent technical staff and with equipment to allow adequate supervision of river basins for ensuring compliance with the regulations. It will always be necessary to provide for the interchange of ideas between the authorities and the users, so that all may be aware of the complexity of the problems and so contribute aiming to their solution. It is in particular necessary that the results of operation are published so that everyone may be satisfied about the justification of the conditions with which he has to comply.

We suggest in this connection that appropriate records should be open to inspection. This may make it possible for river basin authorities to reconsider the conditions they have imposed on industries that are not able to meet their objectives.

The assignment of objectives for the quality of the rivers according to French law is the main tool in the fight against water pollution. At the same time, the registration of nuisance-producing industries by the 'Agences financières de bassin' is making good progress in pursuance of their policy relative to the dues they collect. The apportionment of these dues in connection with the objectives of quality assigned to the water courses is an important inducement to keep within the limiting pollution load permitted to be discharged.

Discussion *by* E.J. Hall
South Africa

(1) *Objective of Water Pollution Control*

The statement by M. Saglio that the aim of a water quality arrangement policy is to maintain or improve the standard of the receiving water, lacks reference to the economic aspects. I suggest the aim is:

To ensure the continued maintenance of the quality of a receiving water such as to ensure its adequacy, as a raw water, for the normal uses to which it is or will be put at the place where it is so used or extracted for use.

This definition of objective implies: (i) Water quality standard should be the minimum required for protection of health or for a raw supply. (ii) The natural regenerative powers of the water should be fully employed; hence the point of use is important.

(2) *Maximum Benefits*

The aim of a Water Pollution Control Programme is to achieve the maximum benefit at the lowest cost. The expected pollutant loads, must therefore be specified as accurately as possible, preferably, as recommended by M. Saglio, by total load and not concentrations, based on the absorbing capacity of the river, and allowing for its powers of renovation.

This point must be emphasised. Rivers are a country's drains and should be used, though not abused, as such. There is no such thing as pure river water. Rivers provide virtually the only practical means of removing wastes from our urban environment economically.

Further, wastewater purification is achieved at a cost which by the law of diminishing returns becomes excessively high for marginally negligible increased plant removal efficiency. Too high removals means costs disproportionate to the gain.

(3) *South African Experience*

M. Saglio suggests that the problem is of such complexity as to be precisely insoluble. A practical compromise is the only answer.

The history of Water Pollution Control in South Africa indicates parallel thinking. The largest settlement, of 2½ million people, is 500 km from the coast and was recently (1896) established on a major watershed with few perennial streams. The consequence is a tradition of wastewater treatment to contemporaneously high standards. As water sources are limited, water conservation or re-use is necessary and this small nation has kept in the van of wastewater treatment. It enacted a Water Act (No. 54 of 1956) which was a not inconsiderable achievement for that time.

This overriding act places full responsibility for water resource and pollution control on one Cabinet Minister. It prescribes enforcement by published standards. The general standard including inter alia, COD 75 mg/l, PV 10 mg/l, NH_3 10 mg/l, SS 25 mg/l, tds 500 mg/l above intake.

A system of permits meets temporary circumstances and special conditions. Sometimes P and NO_3 maxima are specified to avoid eutrophication.

For all its shortcomings, success has been achieved, thanks to enlightened application and excellent co-operation from the cities and industries. Industrial Effluent policies of cities ante dating the Act, have helped materially.

(4) *Industrial Effluent Control in Urban Areas*

Generally sewer tariffs are imposed to fund capital and operating costs. Industrial Effluent tariff charges are added to recover proportionate revenue from Industries and to encourage water conservation and minimize pollution.

Industries benefit as Industrial waste is generally treated more economically in admixture with large volumes of domestic sewage due to the dilution and scale effect. More importantly, charges directly related to volume and PV concentration provide a powerful incentive to industrialists to recycle and recover rather than discharge to waste.

The present charge formula is $3,74 + 0,022 (PV - 80)$ cts/kl which recovers both carriage and treatment costs. Table 1 shows the volumes and loads of total sewage and industrial effluents treated in Johannesburg. The proportionate reduction in the industrial contribution to the tariff.

TABLE NO. I

SEWAGE VOLUMES AND LOADS: JOHANNESBURG

Year	Flow Ml			Load: Kg O_2 x 10^3		
	Total Sewage	Total Industrial	Industrial Total %	Industrial Sewage	Total Industrial	Industrial Total %
1967	93 800	6 160	6,6	8 000	2 360	29,5
1970	109 300	5 905	5,4	8 900	2 540	28,6

Industries contributing large loads are forced to re-think completely their whole process in order to survive. The Yeast Industry is an example. Two factories in Johannesburg produce 70% of the national needs. The waste load discharged to the sewer amounted in 1970 to 8,6% of the total city sewage load though the volume was only 0,16% of the total. (Charge = R230 000 p.a.) Pretreatment systems from anaerobic digestion to incineration were considered, but were too costly. A systems approach by one factory suggests a change in source of raw material. Using refined molasses in place of blackstrap reduction in the strength of effluent promises a saving of 60% on effluent charges.[2]

(5) *Other Industries*

Large Industries treating their own wastes have found it necessary to conserve water by recycling in view of limited supplies. This makes it more difficult to comply with the Act. Table 2 after Heynike[1] gives the trend.

TABLE No. II
WATER USAGE : VAAL INDUSTRIES

Product	Water in Circulation Ml/day	Intake Water	
		Ml/day	% of Total
Steel	860	29,5	3,4
Thermal Power Station	4 530	114	2,5
Oil (from coal)	840	64	7,8
Chemicals	600	18,2	3,6
Total	6 830	225,6	3,3

A consequence is an increase in tds in the river as removal of salts is prohibitively costly at present.

(6) *Subsidies and Incentive*

In South Africa the Government makes no contribution or subsidy to Local Government or Industry to assist them in complying with the effluent standards. The difference between this system and that pertaining in the U.S.A. and other countries has not gone unmarked.

The application of this 'no subsidy' policy is interesting. Because cities and industries must comply, and do so, with their limited resources of Capital and Revenue, they have been forced to refine their process to comply at minimum cost. The very lack of subsidy has proved a powerful incentive. This is a lesson to be learned.

(7) *Total Water Management*

Water Management, means balancing the numerous demands against supply, including purified wastewater at optimum cost. Such a complex mechanism can be balanced at least cost only if incentives are provided. It is suggested a financial penalty and benefit is the best tool available.

Assign a price to water, which is related to its quality, and automatically users seek sources of the minimum standard. An incentive to purify to higher standards is also created because of the higher potential revenue.

REFERENCES

1. HEYNIKE; S.A. Branch, I.W.P.C. : Symposium, Dissolved solid loads in the Vaal Barrage Water System (November 1969).
2. BOLITHO, V. : S.A. Branch, I.W.P.C. : Conference paper (May 1972)

Discussion by Klaus R. Imhoff
"The Potential Role of Effluent Charges
in Water Quality Management"

Water quality management has to satisfy such conditions as adequate quality in receiving water, economical aspects of industries and communities and system flexibility. In this respect it might be of interest to learn about the system which has been applied in West Germany's Ruhr District for more than five decades.

The water authority which is represented by the state minister of food, agriculture and forests makes a study of the different interests in water. If necessary public hearings have to take place. On this basis the concept for water quality is developed.

For 50% of the area of the densely populated state of North-Rhine-Westfalia river associations have been formed to do the practical work. The associations are responsible for entire drainage areas. They are based on special laws. So everybody who has an advantage of the associations' work.

The associations design, supervise construction and operate the plants which are necessary to control water pollution. The design has to get approval by the state authority. Also the financial budget of the associations has to get approval. If the budget is too small the state minister may increase it at his discretion. In such a case the higher expenses have to be covered by the associations' members.

Expenses are paid by subsidies from the government and by effluent charges which are raised from members. Effluent charges are calculated according to quality and strength of the members' discharges. The discharge may either flow into a sewerage system and thus be treated in central works of the associations or it may flow directly into a river.

By this system the river association is able to adapt its policy to pollution increases. It is up to the association to decide at what location the next wastewater treatment plant has to be constructed and what efficiency would be appropriate.

If the effluent charges are calculated in a way that reflects the real construction and treatment costs of the association, such a system is also economical because the decision maker of the factory can make up his mind if it is cheaper to pretreat on the factory's expense or to use central works of the association at the association's costs.

River associations are self-controlling bodies with a board of directors formed from its members. Usually representatives of water works, industries and communities are represented. Thus in general a good balance of interests can be achieved already within the association. If the result does not correspond to the necessary river standards the authority will act and change the program.

Discussion *by* K.R. Imhoff

Effluent calculation charges by two different methods

Due to industrialization and dense population different water associations have been formed at the Ruhr and the lower Rhine.[1] These associations have the legal task of water pollution control within different drainage areas. They design, construct and operate waste water treatment plants for their members. Members are those who load the system like industries and communities, and those who derive benefit from the association's work, like i.e. water works.

The annual budget of the association has to be accepted by a majority of the members. The total amount of money then has to be subdivided according to the advantage of the member or according to quantity and strength of the members' discharges. For the calculation of effluent charges different procedures have been developed. Two methods shall be outlined here.

Dilution Method

The Emschergenossenschaft determines effluent charges on the basis of quantity of flow and its necessary dilution. The basic principles are explained by the dilution formula.[2,3]

$$D = -1 + \frac{S}{SP} + \frac{1}{2}\frac{B}{Bp} + \frac{1}{2}\frac{(P-30)}{Pp} + F$$

D = Dilution factor, S = Settleable Solids (ml/l), Sp = Permitted settleable Solids (mg/l), B = BOD_5 (mg/l) after sedimentation, Bp = Permitted BOD_5 (mg/l), P = Potassium permanganate-consumption (mg/l) after sedimentation, Pp = Permitted $KMnO_4$-consumption (mg/l), F = Toxicity to fish as determined by dilution test.

In each term of the equation the numerator indicates the actual amount of substance found in the effluent sample. The denominator indicates a permitted amount of the substance. The ratio represents a dilution factor. It is important to note that the individually derived dilution factors are held to be additive in character. BOD_5 and $KMnO_4$-consumption have been combined to compensate for differences in degradibility.

Population Equivalent Method

In Ruhrverband a special bio-assay-test is applied to assess industrial waste water and to convert even toxic substances into BOD population equivalents.[3,4,5] It is kept in mind that a given BOD in
In Ruhrverband a special bio-assay-test is applied to assess industrial waste water and to convert even toxic substances into BOD population equivalents.[3,4,5] It is kept in mind that a given BOD in river water results from upstream organic pollution and from selfpurification. Toxic substances may inhibit selfpurification and may cause a higher BOD similar to organic pollution.

Both influences can be compared by a test. For the test two rows of Erlenmeyer flasks are used which contain a standard solution of organic compounds and a small amount of activated sludge. To one row increasing amounts of sewage are added, the BOD_5 of which is simultaneously determined, and to the other row increasing amounts of the effluent to be tested. The flasks are then shaken. After 3 days the potassium permanganate consumption of the filtered contents of the flasks is determined.

If the differences between the $KMnO_4$ consumption of the flasks are plotted over mg BOD/l added to standard solution and over ml/l of effluent to be tested added to standard solution respectively, two curves are obtained. By comparing curve A and curve B it can be concluded that 54 g of BOD added (1 population equivalent) result in the same additional $KMnO_4$ use as 18 liters of industrial effluent. Taking 3 points of the curves into consideration (18 + 34 + 42) : 6 = 15.7 liters of the test effluent are to be valued at one population equivalent.

The main advantage of this procedure is that all factors influencing selfpurification including oxygen demanding and toxic influences are tested simultaneously.

Conclusion

Two methods for the calculation of effluent charges have been shortly described. In practise the methods are modified to cover treatment costs more realistic. KNEESE has pointed out that such approach is most important to get an overall feasibility, including pretreatment or recycling by the factory.

REFERENCES

1. FAIR, G.M.: Pollution Abatement in the Ruhr District, Journal Water Pollution Control Federation 1962, 749.
2. KNOP, E. : Die Schädlichkeitsbewertung von Abwässern, Technisch-Wissenschaftliche Mitteilungen, Heft 4, Juli 1961, Vulkan — Verlag Essen.

3. KNEESE, A.V. : The Economics of Regional Water Quality Management, Johns Hopkins Press, Baltimore 1964.
4. BUCKSTEEG, W. : Teste zur Beurteilung von Abwässern, Städtehygiene, Heft 9, 1961.
5. BUCKSTEEG, W.: Problematik der Bewertung giftiger Inhaltsstoffe im Abwasser und Möglichkeiten zur Schaffung gesicherter Bewertungsunterlagen, Münchner Beiträge zur Abwasser-, Fischerei– und Flußbiologie, Bd. 6, R. Oldenbourg 1959.

Discussion by J.B. Sprague, Canada, of paper by K.R. Imhoff

What kind of numerical value does Dr. Imhoff assign to "F" in his formula (referring to toxicity to fish), and how does he arrive at this number? How does he transform toxicity of effluents to fish to an equivalent BOD value?

Reply by K.K. Imhoff

The dilution figure "F" is determined by a fish test with Red Ides (Idus Idus variatio auratus). The waste water sample has to be diluted by fresh water to that degree that fish survive for 48 hours. During that period the sample is aerated. There is no transformation to an equivalent BOD value.

S.W. Loewenson, Rhodesia

What are the authors views on comprehensive anti-pollution legislation?

R. Savage, Haifa, Israel

In the non-ferrous industry a recent survey showed over 75% of water used as cooling water, could almost entirely be saved by recirculation? Education, official persuasion and financial incentives are necessary to make industries aware of their water mis-use.

M. Neeman, Israel

No standard, regulation or law has been promulgated without using the knowledge and experience of engineers and scientists, often without close cooperation between scientists and politicians. This lack of cooperation may result in poor legal formulations; and the scientists are not less responsible than the politicians.

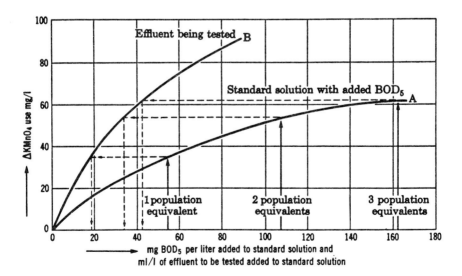

Determination of the population equivalent of industrial effluents.

EDUCATION AND TRAINING FOR WATER-POLLUTION CONTROL

PETER C. G. ISAAC

Professor of Civil and Public Health Engineering and
Head of the Department of Civil Engineering, and Dean of Applied Science,
University of Newcastle upon Tyne, United Kingdom.

Since this short paper is for a Workshop session it is not intended to be a synoptic review, nor does it claim to carry any authority. It is designed to put forward some of my views, based on over a quarter-century of university work in this field. It is hoped that these observations and aphorisms may provide the starting points of a lively, vigorous and useful discussion.

At the outset it must be clear that the paper deals only with the professional or graduate level. It must not be inferred from this that the sub-professional and technician levels are unimportant, and I hope that other participants in the Workshop will take up the training for those levels.

Since words have different meanings to different people I must also define what I mean by 'education' and 'training'. In the control of water pollution we are concerned with the professional realities of an important part of the real world. I see two elements essential in preparing for this, and these I call respectively 'education' and 'training'. These elements may be pursued concurrently or – perhaps more usually – successively, and the whole process must necessarily occupy not less than six years. By 'education' I mean formal instruction in the university, and by 'training' I mean guided experience in applying the formal instruction to real problems. Training may be gained wholly in professional practice, or partly or wholly in project work at university. The division between education and training is far from clear-cut, and my definition is itself very far from adequate, but perhaps may become clearer as I deal with what I believe should be the purpose of each. Just as I believe that education and training cannot be completely separated, so do I believe that they should both continue throughout a professional career.

The clamorous need for the control of water pollution must not cause us to forget that, at its best, a university education endows the men and women who enjoy it with a characteristic intellectual equipment – the capacity to reconcile orthodoxy and dissent. This is a delicate and difficult balance which the demands of the professions too often swing too far towards orthodoxy, so that creativity and the ability to innovate suffer. I have suggested that university education is a dialectic between orthodoxy and dissent, but it is also useful to remember that all education helps the young man or woman to three ends:

1. To survive in his (or her) environment. This emphasizes the importance of orthodoxy, of the acquiring of techniques – of techniques of living together, of finding out, of earning a living etc. However mechanical this may seem – and it is not so – it is a quite legitimate function of education.

2. To make a contribution to his community – in a sense this is the dissent of the university dialectic. Everyone needs a challenge to which to respond, and this need must be provided in the university.

3. To enjoy the fruits of his civilization — the sculpture as well as the drains. This is, perhaps, the most difficult of the three ends of university education, especially in science and engineering departments. It demands that the student should see his subject in the seamless context of his civilization; it demands also an appreciation of the interests of others; and, above all, it calls for a mature judgement which puts proper weight on matters outside the student's own field. (In concrete terms, the efficient professional man must appreciate the functional needs of his work in human, economic and other terms. At the same time he will rise to the heights only if he is, to the limits of his ability, a full man.)

Attention has usually been concentrated on the last of these ends. Indeed, many have advocated a completely non-vocational approach to university education — an attitude in my view both damaging and unhistorical.

I have devoted the opening paragraphs of my paper to these perhaps high-flown ideas, because I believe that it is even more important, than for other scientists and engineers, for those engaged in the control of water pollution to be able to take a broad, informed view of their work. These three ends of university education are its warp; the weft which gives it much of its pattern and colour is, in our case, the science and art of controlling water pollution, and to this I shall devote the rest of my paper.

The control of water pollution is not a single topic, requiring a single professional skill, with its unique point of view. It is achieved only by the activities of at least three groups of specialists. These are:

1. those responsible for the design of sewerage and waste-treatment works;
2. those responsible for the operation of these works; and
3. those responsible for the management of the receiving waters.

Civil and, increasingly, chemical engineers are mainly concerned with (1), and it is a most happy portent that these engineers are increasingly associated with chemists, biochemists, microbiologists and biologists in the earliest stages of their planning and design work. Chemists and, to a much lesser extent, civil engineers are concerned with (2). At present it is doubtful whether all aspects of the receiving waters are being purposefully managed; generally most attention is paid to policing the quality of the receiving water and of effluents flowing into it, and this is primarily the province of the chemist and the biologist.

It is an unhappy commentary on the 'compartmentalization' of life that those in each group too seldom effectively work one with another. The designer often fails to come to grips with the operator and almost always with the pollution-control officer, and those who manage one or another aspect of the receiving water may have only the most indirect means of bringing their skills to bear on design and operation. This could be altered for the better by major administrative changes in our river-basin management, but this is not our concern here. It could also be significantly improved by adequate education — and training — for all those scientists and engineers who eventually enter one of the three groups. At present most of them are educated in traditional fields of science or engineering and acquire, almost accidentally, a knowledge of how this applies to their chosen field of endeavour.

It would clearly be quite inappropriate in a general paper of this nature to attempt to set out programmes of instruction or syllabuses, but it is necessary that I should try to define the needs of, and hence the principles underlying, education and training in this field.

I have already drawn attention to one set of interrelationships — those among the three groups of specialists in the control of water pollution — but these are not the only important interrelationships. There are two other sets which are quite vital to efficient work in this field. The first set arises from the fact that the purpose of sewage treatment has undergone a significant secular change in the last century, more particularly in the highly industrialized countries. At first it was envisaged primarily as protecting public health; soon it was seen as preserving amenity. Now it is clearly a vital part of the overall management of our water resources. The specialists in my group (3) are concerned with much more than water quality, and those specialists in groups (1) and (2) must essentially subserve this wider purpose. Education and training must, therefore, take a sufficiently broad view — of water supply for domestic, industrial and agricultural purpose, of hydroelectric development, of land drainage and flood control, of irrigation, of navigation, etc.

The second important set of interlationships arises from the apparently frequently ignored fact that matter is normally indestructible. Any material, including air and water, which the community uses, must be disposed of. It is not possible — and may indeed be dangerously counterproductive — to isolate water pollution. We must not feel ourselves free to propose a solution to a particular problem of water pollution if it creates, for example, an air-pollution problem, without being sure that, in this way, we shall be producing the maximum net benefits, or, at least, the minimum net damages. This too, demands that the education and training should be broad.

Here, at once, we come face to face with an almost insurmountable problem — how do we reconcile the breadth that I have shown to be desirable with the depth that is essential for any professional man? Part of the answer — but by no means the whole of it — is time. We must have long enough for the young man to cover the fundamentals of his own field of science or engineering, to marry this theory with the practice of his later profession, and to see his work in its wider ramifications. This seems to demand that the specialist education for water-pollution control should be at the postgraduate level. This, in my view, is desirable for another reason. It is possible to envisage an undergraduate course that would make the young man immediately useful, for example, for the operation of a large sewage-treatment works. It is, however, likely that such a course would produce a technologist — or perhaps even only a technician — for 1970, and that he might be quite unable to meet the challenges of the 1980s. We have to educate and train our young men and women to be almost immediately valuable, but also to have a potential for their 30s and 40s and 50s. This is a challenge that most of us in universities recognize, but I wonder how many of us can feel that we are satisfactorily meeting it. The young man or woman should, therefore, read science or engineering at the undergraduate level, taking those options which are particularly relevant to his later career, e.g. freshwater biology, hydraulics, water engineering, water chemistry, etc. He should then receive postgraduate specialist education.

This postgraduate study, which can, in my view, most usefully come after some practical experience, must provide the background for the interrelationships that I have mentioned. For example, the sewage engineer must receive instruction in the engineering of wastewater treatment, in the chemistry and biology of sewage and freshwater, in the relationship of the disposal of waterborne wastes to that of solid and gaseous wastes, in the management of water resources generally etc. In studying sewage engineering it is necessary to consider the unit processes involved, but it is also essential today, as technology becomes more complex, for the young engineer to study whole systems and to have some feeling for the potential of operational research, linear programming etc. All this should have given the young engineer an idea of the importance of harnessing nature rather than resisting it. He must have an appreciation of the social context of his work and must be able to appreciate, if not actually to marshal, the arguments of economics. This is a tall order, but the needs I have outlined apply *mutandis mutatis* to all those concerned with the control of water pollution – scientists as well as engineers. The engineers must have a grasp of all the relevant sciences, and the scientists must have an understanding of the engineering requirements of any situation and of the engineer's special skills and needs.

The kind of professional man whose education and training I have been discussing is very far from being a technician dancing to someone else's tune. He must frequently make important technical decisions, and he must be able to present technical and economic arguments in such a form that the best political decision is taken when a political, not a technical, decision is called for. Each of these situations calls for value judgements, so that our young man or woman must be brought face to face with such judgements during undergraduate and postgraduate education, and, most particularly, during practical training. For this purpose project work in the university can certainly have value, but it seems to me highly desirable – indeed essential – that these judgements and decisions should be made in a real-world context, where the cost of a wrong decision in some way impinges on the young man or woman. It is for this reason that I believe strongly in the value of practical experience (training) coordinated with the university courses. The practising members of the profession have a duty – often not appreciated – of passing on their accumulated experience in an organized way. Too often the senior member of the profession treats his younger colleague merely as convenient cheap labour. The point I am trying to make is probably not relevant to the continental European system of training engineers and applied scientists, but in Britain and the USA there is a complete dichotomy between the two halves of what is essentially one whole process – university education and practical training. To set this right I should like to see a system in which the university has overall responsibility for at least a seven-year programme of education and training: one year in practice as part of the maturing of the young man or woman, followed by a three-year or four-year undergraduate course in science or engineering. This would be succeeded by two years of carefully supervised experience in professional practice and the whole period would be rounded off by a final postgraduate year of specialized instruction. At the end of this long period of apprenticeship the able young man or woman would be able to take his or her place as a fully-fledged member of his profession beginning his career.

I have dealt first – and I hope coherently – with the basic education and training of a professional specialist in water-pollution control. There are two important related matters which I must now briefly look at. These are the role of research in the education of our specialist, and the need for continuing education.

I have emphasized the need for depth as well as breadth in education. Many of my university colleagues see research as the most satisfactory way of pursuing depth. I believe, however, that, especially at the undergraduate level, research is not appropriate for this purpose. Too often the subjects pursued are trivial. It is most important that the young man or woman should, as I have indicated, be brought to face the social and economic realities of our field. This – and the pursuit of depth – can best be achieved by project work of a different kind – design, feasibility study, survey of operation, pollution survey etc., always associated with recommendations for improvement and accompanied by estimates of cost. A research project often – perhaps usually – directs attention away from these 'realities', and is concerned with the pursuit of knowledge for its own sake.

Nevertheless, I strongly believe that research is vital to any university department. In a general sense universities must be close to the boundaries of knowledge, and active and well-directed research ensures this. Even more important is the value of research in re-creating (in its older sense) the university teacher; through research he is kept interested and actively aware of his field. There are also a relatively small number of people whose intellectual gifts will most readily be realized through research, and, for these, the PhD is a valid exercise. Even here, however, I believe that it is of the utmost importance that the research should be directed towards the solution of real problems, and that the PhD student should see his work firmly in this context. Already there is the beginning of a rapprochement between universities and the practising side of our profession. This needs very great strengthening at the research level, and I should like to see many more research projects carried out partly in the university, and partly in the field of practice. It will, then, be essential to reach the delicate balance between intellectual vigour, which is an important end of the PhD, and social, economic and technical needs.

It goes without saying that the able professional man or woman continues to learn throughout his or her career. Much — perhaps most — of this learning will be what I have called 'experience', but the more enthusiastic will also want to keep up with advancing theory and technique in their field. It seems to me, therefore, that healthy development in our field demands well-planned post-experience training courses: It will usually be impossible to release a practising scientist or engineer in our field for more than the shortest period. I envisage, therefore, a coordinated series of one-week courses dealing in depth with a single narrow topic only. These topics should be of immediate value and should be presented in such a way that the information can be immediately applied. Frequently, to keep these courses lively, practical work can intersperse the lectures. These courses should be presented cooperatively by active practitioners, and university and research specialists. The actual organization of the courses may be by university departments, by professional institutions or by government, or other, research laboratories. Each group has a vital contribution to make.

I have no doubt that my paper is excessively coloured by my British experience, but the principles that I have put forward are, I believe, of general application. The expert in the control of water pollution must have a deep knowledge of the fundamentals of his basic science or engineering and of their application to his chosen field. He must be able to appraise his work in its social and economic contexts, and must be able to relate it at all times to the work of related experts in the whole field of water resources. He must have depth and breadth — he must be a whole man. Indeed perhaps we must aim for the impossible — to produce a polymath — if we are to manage our environment for our future well-being or even, indeed, for the continued existence of man.

Development of Instructional Programs
for Wastewater Treatment Plant Operators

by J. Brady, W. Crooks, B. Dendy and K. Kerri

GOALS OF EDUCATION AND TRAINING
(Abbreviated version. Requests for the full text should be made to Dr. K. Kerri (Editor)

Analysis of the education and training needs of wastewater treatment plant operators in the United States today indicates vocational aspects should be given the highest priority. On the basis of current trends in curriculum development, future education of treatment plant operators and technicians in the water pollution control field appears to be advancing towards the three ends of university education as outlined by Isaac:

1. To survive in his (or her) environment.
2. To make a contribution to society.
3. To enjoy the fruits of civilisation.

RESPONSIBILITY OF ADMINISTRATORS

Administrators are charged with the responsibility of insuring proper design, operation, and maintenance of wastewater treatment facilities under their jurisdiction.

An effective education and training curriculum is an essential ingredient in an administrator's program to accomplish his duties. Important to the administrator is a knowledge of available instructional techniques and materials. He should be aware of how they were developed and verified so he can evaluate their applicability to his particular situation.

The remainder of this discussion will review the steps an administrator can follow in developing an education and training program. These steps will be illustrated by a history of how a new field study manual, entitled *Operation of Wastewater Treatment Plants,* was written, applied, evaluated, and rewritten into its current form.

INSTRUCTIONAL OBJECTIVE

Development of capable, trained personnel who can operate wastewater treatment plants efficiently is the objective of any education and training program. Efficiency is defined in terms of producing an effluent quality as intended by the plant designer, meeting established receiving water quality standards, and minimizing costs of operation and maintenance. Accomplishment of the objective can be greatly assisted by providing operators with the knowledge and procedures they need for efficient operation of wastewater treatment facilities.

IDENTIFICATION OF ESSENTIAL KNOWLEDGE AND PROCEDURES

Experienced operators were requested to write the information and procedures they felt an operator needed to know to operate his plant. Site visits to various types of wastewater treatment plants were conducted to observe the situations encountered by operators and their daily operational problems. Special attention was given to plant start-up, daily operational schedule, frequently encountered problems, maintenance, sample collection and analysis, and operator response to lab results.

REVIEW AND SELECTION OF EDUCATIONAL METHODS

A team of experienced operators, consulting engineers, administrators, regulatory personnel, and educators was formed to develop a curriculum capable of accomplishing the objective of this program. How best to convey the desired information and procedures to operators unable to attend conventional classroom programs was the next problem to be solved. For the program to be readily available to as many operators as possible, the cost should be as low as possible without sacrificing the transfer of knowledge. Correspondence schools, home study, and self-study manuals have been used extensively in this type of situation in the past.

Correspondence schools for wastewater treatment plant operators sponsored by International Correspondence School, Clemson University, and University of Arizona and the literature on operator training were examined.

Programmed instruction techniques were selected as the most appropriate means of conveying the desired knowledge because no additional facilities or material other than basic manual would be required.

Strict adherence to either programmed instruction technique requires considerable time and ingenuity to develop the instructional material. Neither approach is familiar to most operators, although they could adapt to the procedure. Presentation of material on the operation of wastewater treatment plants by either technique limits the value of the material for quick and easy reference when plant operational problems occur. To overcome these shortcomings of strict programmed instruction, the material was arranged in the form of a typical textbook, but presented using the techniques of programmed instruction.

Tests by educators and psychologists have indicated that answering questions immediately after reading the material provides immediate reinforcement of knowledge. Research with equivalent groups using programmed instruction indicates that essay and multiple choice questions are equally effective; however, retention is much better one year later in groups using essay questions. The adopted procedure was for the operator to read a short section, write the answer to a few questions, compare his answers with suggested answers at the end of the chapter, and decide whether to reread the short section or continue on to the next section. A major advantage of this approach is that an operator can proceed at his own pace, and operators with considerable variations in education and experience all can use the material.

PREPARATION OF MATERIAL

Information and procedures needed by operators of wastewater treatment plants was divided into 17 chapters on the basis of subject matter, as shown in Table I. Experienced wastewater treatment plant operators wrote the chapters on treatment plant processes with emphasis on the essential knowledge and procedures previously identified that the operator needed to know to perform his duties effectively. Each treatment process chapter followed the format outlined in Table II.

After the writers of each chapter had prepared their material, it was reviewed, edited, and rewritten in the selected programmed instruction format. Following each short section of a particular topic, several questions were prepared for the operator to answer. These questions were designed to reflect problems an operator could encounter and attempted to obtain a solution from the operator that he could apply to his plant. When the operator checked his answers, he found suggested answers that contain a discussion of possible solutions to the questions posed. Numerous sketches, illustrations, photographs, and useful tables were provided to enhance the appearance of the material and facilitate its usefulness.

The material is written to allow an operator to study only selected lessons which apply to his problems, or to proceed through the entire book, whichever he chooses.

TABLE I. TREATMENT PLANT OPERATION COURSE OUTLINE

CHAPTER	TOPIC
1	Introduction
2	Why Treat Wastes?
3	Wastewater Facilities
4	Racks, Screens, Comminutors, and Grit Removal
5	Sedimentation and Flotation
6	Trickling Filters
7	Activated Sludge
8	Sludge Digestion and Handling
9	Waste Treatment Ponds
10	Disinfection and Chlorination
11	Maintenance
12	Plant Safety and Good Housekeeping
13	Sampling Receiving Waters
14	Laboratory Procedures and Chemistry
15	Basic Mathematics and Treatment Plant Problems
16	Analysis and Presentation of Data
17	Records and Report Writing

TABLE II. TYPICAL TREATMENT PROCESS CHAPTER OUTLINE

SECTION	TOPIC
1	Relationship of process to overall plant
2	Purpose and description
3	Plant start-up
4	Daily operational problems
5	Sampling and analysis (includes performance evaluation)
6	Safety
7	Additional useful information

FIELD TESTING

A critical field testing program is essential to evaluate the effectiveness of instructional materials and to identify means of improving the material or method of instruction. Field testing of the material was divided into three phases, with each phase requiring one year. Details of this testing program and results have been reported previously [5].

TYPES OF INSTRUCTIONAL PROGRAMS

Several different types or combinations of techniques may be used to convey to the operator the information and skills he needs to operate and maintain his plant. Conventional classroom instruction is the most common method. Most instructional programs are tending to couple lectures with laboratory instruction and on-the-job training. The field study manual described herein has been used successfully in these situations as well as an individual study manual with or without periodic visits from an itinerant instructor. Another possible use of the manual is in conjunction with educational TV programs.

OTHER INSTRUCTIONAL MEDIA

Considerable research is being conducted to develop better instructional media for the education and training of wastewater treatment plant operators. Of particular interest is the concept of developing instructional modules to allow the operator to work towards certification by completing a prescribed set of lessons plus a specific number of relevant elective lessons on pertinent topics. Each advancing level of certification requires a higher base of modules, plus additional electives. Curriculums have and are being developed for potential and existing operators at the high school and post-high school levels[1].

SUMMARY AND CONCLUSIONS

Procedures followed in the development and use of operator educational materials are outlined in this paper. In summary, the procedures involved are:
1. Defining educational objectives.
2. Identifying knowledge and procedures essential to achieve objectives.
3. Reviewing educational techniques, materials and media.
4. Developing curriculum.
5. Selecting qualified instructors and/or writers.
6. Administering the program.
7. Evaluating the program.
8. Repeating administration and evaluation until material and method of presentation were satisfactory.
9. Providing procedures to revise program when necessary.

A field study manual on the operation of wastewater treatment plants was developed and tested by experienced operators which is capable of providing operators with the information they need to know to operate their plants and solve operational problems. Operators studying alone or enrolled in regular courses who use this manual are provided helpful learning material and a useful reference.

ACKNOWLEDGEMENTS

This project was funded by the Environmental Protection Agency, Office of Water Programs, under Technical Training Grant No. 5TT1-WP-16-03.

The authors wish to express their gratitude to the many operators who participated in the development of this program. Special thanks are due to the men who wrote the chapters and our technical and educational consultants. Illustrations in the manual referred to in the original text were drawn by Martin Garrity. F.J. Ludzack, Chemist, National Training Centre, Office of Water Programs, Environmental Protection Agency, provided many helpful suggestions throughout the entire developmental period.

REFERENCES

1. AUSTIN, John H., *Criteria for the Establishment and Maintenance of Two Year Post High School Wastewater Technology Training Programs,* Vol. I, Program Criteria and Vol. II, Curriculum Guidelines, Clemson University, Clemson, South Carolina, 1971.
2. BRADY, John, CROOKS, William, DENDY, Bill, and KERRI, Kenneth, "Development of a Field Study Program for Wastewater Treatment Plant Operators," *Water and Sewage Works,* April, 1972 (scheduled publication date).
3. Environmental Protection Agency, *A Curriculum Activities Guide to Water Pollution and Environmental Studies,* Volumes One and Two, prepared by the Tilton School Water Pollution Program,

Training Grants Branch, Office of Water Programs, Environmental Protection Agency, Washington, D.C., 1970.
4. KERRI, Kenneth D. and DENDY, Bill, *Operation of Wastewater Treatment Plants – A Field Study Educational Program,* Environmental Protection Agency, Office of Water Programs, distributed by Sacramento State College, Sacramento, California, 1971, 1504 pp.
5. REDEKOPP, A.B., and J.H. Austin, "Systems Approach to Licensing," *Journal American Water Works Association,* Vo. 63, No. 12, p. 743, December 1971.

F.C. Larson, USA

Political authorities should be made aware of the importance of giving incentives to those employees undergoing training and instruction in wastewater treatment processes.

Reply

In the author's opinion, two aspects should be distinguished a. the willingness of the superior officers to let their subordinates follow post-degree courses at the expense of the company; b. the willingness of the subordinates to do so. Provided the courses are of short duration, preferably not longer than one week, the willingness mentioned under a. is readily available in the Netherlands, both with governmental as with private organizations, the employers realizing keenly the benefits for the company thus to be obtained. Technicians and technical engineers in the Netherlands have many opportunities for an earlier or an additional promotion and their willingness to participate in post-degree courses is consequently large. For university graduates on the other hand, the situation is rather different. With the high taxation on additional income, financial motives are nearly absent, while for them promotion often means an additional administrative burden and a correspondingly larger work load. The incentive must now be sought in satisfying their scientific curiosity, to which the subject matter of the course and the way of presentation must be directed.

P.H. Jones, Canada

So far the problem of pollution has been entirely as a technical one. It is suggested that we must ensure a broader approach to the problem. Engineers need to be educated as "whole men" it has been said implying that economists, biologists and lawyers should be involved. Who will help the engineering professor and student understand and be influenced by their colleagues in other disciplines? The problem breaks down into three parts. It is necessary to teach the professors and specialists to listen to and respect their colleagues from other disciplines; to be influenced by their colleagues in other disciplines and to teach engineering students with non-engineers to some extent to ensure learning. To fit into the economic and job structure we must communicate with the employers to determine trends and demands. It was stated that emotional analysis of environmental problems is bad. I suggest that one of the few properties given to man to distinguish him from the beasts of the field is "emotion." To suggest the solution of environmental problems without emotion once more suggests that the environmental problem is a simple technical one with a simple technical solution. This simply is not so.

Reply

Of course, the problem of environmental pollution is not entirely a technical one. Already in a small country like the Netherlands, the abatement of existing air, water and soil pollution alone will require many billions of dollars, while for other environmental problems as noise, housing, racial discrimination, etc, similar costs are involved. Spending this money means that for many years real increases in personal income will be absent or even negative and for the public at large this will be difficult to accept. The help of sociologists, psychologists, public relation officers, etc. is therefore indispensable, while the participation of medical doctors, toxicologists, physicists, chemists and many others is required to define the desired goals. The sanitary engineer should be well aware of these facts and he should be taught to communicate with experts from other disciplines, but he should never try to perform these duties himself.

Emotion is certainly helpful to get things started, for instance the doomsday syndrome created by The Blueprint for Survival and by The Limits to Growth. Sound decisions, however, can only be made by experts with cool heads, basing themselves on well established facts and figures.

Reply

In reply to Dr. KERRI and his colleagues the author stated that it was right to emphasize that education and training were essential at all levels if water-pollution control was to be carried out effectively and economically. He agreed with Dr. ADIN that educational method was as important as content, but did not feel that this specialized workshop was the appropriate venue for its discussion.

He could not, however, agree that there were few opportunities for educators to perform their tasks; such opportunities offered themselves continually to the experienced teacher. He also agreed with both the important points made by Dr. JONES. The engineer in this field must be able to understand, and appreciate the viewpoint of, many other experts; he must himself take a wide view.

A. Adin, Israel

The question of how to educate and how to make fruitful contact with the students is as important as knowing what to teach. There appear to be few opportunities for educators and teachers to perform their tasks.

Reply

The opportunities to study the science of education and to make fruitful contacts with students is as large as the time educators and teachers are willing to spend on these subjects. Indeed much is lacking in this respect, but the time required can easily be found by cutting down on administrative work, streamlining the procedures and above all by abandoning the desire to make all decisions by oneself.

S. Iwai, Japan. Formal Discussion

The reason why environmental pollution is not being actively controlled in Japan and several Asian countries which I visited in the summer of 1971 as an eventual consultant for W.H.O. is probably the lack of public awareness of the problem. Environmental pollution could be generally classified into two categories, the first is related to the hazard to human health and the second to economic and aesthetic values of the human environment. The former results from placing a smaller value on human life, the latter from the policy which gives preference to the rapid promotion of urban and industrial developments on the basis of a faulty estimate of the capacity of this environment to absorb and naturally purify various pollutants.

The first category of pollution which Japan experienced very seriously was not observed except in a few cases in the visited Asian countries, while the second was present in these countries, with a tendency for rapid growth of population and propagation of urban area. However accelerated industrial development in the countries is liable to cause the former type of pollution in the near future. The customs of people and climatological conditions (mostly sub-tropical and tropical) are similar in those countries to those in Japan. This tends to introduce problems which are also comparable. Even in rural areas there is danger of pollution of the former type, due to the misuse of pesticides for agriculture or careless disposal of toxic wastes from mining.

Training and education on water pollution is carried out to some extent in these countries, including Japan, in institutes of health, institutes of public health or environmental sanitation, schools of public health and schools of engineering in universities. For the existing staffs concerned, it would be advisable to practise "on-the-job" training in vocational schools of engineering which have courses in waterworks and sewerage with basic chemistry, biology and public health. Industries must be encouraged to send their employees to such schools or on short training courses. A certificate, a diploma or credits could be awarded upon the successful completion of the courses. Recently, in Japan, such a training course has been initiated in order to qualify the trainees as "a chief pollution control engineer" which is stipulated by a new law. Most of the factories in Japan must have such a responsible engineer for pollution control. Education through research is a very effective means of producing capable scientists and engineers concerned. Many public health schools, sanitary or public health departments of the universities in these countries could profitably co-operate in well-constructed international and regional research programmes. Shortage of teaching staff in this field in higher educational institutions hinders the training and education of the lower levels. This results in less rapid professional growth of the specialists with respect to professional position and salary. General education on pollution would be needed even for the public. Training and education on pollution problems, however, must never be carried out by "a witch-hunting procedure," but on the basis of calm and rational concepts of science and technology.

TRAINING AND EDUCATION IN
WATER POLLUTION CONTROL

PROF. IR. L. HUISMAN
University of Technology, Delft, The Netherlands

1. General

Below the graduate level, opportunities for training and education in the field of water pollution control in the Netherlands are limited to a part time one-year training of artisans (6 years primary and 4 years secondary school) as operators of sewage treatment plants and to a part time two-year post-degree course for technical engineers (6 years primary and 5 years secondary school followed by 4 years Technical College) in the field of water supply engineering.

At the graduate level on the other hand, many opportunities are present. As most important may be mentioned the specialization in sanitary (environmental) engineering in the departments of civil and chemical engineering at the Universities of Technology in Delft, Eindhoven and Twente and in the Department for Environmental Science at the University of Agriculture in Wageningen. In all these cases the pre-university education amounts to 12 years (6 years primary school and 6 years Gymnasium or Atheneum), while at the University the official duration of study equals 5 years, but in reality is much longer with the median value somewhere between 7 and 8 years.

Among the graduate educations in sanitary engineering mentioned above, the one at the Department of Civil Engineering of the University of Technology in Delft is the oldest and the largest. It started in 1950 and at the end of 1971 a total number of 125 civil sanitary engineers will have graduated here. This education will be described in detail in section 2. Post-experience and post-graduate education in the Netherlands has been developed to a high degree, with a large number of participants both on the national as on the international level. This will be discussed in detail in sections 3 and 4. Section 5 will indicate which improvements are necessary to deal with tomorrow's problems.

2. Graduate education in sanitary engineering at the Department of Civil Engineering, University of Technology, Delft.

At this very moment, the Department of Civil Engineering in Delft, the only one in the Netherlands (13 million inhabitants), has 2500 students. The number of freshmen varies between 400 and 450 per year (with 30% having a degree of a Technical College), of whom 40% will never graduate, 30% will do so within 7½ years, while the remaining 30% take much longer, up to 10 and even 15 years. Taking into account the drop-out, these figures mean that eleven student-years are necessary to produce one civil engineer.

At the end of this decade, the number of freshmen will have grown to 500 per year and the total number of civil engineering students to 3500. The number of civil engineers working in practice will then have risen to 4000, while according to the Central Planning Bureau a number of 4800 are necessary. A pleasant and reassuring prospect of today's civil engineering students!

At the Department of Civil Engineering in Delft, the education is aimed directly at the master's level, the bachelor degree being unknown. The traditional civil engineer as jack of all trades has largely disappeared, however, and so has his education. Six specializations are available nowadays, in building, structural, hydraulic, sanitary and traffic engineering, as well as in civil engineering sciences.

The official duration of study of 5 years is subdivided into 3 parts

(a) a propaedeutical study of 2 years, with emphasis on mathematics, physics and mechanics;

(b) a bachelor study of 1½ years, mainly devoted to civil engineering subjects, but with a gradual transition towards the chosen specialization;

(c) an engineer study of 1½ years, fully within the specialization selected.

To obtain the final degree, the above-average student will need 8500 hours, subdivided into 1800 hours of lectures, a little over 3000 hours for laboratory and design work and about 3500 hours for self-study. From this amount of work, the chosen specialization occupies 500 to 550 hours of lectures (30%) and 1200 to 1500 hours of laboratory and design work (40 to 50%), while with the more advanced subjects the amount of self-study will be relatively larger. About 10% and in future perhaps 15% of the students mentioned above choose sanitary engineering as their specialization, meaning the number of graduates increasing from 20 to about 40 per year in this decade. With the basic education in civil engineering, it nearly goes without saying that this sanitary engineering is mainly directed at the subject water. In this field the following lectures are provided, their scope being expressed in semester hours (one semester hour is one hour per week for 13 weeks).

for all civil engineering students

hydrology	1
water and water pollution	1
hydraulics	3
water supply and wastewater engineering	2
sub-total	7

for all sanitary engineering students

hydraulics	2
geology	2
groundwater flow	2
general hygiene and epidemiology	2
chemistry of water and sewage	3
microbiology of water and sewage	2
distribution of drinking water, collection of sewage	2
drinking water purification	3
sewage treatment	3
solid refuse	1
air pollution	1
pumping equipment	1
sanitary engineering design	1
sub-total	25

for sanitary engineering students to choose,
according to their design work, at least   8
from

hydraulics, special subjects	1
hydrology, special subjects	4
water management	3
pollution and self-purification	1
thermal pollution	2
intake and impounding of surface water	1
groundwater recovery	2
artificial groundwater recharge	2
drinking water purification, special subjects	1
tertiary sewage treatment	1
industrial wastewater	2
artificial sludge treatment	1
water treatment for swimming pools	1
process engineering	2
Total	40 hours

Laboratory and design work in the field of sanitary engineering, expressed in full weeks, amount to

under-graduate level

chemistry laboratory	2
microbiology laboratory	1
unit operations laboratory	2
sub-total	5

graduate level

sanitary engineering laboratory	0 − 25	
sanitary engineering design	35 − 10	
sub-total		35
Total		40 weeks

Assuming that 1 semester-hour of lectures conforms with one week of work on the part of the student, the total amount of time devoted to the sanitary engineering speciality adds up to 80 weeks or nearly 2 years, leaving a little over 3 years for the basic civil engineering education.

Next to the formal education described above, the Chair for Sanitary Engineering organizes on a voluntary basis colloquia and excursions. Colloquia are held once or twice a month during a full afternoon, are attended by staff, students and experts from outside the University and are often concluded by a joint drink and meal in one of the students' clubs. During these colloquia teaching, organizational and management problems are straightened out, students give talks about their design and laboratory work and guest speakers lecture on sanitary engineering projects at home and abroad, problems of environmental pollution, career opportunities in and outside the Netherlands, etc. Inland excursions with a duration of ½ to 2 days are held about 10 times a year, while every year there is one excursion abroad of 1 to 1½ weeks. With the Netherlands in the centre of the most densely populated and heavily industrialized part of Western Europe, subjects for those foreign excursions abound with the consequence that most students attend 2 or 3 of them to broaden their views.

3. Post-experience courses in sanitary engineering

In 1959 a start was made with refresher courses in sanitary engineering, meant for engineers, physicists, chemists, biologists, etc., working in this field. Subject matter and ways of approach were based on a graduate education, but this was not a prerequisite for admission, these courses also being open to people without an academic education but with a wide practical experience in the course material to be dealt with. In the beginning various set-ups were tried for these courses, but after a few years the most attractive system emerged: a course devoted to a quite limited subject and held in general, during 5 successive days, in exceptional cases over a slightly longer (6 or 7 days) or a slightly shorter duration (3 or 4 days). On each day 4 to 5 hours of lectures were given, followed by extensive discussions. At the end of the course, the participants were provided with elaborate lecture notes.

Many courses have been held about subjects taken from the field of building or industrial hygiene. As most important courses in the field of water may be mentioned

sewage, 5 courses — collection, mechanical treatment, biological treatment, sludge treatment, design of treatment works — each of one week duration, spread over a period of 2–3 years;

drinking water, 11 courses — chemistry (3 weeks), micro-biology (2 weeks), groundwater recovery, intake and storage of river water, rapid filtration, slow sand filtration, chemical water purification, design of purification works, pumping stations, long-distance transport lines, distribution and storage — mostly of one week duration and spread over a period of about 5 years;

swimming pools, one week duration;

salt and brackish groundwater, 2 courses each of one week.

Interest in these courses is overwhelming. To provide good opportunities for discussion, the number of participants is limited to 40, but this entails that many a course has to be given 3 or 4 times. It also means that whole series have to be repeated after a few years.

Twice a year 1½ day courses (7 lecture hours) are held at the Department of Civil Engineering, University of Technology, Delft, about modern development in the field of water supply (during Christmas recess) and in the field of sewage and industrial wastes (during Easter recess). The courses are meant as eye-openers and enjoy a large audience, for each course attracts somewhere between 200 and 300 participants. The lectures are published, first in the Dutch Journal H_2O and afterwards as a separate booklet, mostly in a more extensive form than actually given. The courses on drinking water started as far back as 1948, meaning that about modern water supply practice a book of 23 volumes is now available. The courses on sewage began only in 1966 and this book is still limited to 6 volumes. The subjects dealt with during the last six years are

drinking water — the biology of water supply; rapid filtration; the technology of water purification; from good to better water; design of water purification plants; quality control in public water supplies.

wastewater — sewage discharge into the sea; sludge treatment; the technology of aeration; recreation and water pollution; wastewater, now and in the future; the oxidation ditch.

4. Post-graduate courses in sanitary engineering

In 1957 the Netherlands University Foundation for International Cooperation (Nuffic) together with the Delft University of Technology founded the 'International Courses in Hydraulic and Sanitary Engineering' in Delft, the Netherlands. These courses have a duration of 11 months, are given in the English language and are meant for graduates in engineering or related fields with at least 3 years of practical experience. The number of participants for each course is limited to about 20, meaning that every year a number of applicants have to be turned down.

During the academic year 1971–1972, a total of 9 courses will be given, 5 in hydraulic engineering (tidal and coastal engineering; reclamation; rivers and navigation works; experimental and computational hydraulics; hydraulic structures), 1 for hydrologists and 3 in sanitary engineering. Next to this a number of seminars of about 5 weeks duration will be held, dealing with management problems in water resources or with highly specialized subjects. The courses in sanitary engineering can be subdivided as follows:

1. International Course in Sanitary Engineering I, dealing with the control of water pollution and the provision of drinking and industrial water in densely populated industrial areas;

2. International Courses in Sanitary Engineering II, dealing with drinking water supply, sanitation and health administration in the agricultural and initial industrial phases of development;

3. International Course in Science and Technology, dealing with the chemical and biological aspects of environmental problems and especially meant to provide chemists and biologists working in this field with a specialized post-graduate environmental training.

At the time of writing this report, the last mentioned course has not yet been given in consequence of the many teaching problems that must still be solved. With the International Courses in Sanitary Engineering on the other hand, much experience has been obtained from which a more or less definite pattern has emerged. Nowadays these courses are composed of:

(a) about 650 hours of lectures, dealing with the following subjects:
Review of sanitary engineering, Principles of human physiology, Determinants of health and disease, Tropical housing,
Public health administration, Water-borne diseases, Environmental chemistry and microbiology, General biology, Hydrobiology, Chemistry and microbiology of water and sewage, Unit operations of water and sewage treatment, Hydrology, Hydraulics, Ground-water flow, recovery and recharge, Intake and impounding of surface water, Pumping stations, Design of water-supply networks, Design of drinking water purifications plants, Design of networks for wastewater collection, Rural water supply, Design of sewage treatment plants, Design of rural waste treatment plants, Marine waste disposal, Selected subjects in water-supply management, Control and automation, Management of water resources, Industrial waste problems, Radio-active wastes, Marine waste disposal, Collection and disposal of solid wastes, Atmospheric pollution, Rural water supply, Corrosion, Statistics, Principles of engineering decisions;

(b) about 180 hours of additional lectures on optional subjects (e.g. hydrology, groundwater flow and groundwater recovery);

(c) a total of 40 half-day sessions of laboratory work in chemistry (9 days), micro-biology (8 days) and unit operations (3 days);

(d) about 40 days of design work, on water distribution, wastewater collection, unit operations and ending with a large design either in the field of water supply or wastewater treatment;

(e) roughly 30 days of excursions, about 10 one-day excursions in the Netherlands and its near surroundings and one excursion of 3 weeks in Western Europe.

During the academic year many examinations are held on separate subjects, while the courses are concluded by a final examination. Those who have fulfilled all requirements and pass the final examinations are awarded a Diploma; otherwise they receive upon request a Certificate of Attendance.

As already mentioned above, the International Courses in Sanitary Engineering command much interest. Up till October 1971 the number of participants amounted to 333 (of whom 257 received a diploma), coming from 80 different countries. The teaching staff is also international in composition, its 10 full-time and 35 part-time members coming from 10 countries.

5. *Future developments*

The Netherlands of today have a little over 13 million inhabitants and many industries, discharging wastewater with a pollutional load equivalent to that of 37 million people. Of the last mentioned figure, however, 25 million inhabitant-equivalent originate from sugar, potato flour and straw board production in the north-eastern part of the country, where the pollution problem must be solved by internal reorganization of the production process, decreasing or eliminating altogether the discharge of liquid wastes. This leaves an amount of 25 million inhabitant-equivalent to be dealt with, of which today

6.4 million receive full biological treatment;

1.2 million receive mechanical treatment;

4.8 million are discharged into the sea.

In the year 1985 the population of the Netherlands will have risen to 16 million people and industrial wastewater production (excluding the north-eastern part of the country) to that of 14 million people, together 30 million inhabitant-equivalent. For the North Sea and inland waters alike, point discharge of raw sewage will no longer be accepted in the affluent society of tomorrow and at least 25 of the 30 million inhabitant-equivalent mentioned above must receive full biological treatment. This means that in the next 15 years treatment plants with a capacity of roughly 18 million inhabitant-equivalent have to be built and operated. For expert design, construction, maintenance and operation of these works, many people are necessary, requiring an extension of existing training and educational opportunities at nearly all levels.

For the basic education of artisans, a large number of so-called lower technical schools are available in the Netherlands, while the specialist education of pipe layers and pipe fitters, machine fitters, plant mechanics and electricians, etc., is obtained by a combination of in-service training and part-time schooling. For this group of people no man-power difficulties are expected for the near future. In the long run, however, the general trend from blue to white collar workers might cause problems, arising from the demand for more respect and a higher remuneration for this category.

At this very moment there is a shortage of technicians in nearly all fields. Experienced draughtsmen and surveyors are very difficult to get and in many cases their work is done by recently graduated technical engineers. With an improvement in educational opportunities and the growing contempt for manual work, this situation will greatly change for the better in the near future.

Without any doubt, technical engineers in the Netherlands are in the shortest supply while their education gives practically no attention to sanitary engineering. For the lack of interest in attending a Technical College, many reasons can be given, the most important being that the official duration of this education over a period of 15 years is only 2 years shorter than that for a M.Sc. in Eng. degree, while the difference in pay is great. In the author's opinion, the best solution for this problem is an integration of Technical Colleges and Universities of Technology into one institute for post-secondary technical education, where after one common year of study participants with a more practical aptitude receive a further 3-year education in know-how (also with respect to sanitary engineering) and those with a more theoretical ability a 4-year education in know-why and research. Both categories, however, receive the same type of degree and will have the same career opportunities. To accomplish this goal, many years are necessary with the result that the shortage of technical engineers will continue for a long time to come, thus forcing many younger graduates to do work for which they have received no training.

As already mentioned, extensive opportunities for graduate education in the field of environmental health and engineering exist in the Netherlands. To offer more diverse programs, cooperation between the various institutes is necessary, preceded by coordination and streamlining of several courses. Some work has already been done in this respect, but notwithstanding the small size of the Netherlands — meaning that all people live close together — it will still take many years before a workable solution is obtained. Career opportunities for civil engineers specialized in sanitary engineering are excellent at the moment and will remain good for the next decade, or even decades while technical engineers and technicians remain in short supply. For chemists, physicists and micro-biologists on the other hand, it is already difficult to find a suitable job, while for the future a sizeable surplus must be expected.

Post-experience courses in the Netherlands satisfy an ever increasing demand and it may be expected that in future they will show a further growth in the number of subjects and participants. To satisfy the needs of the affluent society of tomorrow with its claims for a better environment, they must be supplemented, however, by post-graduate courses on a national level. With the present shortage of expert manpower, it will be extremely difficult if not impossible to obtain the necessary staff and in the opinion of the author, organization of such courses has to wait till the end of this decade. The international post-graduate courses in sanitary engineering will continue to flourish and will be of ever increasing help in satisfying the growing demand for specialists all over the world. The number of participants from the Netherlands, however, is and must remain small so that for the national needs these courses are only of limited value.

Discussion by E. J. Middlebrooks and R. G. Snider

The universities in the United States are still operating under the illusion that it is possible to train a jack-of-all-trades that will be acceptable to practically any area of Civil Engineering. The absurdity of such an assumption has been demonstrated many times by the confusion and disillusionment created when graduates leave a program having been told that their capabilities are unlimited in the field of Civil Engineering. Not only is it necessary for the baccalaureate graduates of Civil Engineering programs to return to the university to obtain a Master's Degree in the speciality area in which he wishes to work, but many employers have become disillusioned with the educational institutions. This is certainly true in the area of Sanitary Engineering education in the United States where very little background is provided for the average undergraduate in Civil Engineering. All knowledgeable organizations hire only graduates of advanced degree programs.

The trend in the United States toward advanced degrees to obtain competence in a specialty area appears to be a logical solution to a very complex subject; however, there are many governmental organizations being limited because of this trend. For example, state water pollution control agencies are notorious for low salary scales, and many were and still are having difficulty in attracting well trained personnel. State agency salaries are not competitive at the bachelor's level, and when an individual invests in another year of education to obtain a Master's Degree, he has eliminated himself from the market that is in greatest need of his services. The inability of state agencies to attract competent personnel results in a weak agency that will continue to be weak because of overall ineffectiveness. In view of the needs of state agencies, a better trained baccalaureate graduate seems appropriate. Another obvious answer to the state agencies' dilemma is to have the states bring salary scales within competitive ranges. However, with the present downward or limited upward trend in state budgets, it is unlikely that any significant changes will be made in salary adjustment. Therefore, if we as educators in the United States are to solve a problem that has plagued us for several decades, it is essential that we think beyond the present system of Civil Engineering education.

The "option" system employed in Civil Engineering programs in the past had significant merit, and the revival of such a system appears very logical. Not only would the baccalaureate graduate have more specialized knowledge, but he would also be far better prepared to enter graduate studies in his specialty area.

It is interesting to note that regardless of specialty area all Civil Engineering students in the programs at the University of Technology, Delft, complete at least as many Sanitary Engineering (Water Supply and Wastewater Engineering) courses as do the supposed jack-of-all-trades Civil Engineering graduate of the United States institutions. If the Sanitary Engineering specialty area is selected at Delft, the student is then exposed to a very substantial program.

The Netherlands program is based on a five year study schedule which results in a Master's Degree, and engineering educators in the United States have suggested that such a program be followed in this country. However, the majority of the educational institutions in the United States have maintained the four year curriculum, and the curriculum has been skeletonized to such an extent that most of the graduates are almost useless upon graduation until they have acquired enough experience to be of service to the employer. Also practically none of the undergraduate students complete a Civil Engineering program in the United States in the prescribed four year period. The majority of the students require an additional year to complete the program because of inadequate preparation prior to entering the program. Many concessions have been made to accommodate unprepared students, and this has further weakened undergraduate education. Therefore, it seems only logical that the educational institutions in the United States would accept the fact that a truly professional engineering degree cannot be obtained in four academic years.

The University at Delft also is to be congratulated on their foresight and recognition of the need for refresher courses in Sanitary Engineering for various personnel that are working in the field. This program originated in 1959 and is one of the earliest, if not the earliest, refresher course in Sanitary Engineering. In the United States it has just become accepted as a necessary part of education to develop extensive postexperience courses in Sanitary Engineering. Within the past five years, there has been a tremendous number of special short courses and refresher type courses offered at many of the American universities; however, very few of these programs have been organized in such a precise manner as presented in the Netherlands. Perhaps this lack of precision in organizing the courses can be attributed to the more geographical situation and needs for water pollution control in the United States than in the Netherlands. Nevertheless, the recognition of the need for such programs certainly indicates a knowledgeable staff and outstanding leadership in the educational field in the Netherlands. Two years prior to the start of short course activity in the Netherlands the International Courses in Hydrology and Sanitary Engineering were also started, and these activities have had a tremendous impact on Sanitary Engineering education throughout the world.

The administrative control of Sanitary Engineering educational activities in the Netherlands is enviable in view of the problems that have occurred in the United States with the recent massive increase in the number of programs offering various water quality and water pollution control

39

programs. This proliferation of programs in the United States has been supported by the Environmental Protection Agency through grants to individual institutions that provide not only stipends for student support but some assistance in hiring faculty and purchasing equipment and supplies for developing a laboratory. Although an attempt has been made to monitor and exercise some control over the quality of these various programs, it has been practically impossible because of the limited staff that are available to the training grants activities in the Environmental Protection Agency. Also the autonomy assumed by various universities makes it very difficult to exercise significant control over these programs except by threat to remove support. The graduates that have resulted from such programs may be numerous enough to have caused a significant dilution in the quality of the graduates now working in water pollution control activities.

The small population and geographical area in the Netherlands certainly lends itself to better control of the educational activities as well as a better guide to the number of graduates that are needed in the water pollution control field. Whereas, in the United States with the responsibility of water pollution control divided among so many agencies, it is very difficult to obtain meaningful information as to the number of graduates that are needed. Although the recent reorganization resulting in the Environmental Protection Agency was supposed to consolidate water pollution activities, administration of the reponsibilities has not improved. However, a concerted effort is being made toward determining manpower needs and establishing a priority for the types of educational programs and activities that should be supported.

Again in a small country such as the Netherlands it is apparently much easier to establish standards and predict the water pollution control activities that will be required to control pollution throughout the country. At the present time in the United States it is very difficult to obtain an opinion, much less a consensus of opinion, as to what will be required in the future for water pollution control in this country. Until such decisions are made it is a very difficult, if not impossible task, to predict the professional manpower requirements in numbers or type in the field of water pollution control in the United States.

WASTEWATER QUALITY CRITERIA

RALPH PORGES

Delaware River Basin Commission

P.O. Box 360, Tenton, New Jersey, USA

INTRODUCTION

It is fitting that we should be discussing wastewater quality criteria here in Jerusalem. One of the earliest and perhaps simplest descriptions of a polluted river may be found in the Bible where is is observed, 'And the fish that were in the river died and the river became foul.'[1] It is interesting to speculate upon the alertness of these early peoples who recorded these events and to wonder about the rationale underlying the observations without the cloak of the then local mysticisms.

Wastes were generally not discharged to sewers until about the middle of the nineteenth century, although it is recognized that earlier in some locations human and kitchen wastes were dumped in gutters and subsequently flushed away. The more common practice was land application or to employ the cesspool, the privy, and the dump. The soil of large cities became foully saturated by the mid-1800s necessitating the development, by mainly unknown and unheralded inventors, of the water-carriage system. Municipal and industrial wastes from the growing technological society soon gave rise to polluted waters. Man's subsequent efforts to control this portion of his environment is the subject of our deliberations.

Stream Standards Versus Effluent Requirements

Stream conditions are intertwined with the discharges that bring about effects on the stream, a logical causal association. Nevertheless, there is a real separation and it is emphasized that this presentation will be directed to *wastewater criteria* as opposed to *stream criteria*.

Stream criteria give direction and policy guidance to the regulatory agency. The objectives as defined by stream criteria usually are based upon protected water uses and the aesthetic or general stream value. Then, regulations governing waste discharges can be tailored to meet the criteria.

Wastewater criteria are the tools for water management by the regulatory agency. Effective control of waste discharges is dependent upon clear and specific limitations. They may be based, in part, upon stream quality objectives, the size of the receiving stream, the accumulative effects upon the environment, aesthetics, and treatment technology. Wastewater criteria are the guidelines that permit municipal and industrial dischargers and their engineers to design process controls and treatment facilities.

New concepts directed towards stream anti-degradation and maximum feasible treatment will tend further to restrict discharge of pollutants and, in many instances, lead virtually to banning of the discharge of certain pollutants.

Background

An extensive review of past information and literature pertaining to wastewater criteria is beyond the scope of this presentation; volumes have been written on the

41

subject, yet reference to some specific source material is indicated. Probably the best known is the excellent State of California, USA compilation – 'Water Quality Criteria'.[2] Another is the Report of the Committee on Water Quality Criteria[3] issued by the U.S. Federal Water Quality Administration, now the Environmental Protection Agency (EPA). Reference is also made to the Section on 'Administration, Stream Standards and Surveys' of the Annual Literature Review,[4] Water Pollution Control Federation, which summarizes annually the current status of standards.

Virtually all nations employ the concept of, or have established, stream criteria, many of which incorporate effluent limits. And many countries have followed along logically to more comprehensive wastewater requirements.

Between waste treatment criteria and subsequent achievement of clean water lies a vast gulf. Universally, there appears to be a breakdown somewhere in the system since abatement has not kept pace with waste discharges even though the understanding and technology have been available. The socio-economic influences and pressures, intertwined amid regulatory agencies and the dischargers of wastes, appear to slow down or even disrupt the planned program to achieve clean waters. Some may call this the politics of waste treatment.

In recent years, particularly in the USA, there has been an awakening of the public which is reflected in a new political stance. A flurry of laws and, at least, gestures towards adequate funding and enforcement is a measurement of our times. Yet waste treatment is not an overnight phenomenon and no matter what laws or funds are available, it takes time to determine treatment needs, develop plans, order equipment, build facilities and operate them efficiently. It also takes persistence, patience, good technical data and often expensive legal preparation to overcome discharger resistance.

DEVELOPMENT OF WASTEWATER CRITERIA

Wastewater criteria evolved from various concepts of which virtually all, to some extent, are currently in use

Most are founded on ways and means whereby wastes could be voided without violating the capacity of the stream to convey and ultimately assimilate the discharged material. Other considerations as aesthetics and treatment technology play a role. It is desirable to review these concepts since problem areas and needed study emanate from past efforts to develop suitable approaches.

Dilution Criteria

One of the most practical applications used early in the science of sanitary engineering was the dilution concept which involved an effluent criterium correlated with stream flow. It provided, in essence, a guideline to limit the amount of BOD discharged based upon available dilution. Probably the first (1887) such use was where for each 1000 persons contributing sewage at the Chicago Drainage Canal Project[5] 3.3 cfs (0.0934M^3/sec) of dilution water was indicated. This was, however, a value designed only to prevent nuisances. Through practice it evolved that generally six cfs (0.1699M^3/sec) would be needed for most streams to handle the waste discharged by 1000 population equivalents. This was a real step forward since the pounds of BOD per 1000 population equivalents could be determined readily providing a mechanism for reasonable control by interposing a treatment plant or by increasing available dilution. At a dissolved oxygen seminar[6] several years ago, it was confessed that any good practicing engineer would apply this rule-of-thumb method against the conclusions reached by the more mathematical approaches. If there was a wide divergence, the assumptions and mathematics were reviewed to ascertain where errors might exist.

Probably the best known effluent requirements are those in the Eighth Report of the Royal Commission of Sewage Disposal, 1913, recommending a BOD not exceeding 20 mg/l and suspended solids not exceeding 30 mg/l based upon normal dilution of English streams.[7] With a 500-fold dilution, crude sewage could be discharged. In recent years, these effluent requirements in some instances are being reduced to 10 mg/l BOD and 10 mg/l suspended solids[8] and there are rumors of five and five.

Assimilative Capacity

The monumental Streeter-Phelps studies[9] along the Ohio River during 1914–1915 gave rise to a better understanding of the stream assimilative capacity for receiving organic wastes. This permitted the engineer to set effluent requirements on BOD that allowed use of the stream capacity similar to any other natural resource. In its application, a required degree of wastewater treatment based upon the raw load was generally specified. The approach exploited, and maybe over-exploited, the available water resources; and by no means is it intended to belittle the economics of waste treatment. Insufficient attention was generally paid to growth, which, with a fixed degree of treatment, resulted in an increasing poundage of pollutants and eventually brought about stream degradation. There appeared to be a built-in pollution mechanism.

Allocations

One can find past reference[10] to allocation of the stream assimilative capacity as an effluent requirement, yet little was done to implement this concept until recently. It appears that the mathematical manipulations had to await the advent of the computer and associated river modeling for practical application when applied to BOD. However, any parameter can be allocated. In the Delaware River Basin[11] an allocation procedure was devised, based on equitable apportionment of the resource to each of the users, which would not permit creeping degradation by population and industrial growth. The allocation in terms of pounds per day of first-stage oxygen demand was given to each discharger in proportion to the raw wastes produced thus requiring equal percent treatment. The allocation is relatively fixed; normal growth must be handled within that allocation. The sum of the allocations plus a reserve for new waste sources or unusual growth of existing sources, equals the assimilative capacity of the stream. When the reserve is exhausted, all allocations will be recalled, a new reserve established and smaller allocations given which will increase the required percentage of treatment. In this fashion the stream objectives will be protected in perpetuity by upgrading wastewater effluents. Full confirmation of this program as measured by stream upgrading must await the passage of time, yet preliminary indications are promising. Generally, there is a paucity of record clearly relating the removal of waste loads to stream improvement.

Negligibles

Efforts to establish standards of purity for effluents has continued for many years; as long ago as 1868, the Rivers Pollution Commission in Great Britain proposed limits for a wide variety of substances.[12] The difficulty of developing wastewater criteria for many substances, let alone a few, is obvious and probably accounts for the lack of resolution of this vexing problem. In an effort to overcome the specifics, most waste criteria include a broad spectrum of prohibitions. Virtually all criteria contain the requirement that only 'negligible' amounts of certain substances may be present in a discharge. These negligibles usually encompass floating material, settleable material, debris, oil, toxic substances, or substances that may produce color, odor, or taint fish. Defining the term 'negligible' in quantitative terms is becoming a necessity in any comprehensive water management program.

Industrial Waste Guides

The United States Public Health Service, in the late 1930s and early 1940s developed guides[13] establishing norms for industrial processing and the waste derived therefrom. These were prepared fundamentally for field engineers in evaluating industrial establishments. Many employed these and other guides for establishing industrial waste effluent requirements. Since these criteria are unrelated to the stream condition, they are of little value where the stream capacity can be exceeded by the treatment norm. In addition, fixed norms would be contrary to improving waste management and waste treatment technology. They are of value in establishing a base line against which to compare housekeeping practices of similar plants and to show and promote what can be accomplished by progressive industrial water management. These guides are being revived by EPA to be used again as a gross measure of industrial waste control.

Economic Criteria

Economic approaches are being recommended as possible methods of controlling pollution.[14] One of these approaches is the so-called effluent charge touted as the most rational approach for the solution of the problem. The intent is to set charges, or financial penalties, on an effluent at such high levels that it becomes more economical to treat wastes than to dump them raw or partially treated to the stream. Another method might be to create 'rights' which would be sold to the highest bidder. These methods might appear merely a means of legalizing discharges, or licensing polluters, because the economics of a given situation could change so precipitously that it would be cheaper to discharge raw waste than to treat. Another danger of effluent charges is that maintenance of suitable stream conditions may be relegated to a secondary role with collection of revenues taking precedent. Also, the establishment of the administrative mechanism seems a real problem.

There are, nevertheless, certain considerations to an effluent charge that merits its use. If there is interposed a regional waste treatment facility between the waste source and the protected river, the waste effluent charge for each establishment can then be set so as to provide for cost of its treatment at the regional facility. This approach is valid and effective. Effluent charges for surveillance, monitoring and administration would also have merit if recognized as a way to support the regulatory agency.

An interesting variation is the approach taken in the American State of Vermont, where effluent charges are assessed during the time it takes to upgrade to meet the standards and while the waste discharges utilize the public resource.[16] When standards are met, the charges cease.

Anti-Degradation

Another restraint on discharges in the USA has been brought about by the anti-degradation concept embodied in the Federal water pollution control law. Here, the assimilative capacity of the stream is not considered available for planned discharge but is reserved for natural or inadvertent demands. Under this concept, the most technically feasible treatment must be imposed and existing high quality streams cannot be degraded unless justified by social and economic reasons. What is technically feasible escapes precise definition although it could conceivably mean in some instances no polluting discharges at all. Even today, some wastewaters are evaporated and ultimately handled as a solid waste. Perhaps the maximum technically feasible degree of pollution control with complete banning of some waste discharges is essential to protect our environment.

DISCUSSION

Wastewater criteria are, or should be, the principle implementing mechanism of the regulatory agency in its water quality management program. Generally, dischargers do not readily accept the restraints of effluent requirements and usually favor imprecise stream criteria or objectives. It is easy to hide behind dilution and it is difficult to prove the effect of one waste on stream quality in the presence of other discharges.

Effluent requirements, once established, should remain relatively firm so as to permit the discharger to function under the requirements. However, in a dynamic society with population growth and constantly increasing industrialization, it is not possible to remain long at a given point. In the near past, the emphasis was on stream criteria with a secondary consideration to effluent requirements. Now, there is need for the establishment of effluent requirements and this is a logical and progressive step. Even more effort will be forthcoming to institute quantitative allocations and other limitations on discharges where numerous waste sources exist, assimilative capacity is limited, stream uses are endangered, or the environment imperiled. For example, the Delaware River Basin Commission in its standards[19] established five direct effluent requirements. The first item required that all wastes shall receive a minimum of secondary treatment defined as 85 percent BOD removal. The second requirement is that wastes containing pathogens shall be effectively chlorinated while the third item states that the discharge shall not create a menace to public health or safety. The fourth requirement is that discharges shall not contain more than the usual 'negligibles'. As the fifth requirement, where the above actions fail to meet the stream quality objectives, the waste assimilative capacity shall be allocated.

The Commission is now defining quantitatively what is meant by 'negligible'. Public hearings have been held on implementation guides significantly limiting discharges of suspended solids; temperature; oil; debris; pH; ammonia; general toxicity in terms of bioassay results; toxic heavy metals; color; odor; phosphates and total dissolved solids. Pesticides are likely to be added to the list.

There continues acute need to define more closely discharge limits and to establish prohibitions of substances that cannot be discharged. More information is required on toxicity, both acute and chronic. Much study is needed on the synergistic interrelationships of toxic substances on the environment and the effects of biological magnification of toxics on man's food chain. Appropriate bioassays will limit those substances having immediate harmful influence on stream environment. Long term toxicity may best be controlled by limiting specific substances. Bioassays of wastes can shed light on toxicity although there is room for a simple, reproducible, bioassay test procedure. The discharge of toxic heavy metals and persistent chlorinated hydrocarbons will be limited more drastically for their effects on life. Insufficient attention has been directed to altering the industrial processes and to seeking substitutes for the more harmful chemicals.

The effect of pathogens, particularly viruses, in effluents needs more attention. There is a paucity of epidemiological evidence linking illness with recreational contact with polluted waters. To chlorinate or not to chlorinate effluents has not been resolved. I have heard mentioned that fish are healthier in streams receiving only chlorinated discharges. These are wide areas for investigations.

Effluent criteria must be enforced equitably by an aggressive regulatory agency. A requisite is up-to-date monitoring. A short time or instantaneous 'BOD' test that can be monitored is highly desirable. Research developments on the total oxygen demand (TOD) and total organic carbon (TOC) appear fruitful.

The establishment of wastewater criteria should consider sampling frequency, variability, and seasonal influence. Should a mixing zone be incorporated into requirements? When setting forth a poundage BOD that can be discharged by a given source, sampling frequency is essential so as to recognize the normal variations that occur. As an occasional, short-term overload to a body of water may not bring about drastic changes, is it unreasonable to utilize weekly or monthly averages?

Another factor is the probability of distribution of a parameter around a given average value. The variability of a waste is axiomatic. It may be desirable to establish a range of values or an average value coupled with a limit which cannot be exceeded.

In the case of some biological waste treatment plants, lower efficiencies due to the cold weather may also require recognition in the criteria. For example, the Delaware River Basin Commission may permit a lowering of its 85 percent BOD reduction requirement to 75 percent during the winter if cold temperatures induce a loss in treatment efficiency. By no means is it implied that treatment facilities should not be operated at a maximum efficiency all year.

The social and political implications of waste treatment are becoming increasingly important. In many countries, the general public appears to appreciate a clean stream. Most people feel that it is their right as part of their well being and general welfare. In the USA, the public now demands clean streams as a right resulting in a political necessity for public officials to support stringent water pollution control programs. Perhaps this is the ingredient required to bridge the gap between wastewater treatment and achieving stream objectives.

Effluent requirements have been geared to stream effects which implies utilization of available dilution, a rational approach until the capacity is exceeded. However, this approach must be considered as an interim step since nonpoint sources of pollution, such as storm drainage, farm run-off and other miscellaneous and inadvertent sources may utilize so much of the stream capacity that future requirements will necessitate virtually complete elimination of controllable pollution.

More and more pressure will be placed upon industry for conservation of resources leading to reuse of materials and process water. This will bring about a reduction in gross water use so that closed-cycle operation will be feasible and economical. One industry in the Delaware Basin plans an 80-fold reduction in water use from 75 to less than 1 mgd. How much further is it to eliminate the discharge completely? Several Basin industries already have closed-cycle operations. Treatment costs may not increase with the reuse concept. By closed-cycle operation, one paper mill saves $40,000 annually by incorporation of the waste solids into the product which provides for any additional costs of the closed system.[17]

Municipal wastes will be scrutinized with the aim of meeting more stringent wastewater criteria and reducing their effect on the environment. Those industries connected to community systems will be expected to undergo the same transition that is visualized for industries discharging directly to water courses. Residual wastes from such industries will be forced to meet higher requirements prior to discharge to the municipal system.

Communities might well consider a dual water system, one of high quality for potable needs, with complete reuse to a safe but lower quality water for uses as bathing, showers, irrigation, lawn watering and fire fighting.[18] This latter effluent could be further recycled or possibly sold for industrial use. In this fashion, virtually complete re-circulation within a community could be safely practiced.

These projections may be visionary, but we are living in dynamic times. There is an awakening to human values and the people are expressing their wishes for a cleaner and healthier environment. Those in engineering and related disciplines have a responsibility in environmental control. It is time the engineer concerned himself more with protection of the whole environment for human needs rather than designing the cheapest disposal system, in a narrow sense, to meet an immediate waste problem.

I would be remiss not to mention the need for an effective administrative mechanism to implement wastewater criteria. Not only must each country and appropriate subdivision be so organized, but each major river basin could benefit from an international, interstate, or river basin agency endowed with appropriate authority to meet its responsibilities.

SUMMARY

Wastewater criteria are the implementing tools for the water management program of regulatory agencies. Wastewater criteria have been based upon dilution, stream assimilation capacity, allocations, discharge of only negligible quantities, treatment technology, waste guides, economic criteria, and anti-degradation.

For the next few decades, more emphasis will be placed upon wastewater criteria which will become more comprehensive and demanding. In addition to the common criteria, exotics as toxicity, radioactivity, heat, nutrients, pesticides and viruses will receive emphasis. Study and research are needed to establish the finite limits. Bioassays will be a must, yet more adaptable and practical bioassays are essential. New procedures and tests to expand current monitoring are needed.

Definitely, the allocation concept will be expanded to encompass a myriad of substances. There will be attempts at economic controls by effluent charges.

Current wastewater criteria must nevertheless be considered only an interim step as future restrictions will ultimately lead to virtual prohibition of controllable, polluting discharges.

To achieve the desired clean streams from the wastewater criteria will require administrative mechanisms adequately endowed with authority and responsibility to carry out the required tasks. Engineers should re-direct their objectives to protection of the human environment rather than to over-emphasis on minimal costs. An awakened public, reflected in a new political stance towards pollution control, may provide the catalyst to bring about implementation of stricter wastewater criteria needed to clean up our waterways.

REFERENCES

1. 'Book of Exodus.' *Bible* Chapter VII, Verse 21.
2. McKee, J. E. and Wolf, H. W., 'Water Quality Criteria.' *State Water Quality Control Board Publication No. 3-A*, Sacramento, Calif. (1963).
3. Federal Water Pollution Control Administration, 'Report of the Committee on Water Quality Criteria.' Federal Water Pollution Control Administration, U.S. Dept. of the Interior, Wash., D.C. (Apr. 1, 1968).
4. Porges, R. and Gross, S., 'Administration, Stream Standards, and Surveys'. Annual Literature Review, *Jour. Water Poll. Control Fed., 43,* 6, 1384 (June 1971).
5. Streeter, H. W., 'Standards of Stream Sanitation'. *Sewage Works Journal, XII,* 1, 115 (Jan. 1949).
6. 'Oxygen Relationships in Streams.' Edited by R. Porges. Proc. of Water Supply and Water Pollution Program of the San. Eng. Center, Oct. 30 – Nov. 1, 1957, Technical Report W58–2, Robert A. Taft San. Eng. Center, Cincinnati, Ohio (Mar. 1958).
7. Schroepfer, G. J., 'An Analysis of Stream Pollution and Stream Standards.' *Sewage Works Jour., XIV,* 5, 1030 (Sept. 1942).
8. Jenkins, S. H., 'British Water Pollution Control.' *Environmental Science & Technology, 4,* 3, 204 (Mar. 1970).
9. Streeter, H. W., 'A Study of the Pollution and Natural Purification of The Ohio River.' U.S. Public Health Service, *Public Health Bull. No. 146,* Wash., D.C. (Feb. 1925).
10. Agar, C. C, "Stream Standards and Their Practical Application.' *Sewage Works Journal, XXI,* 6, 1050 (Nov. 1949).
11. Delaware River Basin Commission, 'Allocation of Stream Capacity for Acceptance of Waste Discharges.' Planning Division Staff Paper No. 111 (July 1971), Trenton, N.J.
12. Lovett, M., 'Standards: Some Pros and Cons.' Control of River Pollution, *Jour. Inst. of Sewage Purif.,* 4, 409 (1958).
13. 'Ohio River Pollution Control.' U. S. Public Health Service, In Two Parts, Supplements to Part II, Wash., D.C. (Aug. 27, 1943).
14. Johnson, E. L., 'A Study in the Economics of Water Quality Management.' *Water Resources Research, 13,* 2, 2nd Quarter 1967.
15. Kneese, A. V., 'Strategies for Environmental Management.' Harvard University Press, Cambridge, Mass. (1971).
16. Arthur D. Little, Inc., 'Economic Incentives in Water Quality Management.' Final Report – Phase One, Submitted to Dept. of Water Resources, Agency of Environmental Conservation, State of Vermont (Dec. 1970).
17. Hammann, C. C., 'Operating Experience With a Total Water Reuse System.' NCASI Middle Atlantic Regional Meeting, Phila., Pa. June 12, 1969.
18. Porges, R., 'Water Pollution and the Potable Water Dilemma.' *Jour. Water Pollution Control Fed., 39,* 10, 1613 (Oct. 1967).
19. Delaware River Basin Commission, 'Administrative Manual – Part III Basin Regulations – Water Quality.' Trenton, N.J., (Nov. 24, 1970).

"Extraction and Identification of Micro Contaminants
of Waters"

Marvin C. Goldberg

The objective of extraction of organic solutes from water is to increase the organic solute concentration. Analytical instruments require about 10 μg, of material to meet the lower limits of gas chromatographs mass spectrometric sensitivity. However, many components of an environmental waterway occur at concentration levels from 1 pg/l to 1 mg/l. A concentration factor of 10^9 is necessary to adequately explore the existence of micro components. The best way to realize this need is to use solvent extraction. Two continuous liquid – liquid extractors have been designed to accomplish that purpose (Fig. 1, Fig. 2). They allow an unlimited input of the water feed solution and choice of any given immiscible solvent as the extraction agent.

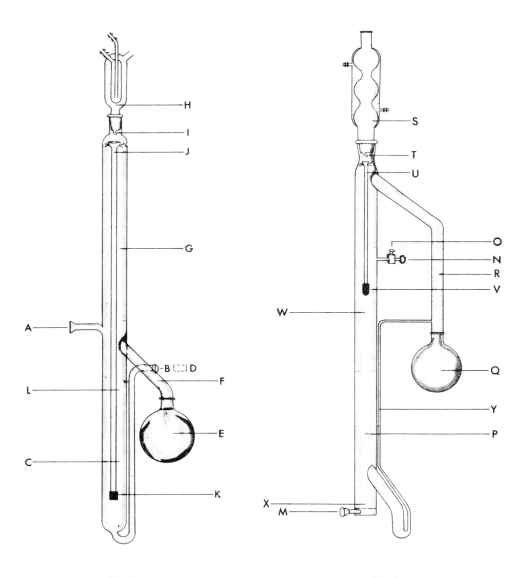

Fig. 1.   Fig. 2.

Operation

The extractor design is a two-cycle system. The water cycle is continuous flow. Water enters at *A* and exits at *B*. In so doing it passes through chamber *C* which is half filled with solvent. A stopcock, *D* can be provided to regulate the water flow rate.

The second cycle is a solvent cycle. This system is closed in that the solvent cycles exclusively in the extractor. The 500 ml bulb *E* contains pure, non-miscible, organic solvent. This solvent is gently boiled and vapor rises in area *F* up through the upper extractor tube *G* into reflux condenser *H*. At this point it is liquified and falls off the drip tang *I* into funnel *J*. The long funnel stem sets up a hydraulic head sufficient to drive a solvent through a porous glass frit at *K*. The frit homogenizes the solvent resulting in fine beadlike particles which form as an emulsion as they rise through the water in chamber *C*. This emulsion extracts organic solutes during the period of water-solvent contact. The emulsion separates in the extractors lower neck *L* and the solvent-solute mixture spills over connection tube *F* into boiling flask *E*.

The closed solvent system cycles fresh solvent from the boiling flask into the extractor. After extracting the organic solutes the "loaded" solvent is returned to the boiling flask thus collecting and concentrating the extracted solutes in flask *E* but always supplying fresh solvent to the extraction chamber at *C*.

Figure 2 depicts the solvent-heavier-than-water, extractor which is similar in principle of operation to the solvent lighter-than-water extractor but somewhat different in design. It operates as follows:

Water enters at *M* and exits at *N*. The stopcock *O* (optional) regulates flow rate. While in the lower neck of the extractor *P* the water flows through the extraction solvent.

The solvent cycle, which is closed to the extractor, starts by being vaporized in bulb *Q*. The vapor rises in arm *R* to condensor *S* where the vapor liquifies and drops from drip tang T to funnel *U*. The solvent under a hydraulic head, is forced through frit *V* where it emulsifies and drops through the upper extractor neck *W*. Extraction of the water takes place at the interface between the emulsified solvent and the water. A stirring bar at *X* (optional) stirs the solvent-water mix. The solvent separates in the lower half of the extractor and flows through tube *Y* into bulb *Q*. Organic solutes extracted from water then concentrate in bulb *Q*.

The extractors vary in efficiency as regards the removal of any given solute from the water system. The mean range of efficiency is from 20 to 100 percent. Factors regulating extractor efficiency can be correlated with the molecular interaction constants such as *a* in the Van der Waals equation. A more convenient index is dipole moment (μ) which is obtained from the Claussius, Mossetti, Debye equation.

$$\frac{(D-1)}{(D+2)} \frac{3M}{4\pi d} = N\alpha + N\frac{\mu^2}{3kT}$$

where D = the dielectric constant of the medium, M is the molecular weight, α is the polarizability, d is the density, N is the number of molecules, k is the boltzman constant and T is the temperature.

The difference between the dipole moment of the solvent and the dipole moment of the solute has been found to correlate with the percent extraction efficiency. As this difference approaches zero the extractor efficiency comes to an optimum. This effect is illustrated by figure 3. Dipole moment calculations for a large number of organic compounds are available in the literature. A common method of approximation using dipole moment values for organic groups can also be used to obtain the dipole moment for molecules where there is a lack of experimental data.

After extraction of this small amount of mixed solute it is necessary to remove as much solvent as possible. Use of Kuderna-Danish concentration apparatus will conveniently reduce the solvent level from several hundred milliliters to about 5 ml. A micro vigreaux still will allow further reduction in concentration to a volume of 0.5 ml. This amount is optimum for analytical purposes.

Analyses are tailored to gas chromatography, coupled to mass spectroscopy where concentrations from 10-100 ng/per component are determined. Several types of mass spectrographs can be used.

A rapid data scanning computer program has been used wherein data of unit resolution between mass numbers 12 and 200 are coded. The computer scans the 200-400 mass range and compares an unknown spectrum to a library of 7000 compounds. The existence of non-existence of each mass peak at a given mass number is matched to the unknown compound. The data received identify the unknown solute, or indicate the nearest molecular configurations in the library and the degree of mismatch.

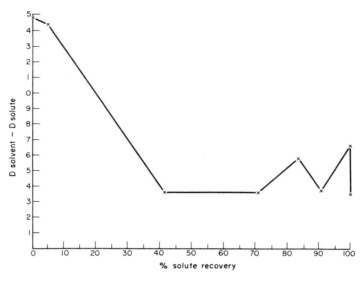

Fig. 3.

Effects attributed to urbanization have been studied in this manner. Protein decomposition products indicative of urban pollution have been traced to determine urban effects. Indole was identified in a river downstream from a large metropolitan area at about .42 mg/kg concentration in the bottom sediment. Another compound put into the river by urban usage is p-cresol. At the same location the concentration range of the sediment was from < .02 to 6.1 mg p-cresol per Kg sediment.

H. Jones, Australia.

What are the author's views on the announced future USA policy concerning zero discharge of pollutants?

Reply

Zero discharge of pollutants could consider high treatment levels to meet this objective, or it could consider closed cycle operation with containment and treatment of any bleed-off or blowdown. It appears to me that inadvertent and natural pollution reaching most of our rivers may utilize much of the assimilative capacity such that virtually complete removal of additional wastes will become essential if the rivers of the world are to be maintained or recovered to serve useful needs. In my contacts with industry, it seems that most industrial plants have embarked or could embark immediately on water conservation programs which would reduce drastically the volume and possibly the strength of waste emanating from the processes. It is only a short although significant step between maximum conservation and completely closed cycle. I believe zero discharge is a goal that can be achieved ultimately by most industrial plants.

J.B. Sprague, Canada.

Free chlorine is very harmful to fish at low concentrations. We found that all concentrations of free chlorine were lethal to trout, down to the lowest concentration we could measure at that time, about 0.02 mg/l although low concentrations took longer to kill the fish, sometimes 4 or 5 weeks.

Recent work in the USA also shows that such trace concentrations of chlorine have lethal and sublethal effects on fish.

It may be desirable to disinfect effluents by some way other than post-chlorination.

Reply

There are indications that the end products of chlorination might also have detrimental effects on people consuming such waters, particularly if those waters have been subjected to pollution. As I have indicated in my presentation, to chlorinate or not to chlorinate has not been resolved and is a subject worthy of much research.

L. Coin, France.

Hydraulic dilution as a factor in controlling discharges ought to be abandoned and replaced by greater reliance on the dissolved oxygen balance. French law now permits quality control on the basis of water usage. This may necessitate complete or partial purification or exceptionally high standards depending on the circumstances. General tests to determine the toxicity of discharges may be used. Disinfection of effluents rather than sterilisation seems to be a practical means of control. Compliance with stated objectives can only result from political decisions.

Reply

It is agreed that effective environmental control can only result from political decisions. I concur fully that dilution must not be considered a factor in meeting effluent requirements. One of the principal problems we face is the determination of what is a process stream, what is excess water and, even, what is cooling water. The effluent requirements specified by the regulatory agencies generally are being applied to the process stream so as to eliminate dilution water as a factor in meeting standards.

C. Lefrou, France.

Having defined one's objectives by a political decision it is necessary to find the best way to achieve them. Arising out of the consideration do the authors think that regulations should be based on flow or on load? In the case of a municipal treatment plant probably concentration is adequate since the flow is known. But if this is applied to an industrial discharge it may lead to disposal by dilution and discourage efficient treatment of effluents and reuse of water.

Reply

The question proposed is an extremely interesting one and something that we have considered in the Delaware River Basin. In the Estuary allocation program, the basic consideration is on the load discharged. Each of the dischargers in this stream reach were provided with an allocation in pounds per day of first stage BOD. This amount had no relation to flow. We encourage water conservation and reuse and generally no industry in the estuary will be penalized for a high concentration of discharge provided (1) it meets percent treatment requirement and (2) it meets its pound allocation. Elsewhere, in high quality stream reaches we do have concentration limitations of 50 mg/l and 100 mg/l BOD_5.

W.F. Lester, UK.

In which other river systems is the allocation concept to control pollution used? In the UK most sewage and industrial effluent is improved by 95% and, in the case of sewage from a higher strength than in the USA to a higher standard than 85% removal of BOD. Would not a greater improvement in quality of river water be achieved more economically by requiring readily treatable effluents to be processed to 95% and 99% purity and accepting a lower percentage improvement for more intractable effluents? Compared with the allocation concept, would river water quality not be better controlled by determining the use of the river and then the water quality criteria necessary for that use and hence the effluent standards necessary to achieve the water quality criteria? He did not consider industry could operate without a discharge. From its "enclosed" systems there must be a purge to the stream and this must carry some contaminants.

Reply

Mr. Lester's inquiry encompasses several questions. First, I am unaware of any other river system besides the Delaware where the allocation concept has been as extensively employed to control a pollution problem. It might be well to indicate the conditions leading to an allocation program. DRBC standards sets forth that protected water uses must be established, stream criteria developed to protect those uses, and effluent requirements specified so that the stream criteria are met. The standards also specify that no less than 85 percent treatment be provided. When the 85 percent treatment will not permit meeting the stream criteria and higher treatment levels are required, the allocation procedure comes into play whereby the capacity of the stream is shared equitably by all dischargers in proportion to the need based on raw load produced. Therefore, in all cases where dischargers are under allocation, treatment is in excess of 85 percent, currently approaching 90 percent, but could go higher if needed.

The original study on the Delaware River evaluated mathematically the least-cost approach which considered high treatment of some wastes and lower treatment for others. There were economic advantages to this method until treatment levels approached the 90-95 percent level for all wastes when the costs were about the same. The approach suggested by Mr. Lester may have merit with a suitable administrative mechanism for implementation.

The method suggested by Mr. Lester of establishing use of the river, stream criteria and effluent standards, is the approach that was taken by the DRBC.

It is indicated that industry must have a discharge. I am aware of several plants that operate on closed cycle as it pertains to liquid wastes. In other instances I feel that with proper treatment and recycling, the volume of purged material necessary to keep the process functioning could be handled by some method, possibly evaporation, so that the final residue would be a solid waste. I am sure that more and more attention will be given to closed system operations and that the handling of the blowdown will need to be faced as astutely pointed out by Mr. Lester.

E.D. Gonzalez, Spain.

What are the criteria for colour in streams? Is a limit for colour used in any river in the USA?

Reply

I am unaware of any color limit specified for streams. Because of the color characteristics of certain natural waters such as streams draining swamp areas, a color standard for streams might well be meaningless. DRBC standards set forth a limit in effluents of 100 color units.

N. Gruener, Israel.

What does the author mean by a Universal toxicological test?

Reply

In my report I indicated the need for a simple, reproducible, bioassay test procedure. Should such a test find universal application it would have additional merit. I have been interested in a substitute organism for fish in the bioassay as well as using an end-point other than death in 96 hours so as to make it more economical, simpler and reproducible. Invertebrates, which can be readily maintained such as amphipods, midge larvae, daphnia, tubifex and others could be considered as alternative test organisms.

S.W. Loewenson, Rhodesia

The authors suggest that criteria should continuously be upgraded with updated information. Industrialists will therefore have to periodically augment their treatment and equipment. This must engender some resistance and non-cooperation from the industrialist and makes difficulties for the administering agency. Is it not better to have high quality standards and allow exemptions for specific periods and specific conditions? Stream quality standards which utilise the dilution aspect and assimulative capacity of the stream are not applicable to rivers in Rhodesia which by climate vary in flow from raging torrents in the rainy season to dry sandy-river beds in the non rainy season. Pollution control must be achieved by effluent quality standards.

Reply

My dealings with the industrial community tend to support the contention that it is better to have high quality standards rather than to seek constant upgrading adjustments. However, should exemptions be universal for either time or conditions, the value of the high standards might be negated. A short while ago, an 85 percent treatment requirement as a minimum was considered an extremely high standard. Likewise, 90 percent treatment resulting from an allocation program was considered extremely high treatment. In the short period of some six years, these requirements are now common objectives and the U.S. Federal Government is targeting on much higher treatment requirements. Upgrading appears essential as the waste load increases and our goals become higher.

I am in agreement that effective pollution control can only be achieved by the establishment of effluent quality standards.

CRITERIA FOR DISCHARGE TO RIVERS, LAKES AND CONFINED COASTAL WATERS

KJELL BAALSRUD

Director, Norwegian Institute for Water Research, Royal Norwegian Council for Scientific and Industrial Research, Gaustadalleen 25, P.O. Box 260 Blindern, Oslo, Norway

The topic of water quality criteria and effluent standards is a highly controversial one. In the author's country, Norway, universally applied standards are not very popular. There is a general desire to examine each lake, river and fjord, as well as each discharge, on its own merits. It is believed that through individual examinations it is easier to arrive at satisfactory conditions than through a rigid system of standards.

However, with the increasing number of discharges, even in a small country like Norway, there is a need for certain guidelines for those who are judging these situations. Whether this can best be achieved through application of standards on quality of effluent discharges or by the use of standardized drainage systems and treatment plants, is an open question.

There are three important sides of the problem to consider. The first is the philosophy of quality criteria for discharge, the second is the scientific basis for criteria and the third one is the management aspect.

PHILOSOPHY

Waste is a product of man and his activities. The human being has always been the source of waste, but until recently the waste was of a type which was acceptable to nature and could be recycled. The present technical and industrial development had led to a number of new types of waste, and in addition the amount of waste has increased tremendously. Handling and deposition of waste has therefore become a major problem in modern countries.

As usually applied, civilization refers to developing and to the developed countries. The process of development has led to much waste and hence waste has become a sign of civilization. On the other hand civilization also means culture and responsibility and calls for a balanced and adjusted behaviour of each individual in the society.

It is, in my mind, obvious that much of the waste is being produced because we have not accepted responsibility for the consequences of modern technical development.

Some people consider the sewer the most destructive of all technical constructions. Through the sewer and sewerage system it has been possible to conduct waste water from isolated houses to places where it does not naturally belong. The desire to get the waste away from the houses is understandable and acceptable. But when this only leads to the same waste being deposited elsewhere, the real achievement is not very great.

In every society there are certain things one is not allowed to do, for practical or moral reasons. The same philosophy should be applied to a polluter. There are certain types of pollution which you should not be allowed to discharge. It ought to be universally accepted that nowhere in nature should it be possible to recognize typical sewage objects, like toilet paper, faeces, etc. Both men and animals are by nature clean: it is odd to consider that through culture and development we have reached a stage of lesser cleanliness than we had before.

Every time a water toilet is installed one should be obliged to contribute towards installing a sewage treatment plant. This obligation is not a difficult one to meet, because in most cases the share of the cost of sewage treatment plant for the owner of the house or appartment is much smaller than the cost for installation of water, water utilizers and the sewerage system. It is only because man has lost touch with his natural behaviour that he has accepted the use of water to transport waste from his home or his shop and discharge it somewhere else.

In most water pollution control considerations, we would like to base recommendations and regulations on a cost benefit evaluation. As far as certain wastes are concerned, for example identifiable objects from domestic wastes, the cost benefit principle should not be applied. For such wastes the proper pollution prevention measures should be taken irrespective of cost.

It would help technicians as well as people in government around the world if we show how to establish a criterion or standard of quality. No discharge should be allowed which makes it possible to recognize that the waste has a domestic origin.

A second standard to be applied universally could be: No waste material present in wastewater should be allowed to enter water if it is not there as a consequence of water utilization. The water power driven saw mills which operated in Norway for centuries and let all the saw dust go to the river, is one example of this principle. A modern example is an oil spill which is flushed away with water.

It is likely that a number of such standards could be set which in reality are prohibitions.

Another philosophy could be that in principle no discharge of polluted water is accepted. The polluter has no right, he is an intruder, and must follow conditions set by others, in this case society.

To have a 'no-pollution-is-accepted' principle is in accordance with recent developments in national and international bodies: a citizen is not allowed to do anything or discharge anything before it can be conclusively shown that the activity or the discharge is acceptable to society or the environment.

Even if it may take some time before such a principle is implemented in all pollution problems, it is important to note this developing trend. Until recently any party or any person who felt offended by a discharge had to prove that the discharge actually did cause damage or nuisance. Soon it will be the discharger or polluter who has to prove that his activity is tolerable.

In addition to being a simple or logical guiding principle this 'no-pollution' philosophy has the tremendous advantage that it makes the polluter responsible for the consequences of his activity.

Some people still believe that pollution control should be executed by new bodies as a new function in society. This is based on an assumption that the polluter has the right of way and that ill effects shall be avoided through some repair-like activity.

No one has proved that water pollution is a necessary consequence of technical development. Admittedly, what has happened for the last one hundred years does indicate that pollution has come to stay. It may, however, as well be that we right now are passing through a transitional phase of maximum pollution, and that during the next few decades, pollution will again become less and less severe.

Another way of expressing the philosophy is: pollution must be stopped at the source.

The source, of course, may be the end of a sewer or pipeline, but it should be within the technical structure which forms a part of the activity or the process.

It is obvious that the 'no-pollution' philosophy cannot be applied all at once. It seems that we have perhaps 10, or at the most 20, years during which we must achieve pollution control. If we do not have success in this regard during these years, and if industrial and technical development continues to go wild, quite likely we will have a lost cause.

SCIENCE

Criteria for discharges must be based on scientific knowledge of the environment, and of how the discharges may influence the health of the environment or create conflicts with other users of the water resources.

Our present knowledge of the composition of liquid waste from domestic, industrial and other sources is very limited indeed. In most cases the composition of the waste is given by a few general parameters. We know that there are vast numbers of different chemical substances in polluted waters. From the practical point of view it would be impossible for any set of criteria to take all these substances into account. The criteria must therefore either be based on certain parameters which are easy to measure, or on certain tests which the effluent water can be subjected to. The necessity for criteria for four important groups of polluting substances is discussed below.

Particulate matter

Wastewater may contain a considerable amount of particulate matter. At times water is used just for the purpose of removing solids, as for instance the washing of potatoes before they are processed. If the particulate matter is biologically inert, it may create few chemical and biological difficulties. The physical effect of changes in the colour of water and of deposition on the bottom are usually easy to cope with.

Sludge accumulation may create only local problems. The volume of particulate material from modern mining activities may, however, be large indeed. Criteria for the amount of particulate matter must be evaluated for each body of receiving water or even for each section of it. As particulate matter in general is rather easy and cheap to remove from effluents, there is no reason to accept such pollution in the environment. And hence it is likely that rather strict criteria would be preferred in most cases.

Organics

The organic matter in sewage was for half a century considered to be the most important pollutant. The well-advanced technology for removing organic matter through biological treatment processes has been the result of the feeling that it was necessary to control organic pollution. Degradable organic matter can only have an adverse effect on local areas, but the extent of the affected area is dependent upon the size of the discharge and the capacity of the receiving water. Even organics derived from the pulp and paper industry in most cases influence only fairly limited areas.

Removal of organic matter from sewage and food processing industries is rather easy and cheap to achieve, and for this reason there is in most places an acceptance of rather strict standards. As a matter of fact, it is possible to meet the requirement that the contents of degradable organics shall be so low that the effluent water without dilution will not lose all of its oxygen over a given period of time.

In those cases where large volumes of diluted water are available, it is possible to have less strict criteria for discharges. It is, however, possible that even in such cases the local demand for clean water will require biological treatment. Non-degradable organics will be mentioned below.

Nutrients

Nutrient elements such as phosphorus and nitrogen are today considered by many to be the most important component of discharges. The nutrients give rise to excessive growth of algae and higher plants in the receiving water, and thereby can change the nature and quality of fairly large water areas. In most cases the effect of eutrophication extends over larger areas than the effects of organic matter. During recent years a technology has been developed for removal of phosphates. We will know in the next few years whether this will give us satisfactory control of eutrophication. It should be borne in mind, however, that there are a number of nutritive substances. Through controlling the amount of one nutrient element only, we are not sure of getting satisfactory results. As phosphates, for chemical reasons, are the easiest element to control, emphasis has so far been put on this component.

The possibility of making eutrophication a benefit instead of a nuisance is being explored. The excessive production of organic matter in water could be used for feeding animals, and could for instance, stimulate the production of fish. Another idea is to harvest the plants and use them for domestic animals. However, it seems to be difficult to change the ecosystem in water and achieve such results. Agriculture in water seems to be much more difficult to handle than on land. It has been tried, but without economic success. It is not likely that we will be able to pollute the high seas with nutrients. But the local enclosed sea areas, as for instance the Baltic, are probably already influenced. We also know that the eutrophication is of major concern in the great lakes of North America. Hence, criteria for nutrients in discharges are highly desirable, and today's technology already makes it possible to apply fairly strict criteria as far as phosphates are concerned.

Substances with adverse biological effect

For the time being this group of polluting substances seems to cause a need for more urgent decisons than any of the others. The increasing release into the environment of substances which are directly toxic, which have adverse effects on organisms, or which are introduced into the food chain and thereby create long-term problems, is creating considerable concern. We may in this regard distinguish between short-term toxic substances and long-term bionegative substances.

Short-term toxicity is in most cases easy to measure, which means that we have the analytical instrument for applying control and setting criteria. Toxins which are easily destroyed in nature can be considered to be only a local problem. Toxins which are resistant to breakdown must be given particular attention and should not be accepted in discharges at all.

The latter group is included in the term micropollutants, which are an ill defined group of substances. As micropollutants we may consider substances which are broken down so slowly that they find a wide distribution in nature, and which in spite of being present in low concentrations may still exert adverse biological effects, either directly or through accumulation via the food chain. In many cases the immediate danger of micropollutants may be rather small, whereas continued release into the environment over many years may create serious problems in wide areas. DDT and PCB are examples of this. DDT has been in use for 3 decades, but it is only recently that it has been considered a danger to environment.

It is highly desirable as soon as possible to agree on substances which should be included in the term micropollutants and to apply rigid criteria for their release into the environment. For some substances, like mercury, cadmium, and lead, we may envisage a total prohibition against their release in waste waters. Although prohibition cannot be applied fully at once, it is likely that the industries could fairly quickly adjust their processes so as to meet this requirement. Anyhow, the presence in the environment of persistent substances with adverse biological effects is highly undesirable and measures against their release must be taken even if it creates some difficulties to industry and costs money.

In this regard it may be of particular importance to follow the activities and recommendations of the GESAMP committee established by several organizations of the UN-family in order to deal with pollution of the marine environment.

MANAGEMENT

Instead of having water as a natural resource of infinite capacity, we find ourselves in a position where water is a limited resource which urgently needs to be managed. It is up to the governments and international agencies as managers to maintain freshwater and seawater as resources useful to us and succeeding generations.

For the management of water a number of guiding principles must be developed and they must be based on international agreements. This is a subject of discussion in OECD's environment committee, and probably will be a topic for discussion at the UN-conference in Stockholm in 1972.

It is in this context of interest to observe the work of the Joint Group of Experts on the Scientific Aspects of Marine Pollution (GESAMP). The GESAMP committee is establishing lists of pollutants according to their importance and impact on man and the environment. It is likely that the criteria for deposition of waste which may reach the sea water, in the future may be based on lists as they are produced by this committee.

Another recent event which deserves attention was concerned with the protection of the North East Atlantic Ocean against pollution. Representatives of 12 countries met in Oslo in October 1971, and agreed on the contents of a regional convention on the control of marine pollution by dumping from ships and aircraft. Hopefully the convention will be signed by the countries involved and in operation by the time this conference takes place in June 1972. Details of the convention will then be available. According to the accepted proposal dumping of certain substances listed in an annex will be prohibited. Other substances listed in another annex shall not be dumped in significant quantities without a special permit from the appropriate national authority.

As international control of water pollution most likely must be based on regional agreements, it seems that the Oslo Convention will be an important milestone in the progress of this work.

Management of water resources must take into account social, economic, scientific and technical aspects. Long lists of items which should be considered in each case have already been established. But management cannot be based on criteria which will give ideal conditions. A way of management must be found which can be used in practice, and still takes as many of the items on these lists as possible into account.

For the management of water resources, establishment of criteria and standards may be of great help. In applying them, two difficulties should be kept in mind. The first difficulty is to establish criteria which are universal. The only substances which lend themselves to universal criteria are those which are stable or are only very slowly degraded in the environment. That means such substances as are called micropollutants in this paper. For all the other categories of polluting material some universal ideas can be applied, but criteria and standards should be worked out for the regions or even smaller areas.

The other difficulty is the temporary value of any set of criteria. The present situation with regard to water pollution is intimately connected with the structure and the activities of modern society. Water pollution abatement and control cannot be reached without a certain impact on the future development. This calls for a stepwise procedure through frequent re-evaluation of guidelines and criteria.

Two or three decades ahead we expect a significant increase in world population and a marked increase in the standard of living in large areas of the world. It is easy to visualize that this will necessitate an almost complete control of pollution. A proper aim of our efforts should therefore be that within 20 years all discharge or polluting material will be stopped. Pollution must be stopped at the source. We know that the impact of man on the environment cannot be avoided and we may not be able to fully preclude the presence of certain substances in water. However, they should be kept at a level and released in such a way that they cannot be considered as water pollution in the conventional meaning of this term.

It follows from this that any criteria of standards that are being set universally or locally should be subject to repeated review and changes. For a number of categories of polluting matter we can start with fairly tolerant criteria and year by year decrease the acceptable levels of various polluting substances. Considering the work which in any case must lie behind establishment of criteria and standards, it is recommended that criteria be used as much as possible because they are easier to change and review than legal standards. This, however, may be different in various areas and where various substances are concerned. We may in the end find that in certain cases we want criteria which are guiding principles and in other cases we want rigid standards.

Discussion *by* Edward J. Cleary
Adjunct Professor of Environmental Health Engineering
University of Cincinnati, Ohio (45219) USA

The question of the administration of a regional pollution-control program involving eight sovereign states, thousands of industries, and the interests of some 20 million inhabitants — almost one-tenth the population of the United States — that occupy the Ohio River Valley have long occupied my attention.

There is growing conviction that public policy for pollution control must be referenced to goals beyond those traditionally identified with sanitation significance. Although difficulties have been experienced in defining specifically what the aim should be, support has gravitated around the notion that the end to be sought should be maintenance of quality conditions in waterways that take into account the satisfaction of a variety of needs.

The means for realization of this objective have embraced the concept of regulation of sewage and industrial wastewater discharges based on a balancing of equities in the use of water resources. A rationale for the balancing of equities stems from ancient legal doctrine which asserts that water resources are common property, the use of which is held in trust by the governing entity, and is to be administered by this entity in the public interest.

From this viewpoint it may be reasoned that an appropriate goal for pollution-control endeavors is the achievement of conditions that maximize public welfare in the use of waterways. The validity of this objective appears to be widely supported. Only the manner of its pursuit poses perplexing questions of how to devise an accommodation with the many conflicting claims as to what is best in the public interest.

The basic decision to be confronted is: Should pollution-control strategy rest on prohibiting the discharge of any amount of any substance in any body of water or should it be based on the conduct of a quality management program that seeks to extract optimum utility from the use of water?

I would suggest that the latter strategy — management of quality — offers the greatest potential for efficient accommodation to the task at hand. However, one of the essential components in development of a management program is the establishment of standards referenced to waterway uses, and this must be regarded as a rigorous undertaking.

There is a distinction to be made between criteria and standards and this deserves some elaboration. Criteria are measures of scientific appraisal for judging the suitability of water for various uses. Standards, on the other hand, are detailed specifications of quality conditions appropriate to the attainment of optimum social use of a particular waterway or stretch thereof.

Standards can be derived only in part from scientific criteria. A standard referenced to optimum use calls for analysis of both tangible and intangible values. This requires the assembly and evaluation of data that will identify the social penalties and benefits associated with alternative choice of standards, along with public hearings to ascertain preferences and the practicability of their attainment. Thus, a decision on standards should be regarded as a public policy issue and not simply a matter of bureaucratic pronouncement.

Implementation of a Management Program

There are certain requirements that may be specified as universally applicable to all waterways that represent the minimum condition that must be met regardless of any other circumstances, and they can be justified solely on the basis of preventing gross and indiscriminate degradation.

There are also supplementary requirements, the determination of which requires detailed study and adaptation to prevailing social and economic circumstances. The virtue of proceeding on a step-by-step basis is that urgent necessities for halting obvious pollution need not be paced to the more painstaking process of promulgating detailed specifications.

This two-stage procedure was employed in the initiation of the Ohio River Valley program. Minimum standards that were established called for all wastewaters to be treated or otherwise modified prior to discharge so as to attain the following conditions in the receiving waters:

Freedom from anything that will settle to form putrescent or otherwise objectionable sludge deposits;

Freedom from floating debris, scum, and other floating materials in amounts sufficient to be unsightly or deleterious;

Freedom from materials producing color or odor in such degree as to create a nuisance; and

Freedom from substances in concentrations or combinations that may be toxic or otherwise harmful to human, animal, or aquatic-life.

Following the promulgation of these basic or minimum standards attention was then given to the determination of supplementary requirements.

See "The ORSANCO Story — Water Quality Management in the Ohio Valley under an Interstate Compact," by Edward J. Cleary and published by The John Hopkins Press, Baltimore, Maryland (21218) USA, in 1967. The Ohio Valley experience has relevance to the comment of Dr. Baalsrud that control of water pollution on international rivers most likely should be based on regional agreements.

The values society has placed on the use of waterways for the disposal of wastes has undergone a vast change during the past half century, and notably so within the last decade. In terms of historical perspective three schools of thought may be identified. 1. Toleration of indifferences about pollution until disease outbreaks occurred or until a nuisance was created. 2. At the other extreme there have been the advocates of pristine purity for water resources. 3. In between these viewpoints on public policy with respect to pollution control — which have ranged from acceptances of foulness to aspirations for recapturing the natural integrity of water resources — we find the emergence of what might be termed a rational accommodation.

This accommodation rests on the concept that social-welfare goals of pollution control endeavors should be to extract maximum utility from the use of waterways.

Achievement of such a goal obviously calls for a rigorous exercise in resources allocation and management. Engineers and economists have provided technology and analytical procedures for the execution of this kind of endeavor. However, there is a lack of appropriate administrative machinery and decision-making apparatus for this purpose.

My conclusion is that accomplishment of water-quality management will reflect ingenuity in interfacing the contributions from the "discoverers of fact" and the "arbiters of value." Discovery of fact as a basis for action falls within the realm of the scientists, physicians, biologists, engineers and economists. The arbitration of values is in the province of the political process, which determines policy and the establishment of priorities, financial resources and institutional arrangements for the achievement of goals.

C. Lefrou, France

Is it practical to control the discharge of micro-pollutants by establishing limits of concentration of these substances?

S.W. Loewenson, Rhodesia.

The authors suggest that criteria should continuously be upgraded with updated information. Industrialists will therefore have to periodically augment their treatment and equipment. This must engender some resistance and non-cooperation from the industrialist and makes difficulties for the administering agency. Is it not better to have high quality standards and allow exemptions for specific periods and specific conditions? Stream quality standards which utilise the dilution aspect and assimulative capacity of the stream are not applicable to rivers in Rhodesia which by climate vary in flow from raging torrents in the rainy season to dry sandy-river beds in the non rainy season. Pollution control must be achieved by effluent quality standards.

L. Coin, France.

Hydraulic dilution as a factor in controlling discharges ought to be abandoned and replaced by greater reliance on the dissolved oxygen balance. French law now permits quality control on the basis of water usage. This may necessitate complete or partial purification or exceptionally high standards depending on the circumstances. General tests to determine the toxicity of discharges may be used. Disinfection of effluents rather than sterilisation seems to be a practical means of control. Compliance with stated objectives can only result from political decisions.

Reply

As a general comment on the floor discussion I would say that criteria can be used or misused and that there is a danger in establishing criteria or standards that are too tolerant. The criteria may then have the opposite effect and permit unnecessary severe pollution. Criteria, when they are set must therefore be strict. This is also the reason why I particularly mentioned sewage objects and other striking examples of pollution as being outside the scope of cost-benefit reasoning.

Mr. Lefrou has raised a most interesting question on micropollutants and I would like to stress that micropollutants are the particular concern in the marine environment. Marine biologists are alert and marine pollution was a very important topic of the recent U.N. conference in Stockholm. The Oslo Convention against dumping is another example of Government concern for marine pollution. Micropollutants should be stopped at the source. It is extremely important that river management takes sea water pollution into account, otherwise a river water quality programme may fail.

TRACE METAL CHARACTERIZATION BY ANODIC STRIPPING VOLTAMMETRY

K. H. MANCY

Professor of Environmental Chemistry, School of Public Health,
The University of Michigan, Ann Arbor, Michigan, 48104

INTRODUCTION

In situ measurement and species characterization are considered to be among the most challenging aspects of water quality analysis. *In situ* measurement, by which the species of interest are determined in their own environment, eliminates the difficult, if not impossible, task of collecting representative and valid samples. Generally speaking, it would be ideal to have submersible sensor systems capable of monitoring water quality parameters both in location and time.

Species characterization, on the other hand, signifies a definition of the chemical form, i.e. in the case of metal analysis this entails a definition of the various forms of metallic species. It is highly significant to provide analytical data in terms of the oxidation state and type of species, i.e. free or complexed, instead of a total metal concentration. This is apparent in view of the fact that chemical reactivity of a given metal is dependent on its form and not necessarily on its total concentration. Furthermore, the availability of metal micronutrients to the biological system or their toxicity effects are dependent on the metal species rather than a total metal content.

Among the various methods of analysis, electroanalytical procedures offer both advantages of *in situ* measurement and species characterization. In addition, electroanalytical techniques offer the advantages of: (a) high sensitivity required for trace analysis, e.g. Anodic Stripping Voltammetry (ASV), (b) suitability for field operations, e.g. portability and ease of operations and, (c) suitability for automation and continuous monitoring purposes. Typical examples include potentiometric membrane electrodes, conventional or pulse polarography, voltammetric membrane electrodes and ASV.

In this paper an attempt is made to illustrate the analytical feasibility of ASV for trace metal characterization in the aquatic environment. Recent modifications of ASV and their application are presented.

PRINCIPLE

ASV involves two consecutive steps: (a) electrolytic separation and concentration of metals to form a deposit or an amalgam on the working electrode, and (b) the dissolution (stripping) of the deposit. The separation step, commonly known as the plating step, may be done quantitatively or arranged to separate a reproducible fraction of the electroactive species. This can be done at controlled potential and time of electrolysis and at reproducible mixing conditions in the test solution.

The stripping step is usually done in an oxygen free, unstirred solution, by applying a

63

potential — either constant or varying linearly with time — of a magnitude sufficient to drive the reverse electrolysis reactions. The resultant current-potential voltammogram provides the analytical information of interest. Figure 1 shows a typical ASV voltammogram obtained with a sea water sample from the Mediterranean. Quantitative measurements are done by integrating the current-time curves (coulometry at controlled potential) or by peak current measurement (chronoamperometry with potential sweep). The peak potential, E_p, is characteristic of the type of metal. Several modifications of the separation and stripping steps have been reported.[1]

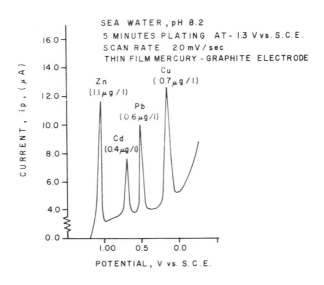

Fig. 1. Typical ASV Voltammogram

ASV measurements are usually done by means of a three-electrode system consisting of an indicator or working electrode, a reference electrode and an auxiliary or counter electrode. The counter electrode serves to perform controlled potential electrolysis and eliminates problems associated with ohmic drop in the test solution.

Different types of working electrodes have been reported, e.g. hanging-drop mercury electrodes by Kemula[2] or Gerisher[3] and thin mercury films on solid electrodes.[4] The latter type of electrode offers higher sensitivity and better resolution.[4,5] Thin mercury film-graphite electrodes have been recently applied for metal analysis in sea water,[6] fresh waters,[5,6] wastewaters[5] and drinking waters.[7]

Metals measurable by ASV include bismuth, copper, lead, indium, cadmium, nickel, cobalt, zinc and others. Complications due to the formation of intermetallic complexes with nickel and cobalt have been noted.[6] Copper, lead and cadmium in natural and wastewaters have been easily determined on a routine basis using ASV techniques.[5] Mercury was determined by ASV techniques using graphite electrodes.[8] There are no data available, however, on the application of this technique for mercury analysis in environmental samples.

Electrode reaction equations for ASV with thin film mercury-graphite electrode have been postulated[4] to be as follows:

$$i_p = nFAl\,v\;\frac{\phi}{e}\;C^o_R \qquad (1)$$

and

$$E_p = E_o + \frac{2.3}{\phi} \log \frac{dlv\phi}{D_o} \qquad (2)$$

where i_p is the peak current and n is the number of electrons transferred in the electrode reaction, F is the Faraday, A is the electrode surface area, l is the thickness of the mercury film, v is the rate of potential sweep, ϕ is equal to nF/RT, C^o_R is the concentration of the reduced metal in the electrode, E_o is the formal electrode potential, d is the thickness of the solution boundary layer and D_o is the diffusivity coefficient of the oxidized species in the test solution.

Equation 1 predicts that i_p is a linear function of C^o_R which represents a reproducible fraction of the concentration of the metal of interest in the test solution. Equation 2 predicts that E_p is equal to the formal electrode potential, E_o, plus a constant term for a given electrode system. In general, regardless of which type of electrode is employed, the analytical results obtained from i_p and E_p measurements are always dependent in some way on the rate of potential sweep and electrode geometry.

DIFFERENTIAL TECHNIQUES

Recent investigations at the Environmental Chemistry Laboratory at the University of Michigan[9] resulted in the development of a differential ASV system, particularly suitable for metal analysis in natural and wastewaters. The differential ASV technique utilizes a four-electrode system which includes two working electrodes instead of one working electrode as previously described. In addition to the two working electrodes, W_1 and W_2, the system included a counter and a reference electrode. Electrode reactions at W_1 and W_2 were followed simultaneously and independently by means of a dual potentiostatic circuit. The differences between current measurement from each of the working electrodes W_1 and W_2 were obtained by means of a summing circuit using an operational amplifier.[9]

Measurement procedures were done by plating the metals of interest on W_1 at the appropriate cathodic potentials (e.g. $- 1.3$ V *vs*. S.C.E.). In the meantime, the second working electrode, W_2, was kept at a more positive potential (e.g. $- 0.1$ V *vs*. S.C.E.). In the stripping step, both W_1 and W_2 were subjected to the same potential sweep. The resultant differential current was then recorded as a function of potential. Since the metals were plated only on W_1, the recorded differential current was the net voltammetric current of the metal dissolution reactions.

An obvious advantage of differential ASV is the increase in sensitivity by eliminating background currents and thus allowing for a greater signal amplification. Perhaps a more significant advantage of differential ASV is the ability to perform measurement in presence of dissolved oxygen. Compensation for background currents of oxygen reduction is afforded by the use of the second working electrode, W_2. Accordingly, it is possible to eliminate the deaeration process usually done by bubbling nitrogen in the test solution.

The ability of differential ASV for metal analysis in presence of dissolved oxygen makes it a more suitable technique for field operations and for *in situ* analysis. Needless to say, differential ASV techniques using four-electrode systems are more easily automated than conventional techniques using three-electrode systems.

An automated, continuous flow, differential ASV system has been described by Schimpff.[9,10]   At present this system is under investigation. To simplify the compilation, storage and analysis of data, the differential current output should be converted to digital form and recorded on a magnetic tape system. By means of a computer program, the recorded data i_p and E_p values can be easily analyzed and printed out in a tabulated form.

SPECIES CHARACTERIZATION

Methods Based on Direct ASV Measurement:

ASV conveniently classifies metallic species into: (a) free hydrated ions, (b) labile metal complexes (MLL) and (c) nonlabile metal complexes (MNL). This classification can be understood upon considering the following metal-ligands interaction (arbitrarily adopting a ligand number of one),

$$yM \quad + \quad xLL \; \rightleftharpoons \; My\,(LL)x \tag{3}$$

or

$$yM \quad + \quad xNL \; \rightleftharpoons \; My\,(NL)x \tag{4}$$

where LL and NL represent labile and nonlabile ligands, respectively. MLL and MNL can be defined in terms of their rates of dissociation,

$$-\frac{d(MLL)}{dt} = k_d^{LL}\,(MLL) \tag{5}$$

and

$$-\frac{d(MNL)}{dt} = k_d^{NL}\,(MNL) \tag{6}$$

where $k_d{}^{LL}$ and $k_d{}^{NL}$ are the dissociation constants for labile and nonlabile complexes. In so far as the electrode system is concerned, the following cathodic reactions may occur during the plating step:

$$M \xrightarrow{\;k_c\;} M - Hg \tag{7}$$

$$MLL \underset{}{\overset{k_d^{LL}}{\rightleftharpoons}} LL \; + \; M \xrightarrow{\;k_c\;} M - Hg \tag{8}$$

$$MNL \underset{}{\overset{k_d^{NL}}{\rightleftharpoons}} NL \; + \; M \xrightarrow{\;k_c\;} M - Hg \tag{9}$$

Equation 7 represents the case of a free metal ion reduced directly to metal amalgam, where k_c is the reaction rate constant or simply the cell constant. This step involves diffusion in a finite boundary layer followed by an electron transfer at the electrode surface. In the case of metal complexes, the complex must undergo dissociation into the free metal and the respective ligand prior to the reduction of the metal on the electrode surface, as shown by equations 8 and 9. It follows then, that labile metal complexes are complexes which dissociate at a faster rate than the rate of plating of free metals, i.e. $k_d^{LL} > k_c$. Under these conditions the rates of plating of free metals and labile metal complexes are equal and ASV current measurement, i_p, cannot distinguish between the two forms.

Nonlabile complexes on the other hand, are defined as metal complexes which dissociate at a slower rate than the metal plating step, i.e. $k_d^{NL} < k_c$. Under these conditions the rate of dissociation of the metal complex is the rate limiting step, and a reduction of i_p is observed. On this basis, it is possible to differentiate between free metals and labile metal complexes on the one hand and nonlabile metal complexes on the other.

The differentiation between free metals and labile metal complexes can be made from the shift in the peak potential, E_p, and the variation of E_p with increasing rates of voltage sweep. This can be seen by considering the anodic reactions during a linear potential sweep. These are essentially the reverse of the plating reactions given previously by equations 7 and 8 and can be represented as follows:

$$M - Hg \xrightarrow{k_a} M \tag{10}$$

or
$$M - Hg \xrightarrow{k_a} M + LL \underset{}{\overset{k_f^{LL}}{\rightleftharpoons}} MLL \tag{11}$$

Equation 10 gives the stripping reaction which involves a charge transfer followed by diffusion of the oxidized free metal, and k_a signifies the reaction rate constant. If labile ligands are present in the test solution, they will bind the metals as they are stripped off the electrode surface. This is shown in equation 11 where k_f^{LL} is the reaction rate constant. If k_f^{LL} is large, a cathodic shift in E_p will occur and if k_f^{LL} is small, E_p will occur at the same potential of metal stripping in the absence of ligands.

The variations of i_p, E_p and width of current peaks with increasing rates of voltage sweep are used to give further information on the concentration, distribution and rates of formation of metal complexes. Further disscussions on the theory and interpretation of these electrode reactions including the effects of ligand adsorption or direct reduction of metal complexes on the electrode surface, can be found elsewhere.[6, 11, 12, 13]

Methods Based on ASV Titration Techniques:

In addition to the above procedures, ASV can be used to provide qualitative and quantitative information on metal-ligand interactions using titration techniques. This involves the addition of spikes of metal ions detectable by ASV (e.g. Cu, Pb, Cd, etc.), metal ions nondetectable by ASV (e.g. Ca, Mg, etc.), or strong acid (H^+). When a spike of metal ions is added to a water sample, it may displace other metals from their complexes; slowly be taken up by a nonlabile complex, or simply increase the concentration of free and labile metal complexes in the test solution. This technique can be used to study competitive metal-ligand interactions in the aqueous phase or metal exchange between water and sediments.

The technique is based on first determining the ASV voltammogram of the original sample. Short plating times, e.g. 5 minutes, are employed in order not to disturb equilibrium conditions in the sample. This is followed by adding a spike of the metal titrant after which sequential ASV runs are performed in order to determine the rate of attainment of equilibrium. Sequential spikes of increasing magnitude are then added, allowed to equilibrate and measured until an increase in signal, linear with the spike concentration, is observed. A typical example, reported by Matson,[6] is shown in Figure 2,

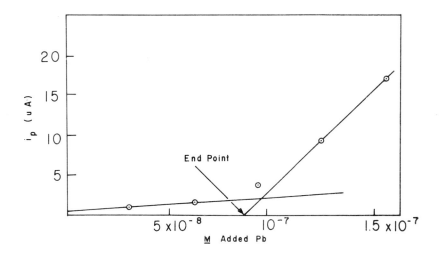

Fig. 2. Titration of Charles River water sample with lead

in which a sample of Charles River water was titrated with lead ions. The extrapolation of the linear portion of the spike concentration versus signal curve to zero signal will give the metal binding capacity of the test solution. This is essentially an estimate of the nonlabile ligands equivalent for that particular metal titrant in the sample. The curvature around the end point has been explained by the author[6] as due to the displacement of other metals from the ligands rather than a simple partition of the titrant metal between labile and nonlabile complexes. It is believed that this simple technique offers a rapid method for the estimation of metal binding capacity of natural and wastewaters.

Identification of the type of ligands in the test solution, responsible for metal complexation can be made by gel chromatographic techniques.[14] The procedure is based on the addition of a spike of metal ions to the water sample under investigation. After allowing enough time to reach equilibrium, the sample is fractionated by gel permeation techniques using Saphadex gels. Individual fractions are then acidified and the amount of metal in each fraction is determined by ASV.

This technique has been used to study copper binding capacity of secondary sewage effluent from Ann Arbor, Michigan.[13] The fractionation pattern obtained with SX100 cm column and G-50 medium revealed three major molecular weight fractions which exhibited strong affinity for copper. These fractions occurred in the column effluent at a high molecular weight fraction at the breakthrough volume of 300–400 ml, an intermediate molecular weight fraction at 500–700 ml and a low molecular weight fraction at 800–1000 ml.

COMPETITIVE METAL INTERACTIONS

One of the most controversial issues in the subject of aquatic metal distribution is how trace organic matter specifically can interact with individual metals, e.g. Cu, Cd, Pb, Co, Zn, etc. Furthermore, if this interaction occurs, does not the presence of Ca and Mg, naturally present in great excess, blur any complex-forming tendency of the organic functional groups? Recent research findings by O'Shea[12] give a rational answer to this question.

Figures 3 and 4 show typical titration curves illustrating the competitive interactions between Cu or Cd with Ca for humic acid ligands. ASV measurements were done using thin film mercury-graphite electrodes at controlled pH condition (maintained by bubbling appropriate partial pressure of CO_2). Water samples containing 5 ppb humic acid were titrated with Cd^{2+} or Cu^{2+} in presence and absence of Ca^{2+}. The calcium concentration was 40 ppm.

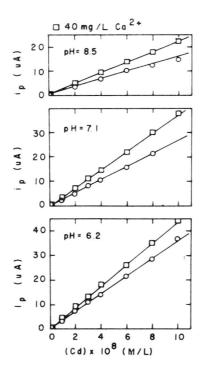

Fig. 3. Effect of Calcium on Cadmium – Humic Acid Interactions

Figure 3 shows the effect of calcium on the titration of humic acid with Cd^{2+} at three different pH values. The linearity of the i_p – *vs.* metal concentration plots is an indication of the presence of excess labile ligands or free metal ions. The latter possibility was discarded in view of characteristic cathodic shifts in E_p. Thus cadmium-humic acid system consists primarily of labile complex formation. Figure 3 also shows increased sensitivity (i_p) in the presence of calcium. This is an indication that calcium interacts with humic acid in such a way as to free cadmium from its complexes. This leads to the conclusion that cadmium forms essentially labile complexes with humic acid and calcium successfully competes for the same binding sites.

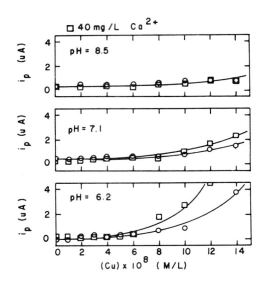

Fig. 4. Effect of Calcium on Copper – Humic Acid Interactions

The behavior of the cadmium-humic acid system is one extreme in the overall picture of trace metal-organic interactions.[12] The other extreme is shown by copper-humic acid system in Figure 4. The copper-humic acid titration curve is typical of metal-nonlabile ligand interaction. The lack of i_p signal at low copper concentrations is evidence of the formation of strong, nonlabile complexes. It is also noticed that nonlabile complex formation increases with increasing pH. After the nonlabile ligand is saturated, there is an increase in i_p signal with increasing metal concentrations. Contrary to the previous case, the presence of calcium exhibits no effect on copper-humic acid interactions. This is an indication that copper competes favorably with calcium for the formation of a nonlabile complex with humic acid.

ANALYTICAL SCHEME

A comprehensive definition of trace metals in the aquatic environment should account for the existence of the following forms:

1. Soluble species –
   (a) Free hydrated ions
   (b) Metal complexes with inorganic ligands (e.g. OH^-, CO_3^{2-}) or organic ligands (amines, proteins, vitamins, humic acids)
   (c) Metal-ligand competitive interactions with H^+, alkaline earth metals and heavy metals

2. Insoluble particulates —
   (a) Colloidal particulates of metal complexes or aggregates of hydrous metal oxides, or
   (b) Metal complexes adsorbed on suspended particles

The existence of metals in one form or another is highly dependent on temperature, pH, pE and ionic strength. Transformations from one form to another and accumulation and release by bottom sediments can be either chemically or biochemically mediated.

The following analytical scheme has been used for trace metal characterization in natural wastewaters:[5]

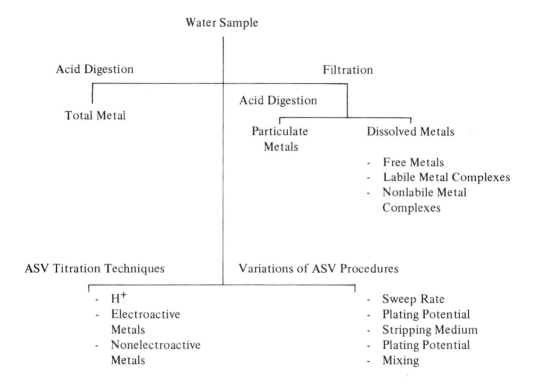

CONCLUSIONS

The above examples illustrate the analytical feasibility of ASV for trace metal characterization in the aquatic environment. The arbitrary classification of metals into 'free', 'labile complexes' and 'nonlabile complexes' cannot be considered as a true metal speciation. Yet it can be used to an advantage to give an insight into whether the metals of interest are present in free or complexed forms and what are the rates and magnitudes of metal complex formation. This is particularly useful in estimating the rates and amounts of metal binding by certain organics in natural and wastewaters and the competitive interactions between metals for organic complex formation.

The use of differential ASV techniques with four-electrode systems makes the techniques more adaptable for environment measurement. The ability of measurement in presence of dissolved oxygen and background current compensation extends the use of ASV techniques to *in situ* analysis and continuous monitoring applications.

REFERENCES

1. Farendicht E., *Electroanal. Chem.*, Bard, A. J., Editor 2, 53, Marcel Dekker, Inc. (1967).
2. Kemula W. and Kubik Z., *Anal. Chem. Acta.*, 18, 104 (1958).
3. Gerischer H., *Z. Phys. Chem.* (Leipzig), 202, 302 (1953).
4. Matson W. R., Roe D. K. and Carritt D. E., *Anal. Chem.*, 37, 1594 (1967).
5. Allen H. E., Matson W. R. and Mancy K. H., *J. Water Pollution Control Federation*, 42, 573 (1970).
6. Matson W. R., Ph.D. Dissertation, M.I.T., Cambridge, Massachusetts (1968).
7. Peterson T. L., Brant D. O. and Mancy K. H., Proc. 160th National Meeting, American Chemical Society (1970).
8. Perone S. and Kretlow, *Anal. Chem.*, 37, 968 (1965).
9. Schimpff W., Ph.D. Dissertation, The University of Michigan (1971).
10. Allen H. E. and Mancy K. H., Report in Press, the Environmental Chemistry Laboratory, School of Public Health, University of Michigan (1972).
11. Nicholson R. and Shain I., 'Theory of Stationary Electrode Polarography', *Anal. Chem.*, 36, 706 (1964).
12. O'Shea T., Ph.D. Dissertation, the University of Michigan, Ann Arbor, Michigan (1971).
13. Wopschall R. H. and Shain I., *Anal. Chem.*, 39, 1514 (1967).
14. Nolan M., School of Public Health, the University of Michigan, Ann Arbor, Michigan (Unpublished 1971).

SOME ANALYTICAL ASPECTS
OF THE BEHAVIOUR OF CERTAIN
POLLUTANTS IN AQUEOUS SYSTEMS

H. A. C. MONTGOMERY
Water Pollution Research Laboratory of the Department of the Environment,
Stevenage, Herts, SG1 1TH, England.

There is an increasing demand for information not merely about total concentrations of various pollutants, but also about the chemical and physical states in which they occur and about the ways in which the relative concentrations of the different states vary with time. An obvious example is the differentiation of inorganic from methyl-mercury, and another is the calculation of concentrations of free, un-ionized ammonia, for which a nomograph has recently been prepared (Montgomery and Stiff, in the press).

This paper consists mostly of a review and discussion of recent work at the Water Pollution Research Laboratory (WPRL) in which attempts have been made to differentiate the forms in which pollutants occur. Two of the projects discussed, the differentiation of cyanide and copper species, were undertaken in connection with studies of the toxicity of the pollutants to fish, and the third, an empirical assessment of concentrations of dissolved heavy metals in digesting sludge, is concerned with the toxicity of metals to bacteria. The projects are discussed together, principles and points of technique being illustrated by reference to all three.

Equilibrium concepts

The nomograph for free ammonia already referred to illustrates the principle that application of the laws of chemical equilibrium is desirable in the study of many systems, both to define the probable situation and to supplement chemical analysis. Equilibrium concepts and their applications in aqueous media are discussed in detail in a recent book by Stumm and Morgan (1970).

Equilibrium calculations may be useful in situations where the information is difficult or impossible to obtain by present analytical techniques. Examples are the calculation of the concentration of 'complex' copper carbonate from a knowledge of its formation constant (Stiff, 1971a) and of the concentration of cupric ion (Stiff, 1971b), and more particularly the calculation of the concentrations of dissolved heavy metals and of carbonate in digesting sludge (Mosey, 1971; Mosey et al, 1971; Department of the Environment, in the press). The latter system is too complex to be described in detail here, but it appears that the toxicity of heavy metals to sludge digestion depends on the concentration of metal in solution and that the following equilibria must be considered:-

Solubilities of metal sulphides
Solubilities of metal carbonates or basic carbonates
Dissociation of hydrogen sulphide
Dissociation of carbonic acid
Solubility of hydrogen sulphide
Solubility of carbon dioxide.

The investigation has shown that the activity of sulphide ion, S^{2-}, is the main factor of practical importance in controlling the inhibition of digestion by toxic metals, and that determination of the *total* concentration of any metal is of little use in diagnosing the cause of inhibition; the application of a sulphide-specific electrode is discussed on a later page.

The equilibrium between hydrogen cyanide (HCN), which is very toxic to fish, cyanide ion, and the complex cyanide ions of zinc, cadmium, copper, and nickel, may be evaluated by calculation (Blaha, 1967; Montgomery and Stiff, in the press). The existence and importance of these equilibria, as well as the fact that HCN predominates over cyanide ion at ordinary pH values, is often overlooked — even by such a source as the latest edition of 'Standard Methods'. In the case of nickel, combining equations for cyanide balance (1), nickel balance (2), and for the dissociation of nickelocyanide and of HCN leads to a fifth-order equation (3) in [HCN] which may be solved graphically or by trial and error for any specified value of [H+]. Use of 6.17×10^{-10} as the dissociation constant of HCN and 1×10^{-30} as the dissociation constant of nickelocyanide (Christensen *et al*, 1963) gives the equations shown:

$$[\text{total cyanide}] = 4[\text{Ni(CN)}_4{}^{2-}] + [\text{HCN}] + [\text{CN}^-] \tag{1}$$

$$[\text{total nickel}] = [\text{Ni}^{2+}] + [\text{Ni(CN)}_4{}^{2-}] \tag{2}$$

$$\left\{3.62 \times 10^{-3} + \frac{2.23 \times 10^{-12}}{[\text{H}^+]}\right\}[\text{HCN}]^5 + \left\{2.5 \times 10^4[\text{H}^+]^4 + 1.54 \times 10^{-5}[\text{H}^+]^3\right\}[\text{HCN}] = [\text{H}^+]^4. \tag{3}$$

These yield concentrations of HCN which have been plotted in Fig. 1 for a dilute solution which is nominally 10^{-5} M in nickelocyanide; the corresponding concentrations of HCN in the absence of nickel are shown for comparison. The validity of Fig. 1 has been confirmed, within experimental error, by chemical analysis (Montgomery *et al*, 1969).

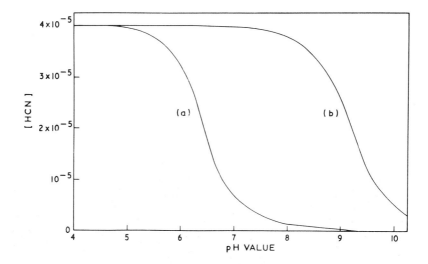

Fig. 1.   Free HCN in equilibrium at 25°C with (a) 10^{-5} M nickel and (b) CN⁻.
Total cyanide, 4×10^{-5} M; no correction for ionic strength; K_a for HCN = 6.17×10^{-10}; K for Ni(CN)$_4{}^{2-}$ = 10^{-30}

The correct application of equilibrium equations depends, of course, on the reactions being known correctly, on their constants being available with sufficient accuracy, and on allowance being made as necessary for the effects of ionic strength. Equilibrium calculations are not applicable in the case of reactions which are slow in relation to the velocity with which the reactants produce their effects; supersaturation, for example, has been shown to persist for several days in bicarbonate solutions of copper, cadmium and zinc under quiescent conditions (Department of the Environment, in the press) and calculations of the equilibrium solubility for such solutions would be highly misleading if applied in a situation where supersaturation occurred. Similarly, copper appears to form one type of carbonate complex rapidly and another type slowly (Stiff, 1971a). Differences in rates of approach to equilibrium may, however, be exploited in chemical analysis as in the identification of a particular type of organic complex of copper by its characteristically slow reaction with a reagent forming a coloured copper derivative (Stiff, 1971b). The velocities of chemical reactions (including adsorption) involving pollutants may be expected to merit as much study in the future as the velocities of biochemical reactions receive already.

Principles and experimental techniques of differentiating chemical and physical species

Samples should normally be analysed, without preservatives, as soon as they are obtained. In particular, dissolved copper has been shown to come out of solution rapidly from a highly polluted river water. Such waters should be filtered under pressure at the sampling site if a true figure for soluble copper is wanted; similar considerations probably apply to other metals. Other reasons for prompt analysis are to avoid biological changes and losses of carbon dioxide, either of which could invalidate determinations of free HCN or cupric ion, for example.

Another principle to be noted is that states of equilibrium existing in an effluent do not necessarily remain unchanged when the effluent is mixed with a receiving water. Profound changes in an equilibrium can be caused by changes in pH value and by the introduction of new reactants which may be present in the receiving water; generally smaller shifts may also occur as a result of dilution by the receiving water and of changes in temperature or ionic strength. For these reasons the determination of 'free cyanide' in a plating-shop effluent, for example, is entirely valueless when it is desired to know the concentration of HCN resulting from the discharge of the effluent to a river; the concentration of HCN could be predicted from a knowledge of the compositions of the effluent and the receiving water by applying the relevant equilibrium equations, but it would be simpler to determine it by direct chemical analysis.

It is important that the analytical procedures themselves do not disturb the equilibrium existing between different chemical forms of a substance. *In situ* methods which consume only a negligible proportion of the substance sought are ideal. The use of specific-ion electrodes is such a method and is discussed in a later section. Ultra-violet spectrophotometry is another suitable technique, which has found use in the study of copper/amino acid systems and of the chromate/hydrogen chromate equilibrium (Montgomery and Stiff, in the press). Phase separation is yet another technique which can be used and there are two possible modes of application. In one, used in a solvent extraction method for the determination of hydrogen cyanide, the proportion extracted, though known, is too small to disturb the equilibrium appreciably and the method is designed also to avoid pH increases resulting from the extraction of carbon dioxide (Montgomery *et al*, 1969). The method has the further advantage that, as the sample is shaken vigorously, the HCN distribution equilibrium is known to be established between the whole volume of both phases whereas in some methods which depend on the evolution of HCN into a gas phase there is a danger that a true distribution equilibrium will not be set up, because of incomplete mixing of the surface film with the bulk of the solution. In these circumstances it is not possible to rely on extracting a constant proportion of the HCN. In gas evolution methods there is also a likelihood of pH changes resulting from losses of carbon dioxide and moreover such methods have a very large temperature coefficient. A solvent extraction procedure has also been used to determine cupric ion, whose weak thiocyanate complex can be formed without appreciably disturbing the equilibrium of cupric ion with other complexing agents. A small proportion of the thiocyanate complex was extracted by partition with amyl alcohol. The determination was carried out radiochemically, using copper-64, to achieve the necessary sensitivity, but was much less convenient than the specific-electrode method referred to later. The other valid mode of application of phase separation methods is when both the complex species and the excess of complexing agent are extracted together, as in the determination of 'hexanol extractable' copper which is believed to be a complex of copper with 'humic acids' (Stiff 1971b).

Another technique for differentiating complex species is to use a colorimetric reagent, of known affinity for the substance sought, which will dissociate complexes whose stability constants are below a certain value. An example is 3-propyl-5-hydroxy-5-D-*arabino*-tetrahydroxybutyl-3-thiazolidine-2-thione (PHTTT) which appears to react with, and hence permit the composite determination of, all but the most stable copper complexes such as those formed with cyanide and cysteine. A difference in the rates of reaction between PHTTT and different complexes also enables a particular type of copper complex to be determined, as mentioned earlier (Stiff, 1971b).

Polarography appears to be a promising method of investigating complex formation, at least in research applications, and the variant known as anodic stripping voltammetry has been applied by Matson (1968) to the study of complex formation by metals, particularly lead.

The nature of samples imposes certain restrictions on polarographic techniques: the need to avoid disturbing the equilibrium (or the need to avoid altering the tendency to form complexes, in the anodic stripping technique) will usually make supporting electrolytes or adjustment of pH value inadmissible; the partial pressure of carbon dioxide should be the same in the gas used for removing oxygen as in the sample; and mixing should be such that the conditions at the electrode can be related reproducibly to those in the bulk of the solution.

In anodic stripping voltammetry, the working electrode may be a mercury pool or a thin film of mercury supported on a graphite rod whose interstices are filled with paraffin wax. A small quantity of the metal being studied is first plated electrolytically from the sample on to the electrode. The plating stage is followed by a stripping stage in which the current/potential curve is recorded with an increasingly positive potential (increasing linearly with time) applied to the working electrode. The current at any potential is proportional to the rate of formation of ions of the metal at that potential and is therefore a function of the quantity of metal originally plated and of the tendency of the metal to pass into solution as a cation. It may be deduced that the potentials at which peaks occur are related to the complexing properties of the medium. The advantages of the technique are its great sensitivity -10^{-10} M cadmium or zinc, for example, could be detected with apparatus costing only a few hundred pounds (sterling) $-$ and the fact that a supporting electrolyte is not needed.

Anodic stripping voltammetry is in use at the WPRL for studying the chemistry of cadmium in water, but at the time of writing the results have not been fully interpreted.

Ion-selective electrodes

The theory and construction of ion-selective electrodes are familiar, but relatively little has been published regarding their applications in the study and control of water pollution (Andelman, 1968; Whitfield, 1971; Briggs and Melbourne, 1972). At the time of writing the Scientific Instruments Research Association is undertaking an evaluation, sponsored partly by the WPRL, of the performance of these electrodes, and a relatively small amount of evaluation work has also been carried out at the WPRL itself (Briggs and Melbourne, 1972).

Under constant conditions of temperature and stirring, the potential of an ion-selective electrode depends on the logarithm of the activity of the ion being determined and such electrodes are, therefore, well suited to the quantitative differentiation of free ionic from complexed forms of an element or radical, especially because the amount of the ion actually consumed is negligible. A copper-specific electrode has been found to give a Nernstian response to cupric ion down to a concentration of 50 μg/l in the absence of complexing agents, and down to very low concentrations ($<$1 μg/l) when the concentration of total dissolved copper was not less than 50 μg/l (Stiff, 1971b). The method has been used successfully for determining cupric ion in polluted waters (Stiff, 1971b) and in solutions used for studying the effect of the medium on the toxicity of copper to fish (Department of the Environment, in the press). A cadmium specific electrode is currently in use at WPRL in a study of the chemistry of cadmium in water and has been found to give a Nernstian response down to 100 μg/l in the absence of complexing agents.

Andelman (1968) has described experiments using a fluoride electrode which suggest the formation of complex fluorides in water samples and there would seem to be numerous possibilities for applying ion-selective electrodes in studies of complex formation in effluents and natural waters. One application which is sometimes suggested, however, the determination of cyanide in effluents, is unlikely to succeed because cyanide electrodes respond only to cyanide ion and, as they are relatively insensitive, the solution must be made strongly alkaline so as to achieve essentially complete ionization of HCN. However, effluents containing cyanide are seldom likely to be completely free from heavy metals and, as has already been seen (Fig. 1), a shift in the pH value of such solutions can cause a profound change in the equilibrium between free and complex cyanide ions. The result given by the cyanide electrode would, therefore, be valid only for the pH value at which the measurement was made and would probably be much lower than the true value. The only present application of the cyanide electrode in effluent analysis seems to be to replace the usual colorimetric or titrimetric finish in the determination of *total* cyanide in the solutions obtained by distilling HCN from acidified samples. If, however, a cyanide electrode responsive to 10^{-8} M or preferably 10^{-9} M cyanide could be devised, direct measurements could be made without adjustment of the pH value in the range (corresponding to about 10^{-6} to 10^{-5} M HCN) of most interest in relation to the survival of fish.

An interesting application of a sulphide electrode which has recently been proposed is in the diagnosis and control of inhibition of anaerobic digestion by heavy metals (Mosey *et al*, 1971; Department of the Environment, in the press). The investigation is being carried out using cheap, specially made, silver/silver sulphide electrodes. At the time of writing the results suggest that digestion proceeds satisfactorily at pS values around 15 but that as zinc, cadmium, copper, nickel, or lead is added, sulphide is precipitated and pS rises, until digestion is inhibited at a concentration of dissolved metal corresponding to a pS value of about 20. The metal ion in equilibrium with 10^{-20} M sulphide ion is apparently the agent responsible for the inhibition. Carbonate is also involved in the equilibrium in a digester and may itself prevent inhibition by metals; the reader is referred to the original publications for a detailed discussion of the system.

Discussion

The techniques referred to in this short paper were (1) application of equilibrium calculations, (2) direct spectrophotometry, (3) partition followed by spectrophotometry or radioactive counting, (4) total extraction of complex and excess of complexing agent, (5) spectrophotometry with a reagent of known complexing ability, (6) application of a kinetic effect in spectrophotometry, (7) anodic stripping voltammetry, and (8) use of ion-selective electrodes. All of these techniques are capable of further application and it is hoped that members of the audience will discuss additional techniques which can be applied to the differentiation of chemical and physical states of pollutants in aqueous systems.

It is also hoped that the paper will stimulate discussion about the definition of objectives as well as about analytical techniques. In the study of analytical and related topics at the WPRL in recent years, attempts have been made not only to develop improved methods for applying to conventional measures of pollution, but also to clarify the real nature of problems in pollution control and research before designing new methods of analysis. Apart from the work described in the paper, Montgomery and Gardiner (1971) adopted the latter approach in designing a respirometric method for judging the strength of sewages and certain industrial effluents (though the project was not completed). From a consideration of the properties it was desired to determine and of the behaviour of bacterial inocula, it was decided to design an entirely new test rather than one intended to predict 5-day BOD values. More recently, the results of a study of another vital aspect of gaining information about water quality, namely the design of sampling programmes, have emphasized the importance of defining objectives (Department of the Environment, 1971).

Topics which might well involve a study of the differentiation of the chemical and physical states of elements include: (1) the availability of trace elements to algae and bacteria — something is already known about the behaviour of iron in this respect but very little about other elements; and (2) the rates and mechanisms of precipitation, adsorption, and dissolution of heavy metals in relation to (a) their transport along, and ultimate fate in, rivers, (b) exertion of toxicity to aquatic organisms, (c) their ease of removal from surface waters during treatment for potable supply.

REFERENCES

Andelman, J. B. (1968). Ion-selective electrodes — theory and applications in water analysis. *J. Wat. Pollut. Control Fed., 40*, 1844-1860.

Blaha, J. (1967). Zur Frage der Bestimmung und Toxizität von freien und komplexen Cyaniden in Wässern. *Vom Wasser, 34*, 175-195.

Briggs, R. and Melbourne, K. V. (in press). Ion-selective electrodes in water-quality monitoring. Paper presented at Conference of Institution of Mechanical Engineers, London, 12 January 1972.

Christensen, J. J., Izatt, R. M., Hale, J. D., Pack, R. T., and Watt, G. D. (1963). Thermodynamics of metal-cyanide coordination II. ΔG^o, ΔH^o and ΔS^o values for tetracyanoniccolate (II) ion formation in aqueous solution at 25°C. *Inorg. Chem., 2,* 337-339.

Department of the Environment (1971). The design of sampling programmes for river waters and effluents. Notes on Water Pollution No. 54.

Department of the Environment (in press). Water Pollution Research 1971. H.M. Stationery Office, London.

Matson, W. R. (1968). Trace metals, equilibrium and kinetics of trace metal complexes in natural media. Ph.D. Thesis, Massachusetts Institute of Technology.

Montgomery, H. A. C. and Gardiner, D. K. (1971). Experience with a bacterial inoculum for use in respirometric tests for oxygen demand. *Water Res., 5,* 147-163.

Montgomery, H. A. C., Gardiner, D. K., and Gregory, J. G. G. (1969). Determination of free hydrogen cyanide in river water by a solvent-extraction method. *Analyst, Lond., 94,* 284-291.

Montgomery, H. A. C. and Stiff, M. J. (in press). Differentiation of chemical states of toxic species, especially cyanide and copper, in water. Paper presented at International Symposium on Identification and Measurement of Environmental Pollutants, Ottawa, June 1971.

Mosey, F. E. (1971). The toxicity of cadmium to anaerobic digestion: its modification by inorganic anions. *Wat. Pollut. Control, 70,* 584-598.

Mosey, F. E., Swanwick, J. D., and Hughes, D. A. (1971). Factors affecting the availability of heavy metals to inhibit anaerobic digestion. *Wat. Pollut. Control, 70,* 668-680.

Stiff, M. J. (1971a). Copper/bicarbonate equilibria in solutions of bicarbonate ion at concentrations similar to those found in natural water. *Water Res., 5,* 171-176.

Stiff, M. J. (1971b). The chemical states of copper in fresh water and a scheme of analysis to differentiate them. *Water Res., 5,* 585-599.

Stumm, W. and Morgan, J. J. (1970). Aquatic Chemistry. Wiley-Interscience, New York.

Whitfield, M. (1971). Ion-selective electrodes for the analysis of natural waters. Australian Marine Sciences Assoc., Sydney, AMSA Handbook No. 2.

Discussion *by* E.M. Levy
"Identification of the Source of Petroleum Pollutants"

Atlantic Oceanographic Laboratory
Bedford Institute
Dartmouth, Nova Scotia

A relatively simple procedure involving an ultraviolet absorption spectrophotometry has been developed at the Bedford Institute for the identification of the source of oil pollutants in the marine environment. Unlike other techniques, no attempt is made to isolate or identify all, or even a few, of the compounds present in the oil. Instead the contributions from all those compounds which are capable of absorbing ultraviolet light are considered as a group — a group which is indicative of the oil involved. This approach has proven useful in several instances in identifying the source of petroleum pollutants in the marine environment of Eastern Canada.

Since the analytical procedures have been described in some detail in a recent issue of Water Research (Levy, 1972), only the essentials will be repeated here. In principle, the method consists of preparing n-hexane solutions of the oil and comparing their ultraviolet adsorption spectra over the range of 350 to 210 nm with those of samples from the suspected sources.

The ultraviolet absorption spectra of several crude, lubricating, distillate and residual fuel oils, typical of those commonly encountered as marine pollutants are illustrated in Fig. 1. The absorption spectra of one Arabian and two Venezuelan crude oils are shown in Fig. 1a. The prominent features in each case are the strong absorption maxima at approximately 223 nm and the less pronounced peak or shoulder at approximately 256 nm. Although the same general features are also present in the spectra of a variety of residual fuel oils (Fig. 1b), there is an obvious difference in the absorbance at 228 nm relative to that at 256 nm for the two types. Considering the spectra for the residual oils, there is a much more distinct peak at 256 nm in Spectra 1 and 2 than in the others, suggesting a substantial difference in their compositions. Indeed Spectra 1 and 2 belong to residual fuel oils derived from Middle East crudes whereas the others were from crudes of Venezuelan origin. The spectra for the distillate fuel oils (Fig 1c) and for the lubricating oils (Fig. 1d) are again quite different from those for the crude or residual oils. Distillate fuel oils absorb very strongly at 228 nm and compartively weakly at 256 nm. Lubricating oils, on the other hand, exhibit practically no absorption of ultraviolet light above approximately 240 nm. These features are a reflection of differences in the compositions of the various types of oil and, as such, provide a convenient means for distinguishing between them. Unfortunately, however, the spectra of different members of the same general type do not differ sufficiently to provide a criterion for distinguishing between different members within the same group. Thus, while the general features of the absorption spectra suffice to categorize a pollutant, they may not by themselves provide a criterion for identifying the source of a particular spill.

As pointed out earlier, however, the absorbance at 228 nm relative to that at 256 nm differ for the various types of oil. Indeed, as shown in Table 1, the ratios of the absorbance at 228 nm to that at 256 nm not only differ significantly from one general type to another, but also amongst different members of the same group. This feature, then, may often be used as a diagnostic criterion for identifying the source of a spill provided samples from the suspected sources are available for comparison.

As an indication of how the absorbance ratio can be used to identify the source of a spill, three pollution incidents that occurred recently on the east coast of Canada will be discussed briefly. In February, 1970, the tanker ARROW was wrecked in Chedabucto Bay, Nova Scotia (Fig. 2) and most of the cargo of 16,000 tons of Bunker C was lost. Shortly afterwards, oil appeared on the shores of Sable Island about 1800 km away. It was not obvious, however, whether this oil had in fact been part of the cargo of the ARROW or if it had been discharged by some passing vessel or other source. At the same time, there were rumors that oil had been discovered by one of the companies carrying out exploratory drilling on Sable Island and on the adjacent continental shelf. This uncertainty was readily resolved by comparing the absorbance ratios of a number of samples from Sable Island with those for the ARROW. As shown in Table 2, there can be little doubt that the source of the oil on Sable Island was the ARROW spill. It is pertinent also that the absorbance ratio for the arrow Bunker C oil had not changed after a year's exposure to weathering processes in Chedabucto Bay. Apparently, this method of source identification is not as subject as most other techniques to ambiguities which arise from the rapid changes in composition which occur on exposure to the environment.

A similar incident occurred in September, 1970, when the barge, WHALE, carrying a cargo of Bunker C oil sank in the Gulf of St. Lawrence. A few days later a heavy residual oil drifted onto the shores of the Magdalen Islands. In this case (Table 2), the absorbance ratios for the samples ranged from 1.28 to 1.39 (Average 1.33) matching perfectly the average value of 1.33 for the suspected source.

A third opportunity to identify the source of a spill occurred in March, 1971, when several miles of shoreline of Halifax Harbor, mostly bordering residential or recreational areas (Fig. 3), were fouled with oil. At the same time a quantity of Venezuelan crude was lost while a tanker was being unloaded at one of the local refineries. At first there was no doubt concerning the responsibility for the spill, but then rumors appeared that a spill of Bunker C oil was also involved. Samples collected over the course of the month following the spill all had absorbance ratios in the range of 2.15 to 2.37 (Table 2), which is in reasonable agreement with the value of 2.51 for the suspected source. In this case, a decrease in the absorbance ratio was observed during the first few days following the spill as the volatile and soluble constituents escaped. However, there is no indication that a residual fuel oil might also have been present.

In conclusion, it should be pointed out that the absorbance ratio would not be capable of distinguishing between two cargoes from the same source – a task which appears to be beyond the ability of analytical techniques proposed to date. In many cases, however, the absorbance ratio can provide an adequate identification of the source of an oil spill, particularly if it can be supplemented by ancillary information such as ship traffic, etc., or if it is supported by some other analytical technique.

REFERENCE

LEVY, E.M. (1972). The identification of petroleum products in the marine environment by absorption spectrophotometry. *Water Research*, 6, 57-69.

TABLE 1. ABSORBANCE RATIOS FOR SOME TYPICAL MARINE OIL POLLUTANTS

Type of Oil		A_{228}/A_{256}	Average
Crude:			
	Guanipa	2.51	2.4
	Mesa	2.62	
	Arabian light	2.14	
Residual Fuel:			2.4
	Arrow	1.57	
	Whale	1.33	
	Patrick Morris	1.60	
	Sydney Harbor	1.23	
	Intermediate bunker	2.11	
	Marine bunker	1.96	
Distillate Fuel:			6.4
	Low ash	4.52	
	Diesel (a)	6.90	
	Diesel (b)	7.41	
	Diesel (c)	6.75	
Lubricating:			4.2
	Stern-tube	2.56	
	Diesel engine	4.64	
	Turbine (unused)	5.33	
	Turbine (used)	4.13	

TABLE 2. IDENTIFICATION OF THE SOURCE OF SOME OIL SPILLS

Location	Absorbance Ratios	Suspected Source	Absorbance Ratio
Sable Island	1.53 – 1.57	ARROW	1.57
Magdalen Islands	1.31 – 1.36	WHALE	1.33
Halifax Harbor	2.15 – 2.37	GUANIPA CRUDE	2.51
		MESA CRUDE	2.62

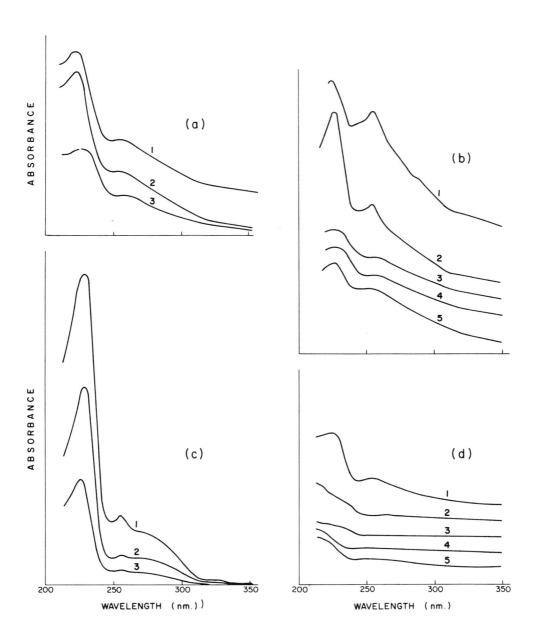

Fig. 1 Ultraviolet absorption spectra of some typical marine oil pollutants.
(a) Crude oils: 1. Guanipa 2. Mesa 3. Arabian light.
(b) Residual, fuel oils: 1. Environmental sample, Sydney Harbor, N.S. 2. Intermediate residual 3. Environmental sample from PATRICK MORRIS 4. Environmental sample from ARROW 5. Environmental sample from WHALE.
(c) Distillate fuel oils: 1. Low ash type 2. Diesel fuel 3. Diesel fuel.
(d) Lubricating oils: 1. Stern tube 2. Diesel 3. Turbine 4. Turbine (used) 5. Diesel (used).

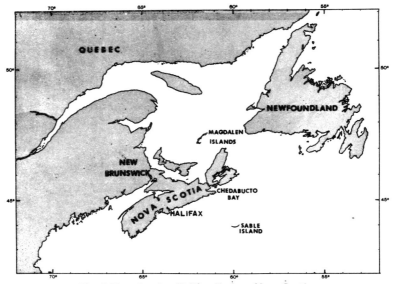

Fig. 2 Map showing Halifax Harbor, Nova Scotia.

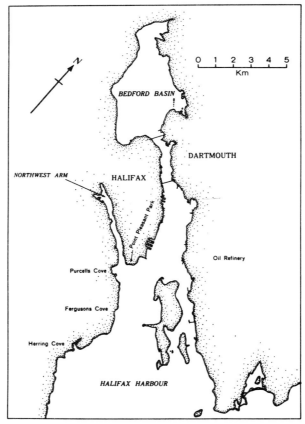

Fig. 3 Map showing a portion of the east coast of Canada and the location of some recent oil pollution incidents.

Ion-selective electrodes can be used to differentiate species, but apparently not at concentrations $<10^{-6}$M. May I suggest a dual technique of liquid-liquid extraction (LLE) for concentration of species and then subsequent ion-selective electrode analysis in the solvent extract. Distribution theory, kinetic data and equilibrium data on LLE of inorganics is available in the atomic absorption spectroscopy literature and may be a helpful starting point. Do you know of any work on the analysis of ionic species by ion-selective electrodes in non-aqueous or partially aqueous solvents?

Reply by H.A.C. Montgomery

The limit of 10^{-6}M mentioned by Professor Suffet would be applicable to total dissolved rather than free ionic concentrations of certain metals, and would have to be checked for each individual metal. I am not aware of any work on the use of ion-selective electrodes in organic solvents. Professor Suffet's suggestion is an interesting one and, I think, worth following up; care would have to be taken to design the solvent extraction methods so as not to disturb the equilibrium in the aqueous sample.

Comment by I. Suffet, USA on discussion by E.M. Levy

Confirmation of the identification of the source of an oil spill is critical. No one technique alone appears capable of confirmation. The pitfalls of one method of identification have been discussed by myself related to organic analysis in water and Kanahara specifically related to oil analysis. The author suggests UV analysis ratios for identification of oil sources. What are his views on more specific tools of infrared, fluorescence and gas chromatography for oil spill identification or on the use of a combination of these methods rather than on the use of one method?

Reply to E.M. Levy to I. Suffet

I agree that no one analytical technique is capable of providing an unequivocal identification of the source of each and every oil spill. In many cases, however, provided the number of potential sources is small, or when ancillary information is available, a single technique may suffice. Nevertheless, there is no doubt that the combined evidence from several analytical techniques provides a considerably higher degree of certainty in the identification. Used in conjunction with gas chromatography, infrared or other technique, the u. v. absorbance ratios provide an additional clue to the identity of the oil. These ratios are particularly useful when dealing with residual fuel oils and others containing appreciable concentrations of aromatic compounds. Since the high-boiling constituents in these oils are not readily amenable to analysis by gas chromatography, an identification by this method must be based on the low-boiling constituents which, unfortunately, tend to escape from the oil or to be susceptible to the effects of weathering processes. As a result, a positive identification by gas chromatography is often difficult. Similarly, infrared and fluorescence spectrophotometry are useful in some instances but not in others. In each case, however, an attempt is made to obtain sufficient information for an unambiguous identification taking advantage of whatever technique or combination of techniques as are required.

Comments by N. Narkis, Israel on discussion by E.M. Levy

How can one distinguish between these two types of oil spills from the upper curve in figure "a" and the second curve in "b" since these seem to be identical? What is the effect of oxidation and weathering on oils in the ocean? Is it not reflected in the U.V. spectra?

Reply by E.M. Levy to N. Narkis

Although the absorption spectra for the crude oils in Fig. 1a and the various residual fuel oils in Fig. 1b are very similar, there are slight but very important differences. These differences become much more apparent when the ratio of the absorbance at 228 nm to that at 256 nm is considered. As shown in Table 1, this ratio often provides a criterion for distinguishing between various possible sources of a spill.

Weathering processes bring about changes in the composition of an exposed oil. Most susceptible are the low boiling and more volatile or soluble constituents. Such changes make the identification difficult by any analytical method which relies upon these constituents. The uv absorbance method, on the other hand, is based on the aromatic and unsaturated compounds, many of which are non-volatile and resistant to chemical or biological degradation. As a result, this technique may provide a positive identification in situations which are difficult to deal with otherwise. In all cases, of course, it is desirable to have evidence from a number of analytical methods.

D.I. Suffet, USA

Many metal complexes (inorganic or organic) can be formed in natural waters. Can these shift the peak potential of the metal and hinder qualitative analysis of the metal? Can the specific species be determined to see what are the controlling equilibrium species?

REASSESSMENT OF THE VIRUS PROBLEM IN SEWAGE AND IN SURFACE AND RENOVATED WATERS

GERALD BERG, PhD.

Chief of Virology, National Environmental Research Center,
Environmental Protection Agency, Cincinnati, Ohio, USA

Man excretes more than 100 different viruses into sewage. Viruses do not multiply in sewage, but decrease in number in this somewhat hostile environment. Since viruses are excreted only by infected persons, and in numbers many orders of magnitude smaller than the numbers of bacteria excreted by all people, the quantities of viruses that occur in sewage and in other waters are never large. The importance of viruses in water resides not in their numbers, however, but in their infectivity. The smallest quantities of viruses that can be detected in susceptible cells in cultures, our most sensitive indicators of infection, are sufficient to produce infection in man (Table 1).[1,2]

TABLE 1

Infective Doses of Viruses for Man

Virus	Dose	Route of inoculation	No. inoculated	% infected
Poliovirus 1	2 PFU	Oral (gelatin capsule)	3	67
Poliovirus 3	1 TCD_{50}	Gavage	10	30

Enteric bacterial pathogens, on the other hand, must often be present in very large numbers to produce infection and disease. Moreover, bacteria are considerably more sensitive to water disinfectants than viruses are. Although in many parts of the world, the bacterial problem is not diminished, water disinfected sufficiently to destroy viruses is likely to be free of bacterial pathogens also.

The Quantitites of Viruses in Sewage

Large amounts of viruses are readily recoverable from sewage. The quantities recoverable may vary considerably from one month to another and from one community to another. In a broad study in Israel, for example, Shuval[3] recovered from five plaque-forming units (PFU) per liter to more than 11,000 PFU per liter in the same city.

Average recoveries per community ranged from about 500 PFU per liter to more than 1600 PFU per liter. Higher recoveries were made in communities of lower socioeconomic levels than in those with higher socioeconomic levels. Seasonal variations in recovery incidence did not occur.

For recovering viruses from sewage, Shuval used the two-phase separation technic. Recent data, however, show that the technic was far from 100% efficient; there must have been much more virus present than he was able to demonstrate — much more than 11,000 PFU per liter, that is, much more than 29 billion PFU per million gallons of sewage.

In the United States, a country of somewhat colder and more fluctuating climate than Israel, the quantities of viruses isolated from sewage have been somewhat smaller, but formidable nonetheless. Table 2 shows the levels of viruses we recovered from sewage in

TABLE 2

Relative Concentrations of Viruses and Fecal Coliforms in Sewage (1970)

Sample	9/2	9/9	9/30	11/11	11/18
Viruses recovered (PFU/100 gallons)	11,900	38,600	40,500	31,800	13,400
Fecal coliforms/100 ml	4,600,000	38,000	160,000	100,000	120,000

1970 in the midwest of this country during the colder months of the year when contact transmission of enteric viruses, and thus the presence of these viruses in sewage, ebbs. Our recoveries at this time ranged from about 12,000 PFU per 100 gallons to more than 40,000 PFU per 100 gallons — more than 400 million PFU per million gallons of sewage. The application of chlorine to primary effluents in amounts up to 3 mg per liter does not always reduce virus levels in sewage, and when virus levels are reduced, they are usually not reduced by much. It is possible, of course, to apply sufficient chlorine to significantly reduce concentrations in primary effluents, but the amounts of chlorine required are generally not economically acceptable and may kill fish and other aquatic forms.

The fecal coliform counts ranged from 38,000 per 100 ml of effluent to 4,600,000 per 100 ml — more than 1.8×10^{14} fecal coliforms per million gallons of effluent.

The Quantities of Viruses in River Waters

Despite the great dilution by the river, viruses are detectable in river waters downstream of sewage outfalls and at water intakes as well. Table 3 shows the amounts of viruses recovered from 50-gallon samples of water along one major United States river. Virus concentrations ranged up to 19 PFU per 50 gallon sample (360,000 PFU per million gallons of water) and this from a water intake. Quantities as great as 1.5×10^6 PFU per million gallons have been recovered from river waters downstream of sewage outfalls.

In samples 4 and 5, viruses were recovered when fecal coliform levels were only 137 and 60 per 100 ml of river water. Sample 6 was also taken at a water intake.

TABLE 3

Recovery of Viruses and Fecal Coliforms Along the Missouri River

Sample	1	2	3	4	5	6
Viruses recovered (PFU/50 gallons)	5	10	4	19*	1	3*
Fecal coliforms/100 ml	5,000	40,000	10,000	137	60	1,900

* Recovered at water intake.

Efficiency of Technics for Detecting Small Amounts of Viruses in Sewage and River Water

The $Al(OH)_3$-protamine sulfate method,[4] which we used for recovering viruses from sewage, is thought to be about as efficient as the two-phase separation technic for recovering viruses in experimental systems, but the effectiveness of these technics with sewage has not been accurately determined.

The polyelectrolyte-silt technic, which we used for recovering viruses from river water, is inefficient. The polyelectrolyte is a Monsanto resin designated PE 60,[5] and the method for recovering viruses from river silt is one developed in our laboratory.[6]

The efficiency of the PE 60 method in laboratory studies ranges widely depending upon the virus present. In one series of studies, when small amounts of viruses were added to distilled water, recoveries of virus ranged from 17% with reovirus 1 to 51% with poliovirus 1 (Table 4). In other studies, recoveries as high as 90% were achieved with poliovirus 1 and as low as 2% were experienced with reovirus 1.

TABLE 4

Recovery of Viruses from Distilled Water by
Adsorption onto Polyelectrolyte PE 60

Virus	Control (PFU)	PE 60 (PFU)	% recovery
Poliovirus 1	75	38	51
Echovirus 7	79	24	30
Reovirus 1	105	33	31
	84	14	17

River silt adsorbs viruses with a modicum of efficiency, but does not relinquish them readily. In a laboratory study, silt adsorbed up to 94% of poliovirus 1 from distilled water, but only 0.3 to 0.6% of the adsorbed virus could be eluted from the silt (Table 5).

TABLE 5

Recovery of Poliovirus 1 from Silt

Grams of silt	Virus mixed with silt (PFU)	Virus recovered from supernate (PFU)	Virus recovered from silt (PFU)	% virus recovered from silt
5	492	78	3	0.6
10	492	28	1	0.3

Nonetheless, viruses were recovered at least as frequently from the silt filtered from river water as from the river water itself (Table 6). Since viruses were not recovered as efficiently from silt as from water, there would appear to be much more virus adsorbed to the silt than in the water itself.

The data in Table 6 represent virus recoveries from water intakes, and are presented in terms of a million gallons of water (a quantity required daily by a community of only 10,000 persons) to convey the magnitude of the problem. The negative findings in four samples reflect the findings in water samples of 100 gallons or less and silt from one of those samples.

The true amounts of viruses in sewage and in river waters must greatly exceed the amounts we have been able to demonstrate. The kinds of cell cultures we used to recover viruses could detect no more than half of the virus types known to occur in sewage, and

TABLE 6

Recovery of Viruses at or Near Water Intakes

	8/28/70	10/1/70	Sample Date 11/12/70 11/19/70 (PFU/million gallons*)		11/19/70	4/1/71
Viruses recovered from water	160,000	40,000	640,000	0	0	0
Viruses recovered from silt	0	20,000	20,000	50,000	20,000	30,000

* Actual samples tested were 50 and 100 gallons and silt from those volumes.

the plaque technic, which we prefer to use for virus detection because it permits the most accurate relative estimate of virus concentrations, yields much lower absolute counts than roller tube cultures.

Moreover, although the aluminum-hydroxide-protamine sulfate technic was believed to be highly efficient, a modified membrane filter procedure described recently by Rao[7] has proved to be considerably more efficient in parallel sample studies.[6] There are indications that the membrane filter technic can be further improved also. So, the amounts of viruses detected in wastewater are diminished all the more.

The polyelectrolyte-silt technic is clearly of low and variable efficiency. If the silt technic is a more efficient detection system than the polyelectrolyte method (Table 6), then, on this basis alone, the amounts of viruses present in river water must be considerably higher than we have demonstrated.

It would seem likely that the amounts of viruses in sewage and river waters exceed by at least two orders of magnitude the amounts yet detected.

Epidemiology of Virus Transmission by Water

Even small amounts of viruses in water are important. Small amounts of ingested viruses may produce infection, but not disease, and go undetected. Disease, however, may be the consequence of subsequent contact transmission by infected carriers excreting large amounts of viruses. Infection and disease spread in this way would give no indication that water was the source of the original infections.[8]

Quite clearly, water treatment processes must be capable of removing virus from our waters. Otherwise, outbreaks of New Delhi magnitude might well be commonplace. Whether treatment processes are always completely effective, however, is a matter the epidemiologist has yet to explore effectively.[8,9]

Certainly, recent studies which show that in some regions of the United States, 20 to 30% of the water supplies have coliform densities exceeding federal government recommendations do not encourage complacency.[9]

Removal of Viruses from Sewage and Water by Treatment Processes

Some sewage and water treatment processes effectively remove viruses, others do not.

Primary treatment. Primary treatment (screening and settling) of wastewater does not efficiently reduce the level of viruses and bacteria in sewage.

Biological treatment. Data are available on a number of biological treatment methods for removing microorganisms from sewage. The one of greatest interest at the moment is activated sludge. In laboratory studies, activated sludge removes more viruses than any other biological treatment process. In the field, however, activated sludge appears erratic. Sometimes, few viruses seem to be removed by this procedure even when good reductions of total organic carbon and turbidity are achieved.

Trickling filters and oxidation ponds appear erratic also, sometimes removing large quantities of viruses from wastewater, but often removing little or none.

Physical-chemical treatment. Coagulation of sewage or surface waters with aluminum, iron, or lime removes much of the virus in those waters. Lime, moreover, may increase pH to a level that is significantly virucidal.

Rapid filtration through sand and multimedia removes viruses only when these processes follow coagulation, probably because small amounts of carryover floc trap or adsorb viruses passing into filter beds.

Carbon adsorption removes some viruses from sewage effluents and from other waters, but adsorbed viruses may eventually be displaced.

Disinfection

Treatment procedures reduce the amounts of viruses and other microorganisms in effluents and water, but never remove all of them. Thus, safe water demands terminal disinfection. Since treatment procedures often remove from sewage and surface waters many substances that interfere with terminal disinfection by HOCl, chlorine is usually adequate for this purpose. Figure 1 shows the rapidity with which HOCl destroys both poliovirus 1 and *Escherichia coli* at 5°C. One-tenth mg per liter of HOCl destroyed 99% of the *E. coli* in about 15 seconds, and 99% of the virus in 100 minutes.[10] Viruses are generally more resistant than bacteria to other water disinfectants also.

When ammonia or other substances that interfere with chlorine disinfection are present, ozone, iodine, or bromine may need to be used. Particularly for sewage effluents, these disinfectants may be especially useful.[4]

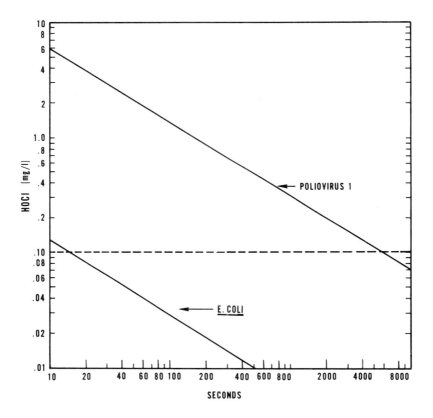

Fig. 1. Time-concentration relationship for 99% inactivation of poliovirus 1 and *Escherichia coli* by HOCl at 5°C

Renovation of Sewage

In many areas of the world, water is in short supply, and renovated sewage may be the most economical new source, and perhaps the only new source, of potable water.

Renovated water is potable water produced directly from sewage, hence the great concern for the microbiological hazard. Yet, renovated water is likely to be carefully produced water with special care taken not only to remove viruses and other pathogens but chemicals as well; since chemicals interfere with disinfection, renovated water may, in the end, be a lesser hazard than the treated surface waters many of us now consume. The renovation process in itself is generally a combination of carefully controlled sewage treatment procedures, and modified water treatment procedures.

Standards for Microbiological Safety of Renovated and Other Potable Waters

Clearly, new microbiological standards are required for renovated and for other potable waters. The following *tentative* standards for microbiological safety are recommended, but they must be subject to revision in the light of developing technology.
1. For renovated and other potable waters to be considered microbiologically safe, these waters should be subjected to disinfection for a dosage-time coupling sufficient to destroy at least 12 log units of reference virus at 5°C. A residual of disinfectant sufficient to destroy, at 5°C, at least two log units of the reference virus per hour should be maintained at the tap.
2. Disinfected renovated, and other potable waters should be tested for viruses at frequent intervals by methods capable of detecting 1 PFU of virus per 100 gallons of water. The permissible amount of virus should be none.
3. Total plate counts for vegetative bacteria should yield no colonies from samples of 100 gallons of renovated or other potable waters.
4. Once on stream, a carefully controlled and continuing epidemiological surveillance should be undertaken among consumers of renovated waters. Since detection of viruses and bacteria can only be achieved long after a water is consumed, prime reliance for microbiological safety of renovated and other potable waters must rest upon accurate continuous measurement of disinfectant residuals. However, the continuous monitoring of viruses and vegetative bacteria will lend important backup that will in time attest to the reliability of the disinfection procedure.

Research Needs

Over the last several years, we have come to see more clearly the extent of the viral hazard in water and the research needs yet to be met. The following are among the more important of these.
1. To obviate today's reliance upon calculations based on theoretical curves for measurement of disinfectant concentrations, methods must be developed to differentiate important chemical species *in situ*.
2. Information must be available for all water disinfectants that will permit us to choose concentration-time couplings capable of disinfecting renovated and other potable waters produced by any treatment train.
3. Continuing efforts must be made to develop and quantify efficient methods for recovering minimal quantities of viruses from sewage and from large volumes of water.
4. Methodology must be devised that will quantitatively detect vegetative bacteria in 100 gallon samples of renovated and other potable waters.
5. Treatment trains for water renovation must be evaluated on a step-by-step basis, before they come on stream, for their capability in removing viruses and other pathogens.

REFERENCES

1. Plotkin S. A. and Katz M., 1967 Minimal infective doses of viruses for man by the oral route. In *Transmission of Viruses by the Water Route,* edited by G. Berg, John Wiley and Sons, New York, New York.
2. Katz M. and Plotkin S. A., 1967 Minimal infective dose of attenuated poliovirus for man. *Amer. Jour. Pub. Health* 57, 1837.
3. Shuval H. I., 1970 Detection and control of enteroviruses in the water environment. In *Developments in Water Quality Research,* edited by H. I. Shuval, Ann Arbor-Humphrey Science Publishers, Ann Arbor, Michigan (London, England).
4. Berg G., 1971 Integrated approach to problem of viruses in water. *Proc. Amer. Soc. Civil Engr.* 97, SA6, 867.
5. Wallis C., Grinstein S., Melnick J. L. and Fields J. E., 1969 Concentration of viruses from sewage and excreta on insoluble polyelectrolytes. *Appl. Microbiol.* 18, 1007.
6. Berg G., Dahling D. R. and Berman D., Viruses in wastewater. Study in progress.
7. Rao V. C., 1971 A simple method for concentrating and detecting viruses in wastewater. Presented at the Second International Congress for Virology, Budapest, Hungary, June 27–July 3, 1971.
8. Berg G., 1966 Virus transmission by the water vehicle. I. Viruses. *Health Lab. Sci.* 3, 86.
9. McCabe L. J., Symons J. M., Lee R. D. and Robeck G. G., 1970 Survey of community water supply systems. *Jour. Amer. Water Works Assn.* 62, 670.
10. Scarpino P. V., Berg G., Chang S. L., Dahling D. R. and Lucas M., A comparative study of the inactivation of viruses in water by chlorine. *Water Research.* Accepted for publication.

INACTIVATION OF VIRUSES

EBBA LUND

The Royal Veterinary and Agricultural University of Copenhagen,
Department of Veterinary Virology and Immunology,
13 Bülowsvej, 1870 Copenhagen V, Denmark

There are different ways of rendering viruses innocuous. A number of chemical and physical treatments may inactivate viruses, which in addition may become harmless spontaneously if kept outside the proper kinds of cells for a sufficiently long period.

Common for most of the inactivation processes is that we know very little about the actual reactions taking place. In only a few cases do we possess definite information of what part of the virus particles which are affected, i.e. if it is the nucleic acid and/or the proteins. We have, however, in a number of cases empirically found rates of reactions as functions of different factors such as temperature, pH, concentration of reactants etc. We know in fact little of what we measure when we find that virus has become inactivated. All we may learn by this is that the virus genome has under the circumstances been unable to take command in a proper cell. The virus particle may well have gained entrance to the cell, but failed to get stripped, or it may not get into the cell at all.

It may also mean, however, that the infectious nucleic acid has truly become inactivated. A number of different situations may thus be covered by this common denominator: Inactivation. I don't think that this is just idle speculation because as long as the nucleic acid is still potentially infectious, but only imprisoned then there may be a potential possibility that it may be let lose. For most practical purposes, we are, in spite of this quite satisfied with the term inactivation, even if we cannot specify it further.

The viruses we are mostly concerned about as water pollutants are enteroviruses, reoviruses and adenoviruses. The human pathogens among them are found quite frequently when we care to look for them in waste waters from cities. In addition to the ones we find we must worry about the unknown virus of infectious hepatitis. Assuming that this virus will prove to have chemical properties not too far from enteroviruses, and knowing that adenoviruses are chemically and physically more labile than the RNA-viruses mentioned, we may concentrate on the small RNA-viruses in our reasoning.

'Spontaneous' inactivation of the small RNA-viruses is a slow process in water at a temperature of 20°C or below: after one or more weeks there is little or no change in virus titer at a neutral pH and with no specific factors present. It is accepted that the more the virus suspension is purified the faster the inactivation. The true nature of the process is unknown; from my own experiments I have concluded that it is not an oxidation process. If it is the same process as the one we call 'thermal inactivation' then the process works faster on the proteins than on the nucleic acids. Heat-inactivated virus may contain infectious nucleic acid. Viruses are inactivated faster at basic pH values. It seems definitely possible that the inactivation process which is working as a function of basic pH values has a similar nature as the thermal inactivation; it may even be the same process. Maybe it has something to do with protein denaturation by the 'splitting and uncoiling' of the proteins.

A number of workers, among these especially the Israeli group, have found that this slow spontaneous inactivation is either speeded up considerably or is supplemented with another process when the water is some kind of sea water. The virus inactivating capacity of sea water seems quite universally bound to sea water, and is not found in fresh water. Most workers point to one or more species of the marine microflora to be the source of this inactivation, which may result in a 3–4 log reduction in titre within a week at 15–20°C. The nature of this process has, in spite of a lot of work, remained essentially obscure so far.

Although we generally believe that there are no specific factors or organisms in fresh water which may render virus inactive, this cannot be quite true, as something is definitely happening in an activated sludge process. The activated sludge is supposedly working through its microflora, not only in the mineralization process in general but also in the virus inactivation. Unfortunately no information is available regarding which specific microorganisms are the active ones. Consequently one cannot just inoculate a waste water treatment plant with a particular organism, but must rely on nature to provide the proper biomasses. It must thus be assumed that the particular microorganisms are also present in the water, although in a quantitatively different way.

The virus inactivating capacity of an activated sludge and a trickling filter plant is considered inadequate by all those of us who have been finding viruses in sewage treatment plant effluent. We find that it depends — too much — on a number of unknown factors in addition to water/sludge ratio, temperature and other things, but we cannot rightly deny that viruses are bound and inactivated to some — and sometimes quite a high — degree. The nature of the process is unknown. It seems that it consists of several steps. At first, as may be found in a primary sludge, virus is 'bound' — may be precipitated — in such a way that some virus is removed from the sewage. Viruses are in fact sometimes more easily detected in primary sludge than in the water, but the very fact that it is so easily found indicates how losely it is bound. Gradually virus is bound more irreversibly, but may often still be demonstrable in the sludge of the final settling. If we had some idea of the nature of the process we might be able to improve upon it.

Chemical precipitation of sewage — it seems especially by means of calcium and aluminium salts — may give a precipitate which to a very high degree may remove viruses from naturally contaminated waters, but the virus is essentially not inactivated, only removed from the water. Consequently it has in some cases with low virus contents not been possible to detect the virus in untreated water or in the effluent, but only in the precipitate. Contrary to what migh be expected virus may be demonstrable even when the precipitate has been obtained at pH 10.5–11.0. This will have to be considered in the disposal of the sludge.

Chemical inactivation of viruses is, in addition to formalin treatment and other processes, of no importance in water treatment, essentially only obtained by oxidation. Among the oxidants of practical importance we have especially the halogens and oxidative, halogenated compounds. In technical terms we have to use 'free chlorine' or 'combined chlorines' in different forms, or ozonization. It is commonly accepted among virologists that under the circumstances usually found in even highly treated wastewaters chlorination does not give a decontaminated wastewater. If, for example, a denitrification process may work on a technical scale then the situation may become very much changed. Then a free residual of chlorine may be obtained, and therefore it is to be expected that a reasonable contact time may result in an effluent without active virus.

In a number of papers I have presented studies on the oxidative inactivation of enteroviruses and adenoviruses. In what seems reasonable agreement both with laws of physical chemistry and empirically found results I found that the rate of inactivation is dependent on the oxidation potential according to $\log K = c_1 E + c_2$, where K is the rate-constant for the inactivation and E the oxidation potential. This has been tested under varied circumstances. The equation

$$E = E_0 + \frac{RT}{nF} \ln \frac{Ox}{Red} - \frac{mRT}{nF} pH$$

is derived from the use of the law of mass action on a redox-process where a reductant, Red, is in equilibrium with the corresponding oxidant, Ox, in a reaction involving m hydrogen ions and n electrons. The R is the gas constant, T the temperature and F the electric equivalent. From this equation a number of things may be read. Thus that the specific oxidizing compound influences E through the normal potential E_0, and its concentration through $[Ox]$. It may be seen, however, that the amount of reducing compounds, $[Red]$, influences the potential, and thus the rate of inactivation just as much. In my opinion this is the simplest way of explaining why virus is inactivated slowly when a lot of dirt is present, in spite of residuals of chlorine. I find it more convenient, too, to regard the effect of pH on the rate of inactivation through its effect on the oxidation's potential, rather then being forced to complicated chemical reasoning.

A very inconvenient thing about the measurements of oxidation potentials is that redox reactions may equilibrate slowly and that electrodes are easily contaminated by chlorine. This may make measurements very difficult. Just as for other inactivation-reactions little is known about what part of the virus particle is taking part in the reaction. Something points at the nucleid, at least no 'reactivation' was ever observed, but this has not been investigated properly.

As pointed out we have a number of natural or man-made ways of rendering viruses innocuous, but the reactions are often much slower than would be desirable and the actual reactions in most cases virtually unknown. In all those situations where waste-water is the raw material for drinking water this is potentially a very difficult situation and work on virus inactivation should be encouraged.

Discussion by
Dean O. Cliver, Ph. D. Associate Professor
Food Research Institutes and Department of Bacteriology
University of Wisconsin Madison WI 53706, U.S.A.

There are fundamental differences between the two papers which I have been asked to discuss. Dr. Berg's approach is extensive, and Dr. Lund's is intensive. Both suffer significantly for lack of solid information regarding virus infection by the oral route. Neither mentions infectious hepatitis, which is the leading (or perhaps the only) documented, water-borne virus *disease*.

Dr. Berg cites reports which indicate that, for attenuated polioviruses, a human oral infectious dose is equivalent to a tissue culture infectious dose. Though I have cited these same references myself at times, I think it important to mention that the viruses used are not necessarily representative of all those shed in human feces, and that the recipients are not necessarily representative of humanity. This is not a criticism of the cited studies: their results simply were not intended to be applied in this way.

One would like to know more of oral infection so as to add a perspective to our view of Dr. Berg's data on virus incidence. One would also like to have some basis for evaluating Dr. Berg's suggestion that an oral pathogenic dose significantly exceeds an oral infectious dose for man. Finally, one would like to be able to deal with the question of whether a virus particle, the coat protein of which has been denatured, can still infect man. This last uncertainty is one cause of Dr. Lund's inability to phrase a rigorous definition of virus inactivation. Granting that a virus is inactivated when it has lost its ability to infect (under some agreed-upon conditions), we still do not know whether all of the inactivation phenomena described by Dr. Lund are relevant to human health.

I shall pass from this point and assume, for the purpose of this discussion, that the viruses which can now be studied in the laboratory are valid models of those which are significant in wastewater. Dr. Berg suggests that virus-free water is likely to be free of bacterial pathogens also. This might mean that selected viruses could be used as primary models in the development of new wastewater treatment strategies. My idea may be unwarranted extrapolation of Dr. Berg's statement, but I find the thought attractive. Rather than being *indicators* of fecal pollution, human enteric viruses are quintessential fecal pollutants. If they are, in fact, the most resistant of these pollutants, then it is time we stopped using them *last* to evaluate wastewater treatments.

Dr. Berg surveys a number of treatment processes with an eye to their effectiveness in removing viruses. Both he and Dr. Lund are interested in the activated sludge process, which sometimes removes large quantities of viruses. Unfortunately, it is said to be unreliable and mysterious.

Both authors arrive at chemical disinfection, principally with chlorine or other halogens or with ozone, as a last resort. This may be unfortunate for at least three reasons. First, I suspect that the cost of disinfectants is relatively high in developing nations whose water safety problems are a good deal farther from solution than those in the United States or in Europe. Second, little effort has yet been made to characterize the products of reaction with these powerful disinfectants. I suspect that when these reaction products are investigated by workers who are principally concerned with environmental mutagens and carcinogens, we shall see chlorine and ozone in danger of being labeled pollutants themselves. Third, if one disregards other available means and bases his virus removal strategy entirely upon chemical disinfection, the result is likely to be the kind of "overkill" advocated by Dr. Berg. I submit that one should use a disinfection process designed for 12 log units of virus inactivation only if the previous treatment steps are entirely ineffective against virus or if the disinfectant and its reaction products can be shown to be free of toxicity for man and animals.

Partial alternatives to disinfection merit more study. Any treatment which can be relied upon to remove some virus under known conditions should be explored and exploited. I described previously (Cliver, 1971) how radio-actively labeled virus could be used for rapid, bench-scale evaluation of flocculation and adsorption treatments. We used this only to determine the virus-removing capabilities of some proposed treatments, but labeled virus could also have been used in developing improved physical-chemical treatment methods.

More important, I hope, is the potential for use of labeled virus in the study of biological inactivation. Studies in our laboratory, reports of which are in press (Cliver and Herrmann) and in preparation, have been concerned with the abilities of bacteria to inactivate viruses. Some laboratory strains of bacteria have been found to dismantle particles of certain enteroviruses ("virolysis") and, in some instances, to use virus coat protein as substrate ("virophagy"). I do not suggest that this proves that biological virus inactivation will be a preeminent part of future treatment processes. I merely wish it known that the means are at hand to answer some of the questions raised by Dr. Lund about biological treatment. It should be possible to survey individual organisms for virucidal capability and to determine optimum conditions on a more rational basis.

Such an approach ought not be expected to solve all of the problems cited by Drs. Berg and Lund, but it should almost inevitably solve some. Certainly biologic wastewater treatment, under optimal conditions, is likely to prove of great value for small-scale applications in "non-affluent" and "non-urban" situations, as well as where "non-human" wastes predominate. The conditions chosen might be those which caused virus to be removed most reliably, rather than just in the greatest quantity, as this would allow decreased use of final disinfectant.

Dr. Lund urges further studies of virus inactivation, and Dr. Berg lists several specific subjects which need more investigation. I shall close by complementing these proposals with three that I have made in the course of this discussion. First, we should be learning a great deal more about how enteric viruses initiate infection by the oral route. Second, we should be developing more efficient flocculation or adsorption methods, while keeping in mind that virus removed in this way from water is not necessarily inactivated and must be handled with great care. Third, we should be seeking means to exploit the natural virucidal capabilities of microorganisms. If these last are not immediately applicable in treating the wastewater of large urban centers, there are certainly a number of other situations in which they might be used to advantage.

REFERENCE

CLIVER D.O. (1971) Viruses in water and wastewater: effects of some treatment methods. In *Proceedings of the Thirteenth Water Quality Conference — Virus and Water Quality: Occurrence and Control* (Edited by SNOEYINK V. and GRIFFIN V.), p. 149. University of Illinois Engineering Publications, Urbana.

L.J. Bollyky, USA

Ozonation deactivates viruses very rapidly in seconds. Electron microscopic examination shows that the viruses appear to have exploded. A dosage of 15 ppm ozone is sufficient to inactivate viruses in an activated sludge treatment effluent. These studies were reported at the Ozone Symposium at the University of Wisconsin in November of 1971.

D.H. Rosenblatt

What information is there on the efficacy of chlorine dioxide (ClO_2) as a virucide?

G. Berg

Such studies have been undertaken by Dr. J.E. Smith at Syracuse University under grant by the Environmental Protection Agency, but he has not yet published data.

L.J. Bollyky, USA

The virus content of discharged effluent should be correlated with occurrences of disease and infection. Are data of this kind available?

G. Berg

Good correlations are not available.

Y. Kott, Israel

In Haifa (Israel) wastewater and in trickling filter effluents approximately the same number of enteric viruses have been found, (100 PFU/100 ml). The numbers in oxidation pond effluents were only 30 — 60 PFU/100 ml. When 8mg/l chlorine was applied to oxidation pond effluents a 30% decrease in viral number was recorded. With 30 — 35 mg/l chlorine no enteric viruses were detected.

L. Coin, France

What quantities of chlorine are necessary to obtain the results given in the section "Disinfection" (Fig. 1) from well purified effluents? The proposed norms (1 PFU per 100 gallons) are very severe. The probability of human contamination under these conditions seems to be minute. The WHO norm is quite adequate.

G. Berg

Dr. Coin's question on chlorine dosage was answered by Dr. Kott's comment. The proposed tentative renovated water standard of less than one PFU of virus in 100 gallons of water is readily attainable in a good quality water by HOCl or O_3 treatment. The studies on HOCl by Dr. P.V. Scarpino of the University of Cincinnati, and those on O_3 by Professor H.I. Shuval and Dr. E. Katzenelson at the Hebrew University clearly show this. Their data demonstrate that we should be able to reduce the virus level of a good quality renovated water to less than 1 PFU per 1000 gallons, or per 10,000 gallons, or to levels much less than that with low concentrations of HOCl or O_3. The limiting factor in the proposed tentative standard, at this time, is our capability to detect small quantities of viruses in large volumes of water. Disinfectant-time couplings which will achieve the amount of virus destruction that I recommended, however, should reduce virus levels to much less than one PFU per 10,000 gallons of water.

Y. Zohar, Israel

Is there any connection between the author's choice of units and the reported infective dose? Do most present municipal drinking waters meet this proposed standard?

G. Berg

I assume that all viruses of human origin are capable of infecting man.

Many municipal drinking water supplies do not meet the proposed tentative standards, but in the interests of the people they serve, they should.

Reply by G. Berg to Discussion by D. O. Cliver

Dr. Cliver is correct in his assessment that we need to know more about the minimal infective dose and the relationship of the minimal infective dose to the disease-producing dose. He is well aware of the efforts we have made since 1965 to obtain funding for such research. Unfortunately, even when funds become available, such studies in man will present us with important and obvious moral and ethical problems. For the time, we must be content with the data that we do have which show clearly that the smallest amounts of some viruses that infect cell culture are sufficient to infect man.

I do not share Dr. Cliver's concern with the cost of disinfectants to developing countries, because disinfectant costs are usually very low compared to the costs of the treatment processes that must precede their use and upon which Dr. Cliver wishes to rely. Moreover, a good disinfectant is not only the more effective destroyer of viruses, and by many orders of magnitude, than the best treatment processes we know today, but it is more consistent and reliable than the treatment processes. A disinfectant, if it is present can be counted on to do some killing of pathogens when treatment process operating marginally may remove few or none.

Dr. Cliver is correct that chlorine and ozone may react with pollutants in water to form carcinogens or toxicants. It would have been folly, however, had we chosen sixty years ago to await such data before initiating chlorination to control the waterborne enteric pathogens. We have, of course, been slow to recognize the problems with carcinogenic and toxic by-products of disinfection (and other chemical treatments as well), and we must deal with them. We already know we must neutralize chloramines which are toxic to fish and smaller living things. We must deal with other such problems quickly as they arise, experimentally and before massive field applications, when possible.

I have dealt with Dr. Cliver's concern with overkill in my reply to Dr. Coin.

J. L. Melnick

It is the better part of wisdom to assume that one tissue culture dose is enough to infect a human being, and therefore detectable levels of human pathogenic viruses in drinking water should not be tolerated.

I feel that attention should be directed towards methods that destroy the nucleic acid core — the genome of the virus. We know that virus supposedly inactivated by heat may still contain recoverable infections nucleic acid. It is conceivable that such virus may be reactivated in macrophages in vivo. More work should be carried out on physical removal of viruses from water. In connection with carbon adsorption Wallis has recently found that carbon may be treated so that it binds viruses in an irreversible manner, and we are examining the use of such specially activated carbons for field use.

CONSIDERATIONS ON WATER POLLUTION PROBLEMS IN DEVELOPING COUNTRIES

R. PAVANELLO
Chief, Environmental Pollution Division of Environmental Health,
World Health Organization, Geneva, Switzerland.

G. J. MOHANRAO
Central Public Health Engineering Research Institute,
Nagpur, India.

A definition of water pollution* which has received general agreement links together three essential concepts: first, the action of man, excluding therefore purely natural changes in composition or condition; secondly, the introduction of any kind of waste matter, other foreign substances, or energy, in concentrations or quantities such as to impair the use of the water in question; and, lastly, the use specified by law or established by custom for any particular body of water. Thus, for instance, the abstraction of substantial amounts of water from a river may be considered as a form of pollution since this reduces its dilution capacity. In countries with high population density and highly developed industry, this definition applies almost generally; it may in fact be said to have been derived from experience and from study of the effects of waste discharges of all kinds into water bodies, and of the measures that have been, or are being taken, to reduce or control the steady degradation of water quality in these countries.

Developing countries, however, where municipal water supplies are far from being available to every urban dweller and where the sewers serve only a small proportion of the towns — and then only in part, where industrial development is only beginning, and agricultural practices do not yet include extensive use of irrigation or of chemical fertilizers and pesticides, present a somewhat different picture. In many of these countries several other factors lead us to consider existing or potential water pollution problems from a different point of view. An analysis of these factors based on information collected by special surveys, meetings and various studies, may perhaps help in appraising the magnitude and severity of the problem in these countries.

Population – Urbanization – Industrial development

A WHO survey (1964) concluded that the growth of population in 75 developing countries was 40% greater than the average for the world as a whole and that the rate of urbanization might be even greater.[1] Even assuming a modest 5% per year increase in the needs for water supply and sewerage (the United Nations Second Development Decade anticipates that by 1980 another 420 millions will have been supplied with water by house connexions or standpipes), this will mean that in little over a decade the total requirements for sewerage and water pollution control facilities will have doubled and this, of course, does not take into account the present backlog, the extent of which is outlined in another section of this paper. Industrial growth is proceeding at an ever faster pace. These last two decades have witnessed an expansion of industrial production of ten or twenty times that of 1950 — in some sectors the rate of growth has been even higher (see Table I – Rate of growth in population and some industrial sectors influencing environmental pollution in India). What is important is not so much the short time at our

---

* Water is considered polluted when it is altered in composition or condition, directly or indirectly, as a result of the activities of man, so that it becomes unsuitable or less suitable for any or all of the functions or purposes for which it would be suitable in its natural state.

TABLE I — Rate of growth in population and some industrial

sectors influencing environmental pollution in India

		Units	1950–51	1960–61	1965–66	1968–69	1978–79	1985–86
1.	Popn — total	millions	359	435	490	527	666	740
2.	Popn — urban	millions	62	78	93	103	144	166
	Mineral production							
3.	Coal	mill. tons	32.6	52.6	67.7	69.5	130	180
4.	Coke	mill. tons	N/A	2.5	4.8	7.9	17.4	27
5.	Crude oil	mill. tons	0.2	0.5	3.5	6.1	11.0	15
6.	Iron ore	mill. tons	3.0	10.7	24.5	28.6	71	107
7.	Bauxite	1000 x tons	65.0	387.0	680.0	908.0	3000	5100
8.	Limestone	mill. tons	1.5	12.7	20.0	21.3	50	81
	Industrial production							
9.	Steel ingots	mill. tons	1.5	3.4	6.5	6.5	16.8	32
10.	Sugar	mill.tons	1.13	3.03	3.54	3.55	6.50	100
11.	Paper and paper board	1000 x tons	116.0	350.0	558.0	647.0	1250.0	2100
12.	Newsprint	1000 x tons	–	23.0	30.3	31.0	300	600
13.	Chemical pulp	1000 x tons	–	–	48.0	55.0	240	500
14.	Cotton yarn	1000 x tons	534.0	801.0	907.0	959.0	1500.0	2000
15.	Jute	1000 x tons	837.0	1097.0	1302.0	998.0	1900.0	2400
16.	DDT	1000 x tons	–	2.8	2.7	3.0	12.0	20
17.	Alcohol	1000 x tons	20.0	91.0	190.0	154.0	400.0	600
18.	Petroleum products	mill. tons	0.2	5.8	9.4	15.4	39.0	70
19.	Fuel oils	mill. tons	–	1.5	2.8	4.2	8.2	12
20.	Tannery (hides)	thousands	2700.0	3500.0	4283.0	4500.0	6000.0	7500
21.	Nitrogenous fertilizers	1000 x tons	9.0	101.0	232.0	541.0	5200.0	8000
22.	Phosphate fertilizers	1000 x tons	9	53	123	210	2100	3000
23.	Commercial vehicles	thousands	116	225	332	386	1025	2100
24.	Cars, jeeps, etc.	thousands	169	310	456	560	1110	2200
25.	Scooters, motor-cyles, etc.	thousands	27	95	243	370	1800	3400
26.	Electricity, total install-ed capacity	mill. kw.	2.5	5.7	10.2	14.3	39	65
27.	Thermal	mill. kw.	1.7	3.7	6.1	7.9	19.7	30

disposal to meet these requirements, but the fact that we are dealing with exponential trends and that, if these continue, the possibility of any country being able to maintain the existing quality of water resources is inconceivable at the present time.

In highly industrialized countries the present concern about the conservation of water resources and of the marine environment centres largely on chemical contamination. Although pollution by putrescible organic matter is increasing, these countries rely on the 'assimilation capacity' of water bodies, as well as on the general application of conventional technological wastewater treatment, which can remove more than 90% of the organic pollution load. From the point of view of domestic use, they likewise rely on the fact that normally a large percentage of the population is supplied with piped water treated and disinfected so as to be safe for the consumer. Sanitary control of the supply has long been established as a routine practice thanks to experienced personnel and modern laboratory equipment.

In most developing countries the situation is vastly different.[2] First, many of them are situated in or include areas where water resources are scarce; others may have such great seasonal variations in precipitation that streams run dry for several months of the year. Secondly the rates of population growth and of urbanization are greater than those of industrialized countries. Taking into account a WHO estimate that only a small percentage of urban dwellers are supplied with reasonably safe piped water, the vast majority of the population in such countries have to use sources that are quite unprotected from pollution (see Table II).

Thirdly, in many countries water-borne diseases are still endemic and the prevailing insanitary conditions often result in epidemic outbreaks. It has been reported that in the developing countries about 5 million die each year from enteric diseases alone and 500 million suffer from such diseases.[3] In some of these countries certain communicable diseases have shown a downward trend as a consequence of improvements in standards of living, better public health and medical services, and of general economic development. However, the need to grow more food and for industrial development is leading to more intensive use of water resources often resulting in the spread of some water-borne diseases, such as malaria, filariasis and schistosomiasis (i.e. some of the most debilitating chronic diseases) to new areas.

TABLE II. Population served with water
supply and sewerage in selected countries

Country	Urban population		Total population	
	millions	% of total	% with water supply	% with sewerage
India	80	13	12	5
Japan	98	68	69	11
Lebanon	2.4	32	85	?
Philippines	35	17	30	6
Thailand	26	15	3	Nil
Ethiopia	22	8	8	Nil
UAR	30	40	40	?

Examples of the existing situation

There is no acceptable index which could give a measure of the increase of water pollution apart from figures showing population density and industrial production. Some very rough formulae based on economic and population growth, or associating these with population density, can give only a very approximate idea of the extent of the problem. Perhaps illustrations from large developing countries such as India, and regions such as Latin America, give a better perspective of the trends in developing countries.

TABLE III. Status of sewage disposal in some of the major cities in India

City	Approximate population (1961 census) millions	Present practice of sewage disposal
1. Greater Calcutta	7,500,000	Only a part of the city is sewered and primary treatment provided. Part used for farming; balance to river in untreated condition.
2. Greater Bombay	5,500,000	Only 10% of city sewage is treated partly with secondary treatment. Balance goes untreated to the sea, and an odour problem exists in some areas.
3. Delhi	2,400,000	Two modern plants with primary and secondary treatment.
4. Madras	1,800,000	Part of sewage used for irrigation and part goes direct to the sea.
5. Kanpur	1.500,000	No sewage treatment plant. Part for irrigation.
6. Hyderbad	1,300,000	Screens, grit and balancing tank followed by irrigation
7. Bangalore	1,250,000	No sewage treatment plant.
8. Ahmedabad	1,250,000	Two modern sewage treatment plants. Also irrigation.
9. Poona	750,000 ⎫	
10. Nagpur	700,000	
11. Lucknow	675,000	
12. Agra	550,000 ⎬ No treatment. Raw effluents used for	
13. Varanasi	500,000	farming.
14. Madurai	450,000	
15. Allahabad	450,000 ⎭	
16. Amritsar	400,000 ⎫	
17. Indore	400,000	
18. Jaipur	400,000 ⎬ Secondary treatment plants. Effluents	
19. Sholapur	350,000	used for farming.
20. Patna	375,000 ⎭	

The population of India (1971 census) is 547 millions of which 109 (20%) live in urban areas (i.e. centres with more than 5000 inhabitants). Only 7% of the 2430 towns and cities are sewered — in terms of population, approximately 30% or 33 millions are served with sewers. This wholly inadequate situation is in fact even worse than first appears, when one considers that many of the cities with a sewer-network have in recent years undergone rapid industrial development and that industrial wastes are also produced in increased amounts, while treatment facilities, as shown in Table III, are adequate only in a few instances.

In West Bengal the river Domador is the main source of water for millions of people. The industrial development that has taken place in the last decades in this basin (and this development continues), according to the Central Public Health Engineering Research Institute, now produces 100,000 m^3 of industrial wastewaters per day containing 43,100 kg of BOD, in addition to toxic compounds such as cyanides, ammonia and large amounts of untreated sewage. The concentrations found in the river water reach 30 mg/l of BOD, 3 mg/l phenols and 8 mg/l iron.

Also in India, the city of Kanpur on the Ganges discharges into the river the untreated sewage and drainage waters of more than 1.5 million people along with the effluents from tanneries, textile factories, jute mills, and a number of chemical and pharmaceutical plants with a population equivalent of 15 million people. The river Kalu, a tributary of the Ulnas, near Bombay, also receives practically untreated wastewaters from a variety of chemical plants. The resulting disastrous effects can be seen not only in the two rivers, but also in the destruction of marine life in the estuary of the Ulnas in the Arabian Sea. The river Chaliyar near Calicut, once of good quality, has since the construction of a rayon pulp mill which discharges 57,000 kg BOD per day, becomes useless but for navigation and waste disposal. The waters of the Mahi Sagar — into which, among other wastes, the effluents from a nitrogenous fertilizer plant containing concentrations of ammonia and arsenic of 500 mg/l and 20 mg/l respectively are discharged — have been reported to kill fish, cattle — and elephants.

The river Chiliwong flowing through Djakarta is now so polluted that septic conditions have developed giving rise to objectionable odours. In Nepal the discharge of wastes from the sugar, jute and tanning industries into the river flowing through Katmandu makes its waters unsuitable for any other use downstream, and during dry-weather flow stretches of the river become septic. Similar conditions prevail along some stretches of the Mekong in Thailand, while there are even more serious problems in the southern part of the country from the discharge into ditches and canals of untreated wastewaters from fish-meal factories and other plants processing agricultural products. In some areas pollution is already interfering with rice-growing and in many localities gross pollution has practically excluded any use of the water and there is a great odour nuisance.[4]

In a group of East African countries surveyed in 1970 a number of water pollution control activities have been reported.[5] In Kenya plans for a paper-pulp factory on the river Nzoia include provision for maximum in-factory recirculation of process water, removal of fibres by flotation and treatment of the dissolved organic matter in aerated lagoons. In the copper belt area of Zambia a water quality sampling and measurement programme has been initiated under the coordination of a national water pollution committee. The existing sewage treatment plant in Livingstone is not in operation, and outbreaks of waterborne diseases are reported from Mpulungu. In Tanzania wastewaters from coffee and sisal processing create local pollution problems, but in each of the 17 regions of the country water advisory boards have been set up to advise the water officer on applications for the establishment of new industries. In Uganda investigations are being carried out on sewage treatment by means of lagoons. The Government, with the assistance of the Special Fund of the United Nations Development Programme and WHO, is preparing a master plan and feasibility study for water supply and sewerage for the metropolitan areas of Kampala and Jinja, the two largest population and industrial centres in Uganda.

The river Pasing in the Philippines, once a source of household water and fish for the thousands of people living on its banks, can only be classed now as an open sewer serving as a receptacle for most of the untreated domestic and industrial wastes generated in Manila.

In 1961 with the agreement reached at Punta del Este, Latin American countries set goals to provide, within 10 years, water and sewerage installations to serve 70% of the urban population and 50% of the rural population.[6] Although specific figures were not available at the time of writing this paper, it is known that the programme launched in 1961 has achieved extraordinary success, as regards the provision of water supplies in many countries. Sewer construction has, however, continued to lag behind. Pollution caused by sewage in urban communities in Latin America is summarized in Table IV. This table does not indicate pollution by industrial wastes — organic or chemical. It is, however, known that industry has been developing very rapidly and is concentrated in and around large metropolitan areas. The annual growth of the fish-meal industry is 23%, of the chemical and wood-pulp industries 18%, steel processing 13%, petroleum products 12%, this in addition to other industries with a high pollution potential such as meat-packing, canning and freezing of food products, distilling, milk processing, and the plastic, rubber, sugar and textile industries. The need for increasing food production for a population growing at a rate of about 3% per year has brought about, as elsewhere in the world, an expansion in the production and use of fertilizers and pesticides. However, the worst water pollution problems are to be found around the large metropolitan areas which, because of their size, produce enormous amounts of wastewater which can often be seen flowing from open drains into rivers or coastal areas. Information from 1962 indicates that less than 10% of the population provided with sewers also benefitted from

TABLE IV. Pollution in urban communities in Latin America

Facilities	Urban popn millions	Degree of pollution
sewers and some form of treatment	5.7	partial control
sewers but no treatment	51.7	severe to moderate pollution of streams and coastal areas
house water-connexions but no sewers	30.0	diffused pollution of soil and streams
easy access to piped water but no sewers	19.0	some degree of land and stream pollution
no access to piped water	39.0	little or no water pollution problems

some degree of sewage treatment, and this situation is not considered to have changed much during the last 10 years.

These are but a few examples. Detailed surveys may well bring out a vast number of situations where the uncontrolled discharge of wastes into streams, lakes and coastal areas has not simply altered the appearance, biology and biochemistry of these water bodies, but has had catastrophic consequences in rendering useless (except as a receptacle for wastes) water resources that are already scarce.

Discussion and conclusions

It is to be hoped that the situation as regards water pollution (fresh and coastal) in the developing countries will be clarified with the additional information that the countries concerned are collecting for, and the discussions expected to take place at, the UN Conference on the Human Environment on this matter. At the moment only general provisional conclusions can be derived from the information at hand, some of which has been summarized above.

1. The first observation is that in the large majority of the countries considered there is a lack of general awareness of the importance, for the health and the social and economic development of the population, of water resources conservation. This has led in turn to the absence of a water quality policy as an essential part of a more general policy on environmental quality conservation or restoration. Most of these countries, even where state or local authorities have initiated water pollution control programmes, do not have adequate national legislation. Although many now recognize the need for such legislation, or are in the process of drafting national laws, the complexity of the question, conflicting interests and, in general, the lack of precise data, may delay the passage of such laws and hence their enforcement for many years. While there is no doubt about the basic principle that, as a community responsibility, water pollution control should be handled by law through governmental or public authorities, many still question whether the great cost entailed by the application of control measures should be borne directly by the polluter or by the community benefitting from their application. Experience from the advanced countries confirms that the problem is a difficult one and that even the most careful compromises hardly make for equitable application.

2. Another observation, more pertinent to that group of countries where sporadic efforts are made to control pollution, concerns the approach used. No generalization can be made, but it is clear that some countries have enacted legislation based on a classification of water uses and have set basic quality criteria for the various classes of water, but have failed to classify the water bodies in the country accordingly. In the absence of specific regulations, control becomes difficult or impossible. Still other countries have only 'nuisance control' legislation which can be applied only when water bodies become so degraded that the cost of restoring them is out of reach of the community in question. In still other countries, legislation is so strict that it has fallen into disrepute and is quite impossible to enforce. In almost no case does there appear to be a rationale behind the action taken with a view to practicability or to preventing further deterioration on a national scale.

3. It is clear that in many countries the major pollution problem is associated with the discharge of untreated municipal wastewaters. It is also clear that a very small proportion of urban areas are adequately equipped for hygienic disposal of sewage. Priority should therefore be given to the acceleration of programmes aimed at the proper collection, treatment and disposal of domestic sewage (including sullage). As existing water supplies

for urban areas are extended, or new ones installed, the backlog of sewerage is likely to increase rather than decrease, and the pollution of the soil, and of ground, surface and coastal waters is bound to spread further especially around densely populated areas and regions. A major change in policy is needed in many countries if this trend is to be reversed.

4. As industrial development progresses, more severe — even if localized — problems are likely to arise. Pollution by organic and inorganic chemical wastes is bound to increase faster than that due to domestic wastewaters. Developing countries have, however, the possibility, through sound planning, adequate siting of industrial development schemes, choice of processes and especially by regulatory action, to curb or minimize the inevitable deterioration of water quality brought about by economic development. Regional and international cooperation may help in surmounting present difficulties of finance, technical 'know-how' and equipment.

5. In some countries attempts have been made to integrate water resources planning into national and regional planning for economic development. Unfortunately, little or no attention has, however, been paid to water quality, for by the judicious siting of urban and industrial development areas, the re-cycling or re-use of wastewater (especially by incorporating these measures when planning water pollution control), impairment of the quality of essential water resources could be contained within acceptable limits. And though, during the first years or decades of development, restrictions on waste discharges in order to maintain a high water quality in streams and lakes may, understandably seem to be an unnecessary luxury, experience in the advanced countries shows that once un-controlled discharge becomes the established practice, it is difficult, if not impossible, to change.

6. As is being done as regards quantitative measurements by the hydrological services in almost all countries, water quality data should also be collected systematically (for surface, ground and coastal waters), as a basis for planning and, wherever possible, for determining priorities of intervention with suitable control measures, or to assess the efficacy of measures put into operation. A programme in this direction need not be too elaborate. Initially, quality measurements could be limited to critical places such as intakes at waterworks, major fishing areas or where irrigation water is extensively used, important resorts, and only essential parameters to assess significant variations (pH, temperature, suspended solids), organic content (DO and BOD) need be applied. In special cases parameters of health significance (coliform bacteria, toxic matter) or for agricultural use (TDS, salinity) need to be measured periodically. As the programme gets under way, the necessary organization and facilities established, and personnel adequately trained, the number of sampling points could be expanded, as well as the frequency and the analysis of other parameters and polluting substances, on the basis of the number and kind of waste discharges and on the actual or planned uses of water.

Most countries where water pollution control programmes have been under way for years are still struggling to find a practical as well as a rational approach which may be applicable to the whole territory. Many have set up classifications for various water uses, the criteria defining these classifications and, in some cases, plans for implementation. However, because of the very nature of the water pollution problem a great degree of flexibility in interpreting and enforcing the national legislation has to be left with state, provincial or local authorities. This difficulty is felt much more acutely in developing countries. If national legislation prescribes certain standards for water quality, measured in terms of defined parameters, such quality will be maintained only by controlling the discharge of wastewater at the source. This in turn implies that the regulating authority must be able and equipped to license every source of pollution and to decide in each case on the concentrations and amounts of polluting matter that can be discharged along a river or into a certain water body, so as to comply with the national classification and with the established use of that river or water body. This approach may well be beyond the possibilities of many of the developing countries, in terms of the type of organization needed, its cost, the availability of trained personnel and supporting services. In addition, at the early stage of economic development the authorities concerned may well decide that the very limited resources available should be entirely devoted to the development of agriculture, industry, communications, education, housing, and other essential projects. Even in this case, however, it is necessary to set up limited monitoring schemes, so that significant changes in water quality may be detected or predicted in time, and so that possible effects on further development — agricultural, industrial and urban — may be assessed.

REFERENCES

1. Water Pollution Control in Developing Countries — Report of a WHO Expert Committee. *Wld Hlth Org. Techn. Rep. Ser.,* 404, Geneva, 1968.
2. Report on a WHO Inter-Regional Seminar on Water Pollution Control — New Delhi, India, November 1967. World Health Organization, Geneva, 1968 (Document EP.68/4).
3. Summary Report of the Regional Development Plan for the South-East Resource Region. Joint Planning Board for the SERR, Ministry of Works, Health & Urban Development, New Delhi, June 1971.
4. Report on a Regional Seminar on Water Pollution Control, WHO Regional Office for South-East Asia, New Delhi, March 1971 (Document SEA/Env.San/96).
5. Survey of Inland Water Pollution in Uganda, Kenya, Zambia & Tanzania Food & Agriculture Organization, Rome, 1971 (FAO document RP.11).
6. World Health Organization/Pan-American Health Organization Symposium on PAHO document ES/WPS/1, 1970.

Discussion *by* Leonard B. Dworsky,
Director of the Water Resources
and Marine Sciences Center,
and Professor of Civil Engineering;
Cornell University, Ithaca, New York.

In connection with the authors' conclusion should we be concerned with the lack of general awareness? Should we not be concerned more specifically with lack of awareness by political and economic planning leaders, for example, that may be much more instrumental in water and environmental policy development? What is the role of this Association in this matter?

As to their Conclusion 2 it is essential that a national program be based on some kind of national strategy. It is not clear what the nature of that strategy should be. It is interesting to note that the authors suggest the need — not for a rational strategy — but rather for a rationale (a set of reasons) "with a view to practicality." What is needed is an effective, working strategy.

National strategies in the United States are under criticism and revision. For example, the formula by which funds are allocated by the Federal Government has not yet effectively controlled pollution on any single U.S. stream over a sixteen year period. Stream classification, initiated in the 1965 Water Pollution Control Act is now being challenged by the development of National Effluent Standards.

With regard to their Conclusion 3 what shall the nature of this change in policy be? If the data leads us to the conclusion "we are dealing with exponential trends and that, if these continue, the possibility of any country being able to maintain the existing quality of water resources is inconceivable at the present time" as suggested by the authors, then we had better assess early and rapidly the character of the technology we want to apply.

In the U.S. major policy changes are being explored by the Army Corps of Engineers in cooperation with the Federal E.P.A. and State and local authorities at five test sites (San Francisco; Detroit; Chicago; Cleveland; Merrimac River Basin in the N.E.). These and other innovations should be demonstrated in a wide variety of places in the next few years to facilitate the " "marketing" of new technology.

Concerning Conclusion 4 in some aspects of water resources planning — as in irrigated agriculture — water planning and land planning have, more or less, been considered together. Increasingly it should be recognized that water quality planning and land planning must be an integrated process.

Proposed land planning legislation in the United States may assist in developing a stronger land planning program, sorely needed if water quality planning is to be effective. A large degree of responsibility for this present lack lies with water pollution control authorities who neither sense this relationship nor are willing to provide the leadership to facilitate its development.

Water resources planning and water quality planning, referred to in Conclusion 5, while growing stronger, have proceeded along parallel lines and, thus, in few places do we have "integrated" planning.

We have long been aware of the need for integrated planning. Unfortunately, authority is still divided among, in the U.S., the E.P.A. and resource development agencies as the Corps of Engineers. It will be necessary for Congress to provide for this integration, since it is unlikely that the executive agencies themselves can solve this problem.

The matter raised in Conclusion 6 is of extreme importance if we are to determine our future strategies. Only if we can assess the value of programs in time can we change as needed.

The most important element in a water quality monitoring program, however, is the analysis and interpretation of the data collected. An adequate interpretive program is still not available in the United States despite a decade of water quality monitoring.

I believe this paper implies three important points, none of which are explicitly stated:

1. It should be clear that progress will be generated only when a single individual — and then other single individuals — develop a recognition and understanding of the water pollution problem to the health, economic and cultural qualities of the country concerned, and then to the region and world scene.

2. It is abundantly clear that this Association, together with comparable national Associations, is an effective force for the stimulation and exchange of technology. It is equally clear that, based upon a review of the program of all six conferences, and with the exception of the San Francisco program, the opportunities provided by the Association to contribute to the effective transfer of technical information to those responsible for public policies or private sector policies is essentially non-existent.

3. that two major challenges face the Association (i) to recognize the very great need for a broader outlook, acknowledging that the absence of technology *is not* the limiting factor in solving national, regional and global problems of water pollution, but rather that the social and economic, including governmental and political, problems are the major obstacles to solutions in water quality management. (ii) to recognize the opportunities available to the Associaton to fill an important gap in the transfer of technical information among not only scientists and engineers; but among those who develop knowledge and are trained to apply it, and those who carry *the public responsibility* to manage and carry out the goals of society.

Dr. Wolman, in his remarks, noted the absence of economic and social concerns in the programs of the Association. I have stated, in terms somewhat more blunt, and broader, the same idea he hinted at.

President Stander asked for help from the delegates as the Association moves along the critical path of developing a viable program.

I hope he, and the Association Governing Board, will seek to use the full potential of the Association in moving into the new future designed during the past weeks in Stockholm.

John Pickford, Loughborough, England

I disagree with Professor Dworsky's claim of similarity between developing countries and the United States; there are vast differences. For example many cities in developing countries need to construct sewers, where there none at present, before they can provide even the lowest cost treatment. Even when a new sewerage system is developed, there is a danger of its benefiting areas which have least need of it. It is easy to construct sewers through low-density residential areas already provided with waterborne sanitation. Septic tanks are replaced by a sewerage and sewage treatment system, but high-density districts served by bucket latrines are left because of problems of sewer construction and low rate revenue.

Proper operation and maintenance in countries adopting water-born sanitation for the first time often receive insufficient attention. In Africa I recently visited a country's only substantial sewage treatment plant, a well-designed waste stabilisation pond system. Virtually no flow was coming in. Following the sewer upstream led to a shattered pipe from which the whole flow, including that from a hospital, poured into a ditch. The accident had occurred sometime before, but no one had bothered to do anything about it.

A major reason for the difficulties in such places is undoubtedly that given in Mr. Pavanello's first conclusion a lack of general awareness of the importance of water quality. It might be better to spend international agencies' funds on vigorous public health education rather than paying for the numerous feasibility studies which produce such a variety of conflicting solutions to developing countries' problems.

Z. Jalon, Israel

The authors' figures show a widening of the gap between population growth and water supplied to the population. From personal experiences encountered as a consultant in developing countries I find there is a feeling that U.N. organizations should act more in policy guidance to governments. Governments should be directed into spending limited budgets available on small schemes rather that on big, long range water supply schemes and thereby ensure immediately a supply of say 5-10 g.p.d. of safe potable water to more people. Agencies like W.H.O. should act more in this respect at government level and on IBRD policy making. This is also true with water pollution projects. Big sewage treatment plant involving high capital cost and high maintenance costs should be postponed for future stages to be preceded by less costly and easy to maintain works.

SANITATION AND LOW COST HOUSING

G. v.R. MARAIS
Professor of Water Resources and Public Health Engineering,
Department of Civil Engineering, University of Cape Town,
Rondebosch, Cape Town, South Africa

INTRODUCTION

Health is an essential component of the infrastructure of a country. Training and acquisition of skills take time; if life expectation is low, training takes a disproportionate part of the active life of the trainee and the country does not reap in full the investment in training. In the developing countries the State is interested in the most rapid development directly productive in wealth and desires to commit as much capital for this purpose as possible. In this situation, in what measure must the State assume responsibility in supplying those components regarded as conducive to good environmental health, i.e. housing, water supply, sanitation and other services.

The focal point of the problem hinges round the concept 'Housing'. Housing can be defined as the environment to promote the physical, sociological and mental well-being of man within the family unit. Well-being, however, is a relative term and one man's well-being is not another's. In tropical developing countries having a high proportion of the population with low productivity and low income, with poor and non-existent health education, the relative concept of well-being as a measure of success of Housing does not provide a practical criterion for decision making. It is necessary rather to define the absolute basic requirements which Housing must satisfy. Three can be distinguished, viz.:

(1)   Access to work
(2)   Water supply and sanitation
(3)   Shelter.

The order of these three is not immaterial and 'shelter' is consciously relegated to third position. It is the thesis of this paper that in 'Housing' the 'house' is the least important.

Access to work

Without access to work there appears to be no justification for constructing a house. In a rural society habitation arises naturally near the place of work. In an urban society this is equally true but the location is more flexible depending on the speed, efficiency and cost of transportation. If access to work is difficult or expensive the journey to and from work is a constant drain on the energy, time, finances and efficiency of the low income groups. Due to the often dismal aspect of badly designed low cost housing areas, there is pressure on the planner from the more affluent (and influential) urban dwellers to site low cost housing away from their areas. This is a problem that presents itself sooner or later in all urban communities.

Water supply and sanitation

These two aspects of Housing go together for they are indivisible. The provision of a potable water supply is a priority which cannot be denied. What is not so obvious is that the provision of a satisfactory form of sanitation is equally vital.

Over the past ten years there has been a drive by the World Health Organization to provide potable water supplies to communities in the developing countries. Where implemented, it has significantly reduced the incidence of diseases such as typhoid, dysentery and cholera. This success has to a degree obscured the fact that increasing the water supply also increases the problem of its disposal. This point is well appreciated in the following quotation from India:[1]

'In many urban towns where water supply was long since installed and a sewer system held up for want of adequate "subsidy", the delay is costing the communities dear. Insanitation and mosquito nuisance have taken root and filariasis is becoming endemic over an ever widening urban area in the entire country. This is hardly a comforting thought. Ironically enough filariasis control is fast assuming an increasing importance as a health measure with prophylactics pressed into service. This is but fighting the shadow and not the substance.'

Thus installation of water supply without waste water disposal merely results in a shift from one dominant set of diseases to another. Such a shift develops only with time; in the interim it engenders a feeling of accomplishment in the unwary so that the true consequences may never be appreciated.

With limited financial resources should water supply with sanitation be provided for a smaller, or water supply only be provided for a larger number of people? In this form the question is unreal for it is not a simple situation of have or have not but a graduation of needs. Rather one should commence by circumscribing the needs to identify the situations where the needs justify priority for assistance.

In a rural society the very dispersal of inhabitants is a powerful factor in reducing incidence of diseases due to insanitation. To achieve a tolerable standard of sanitation and water supply in this situation may require provision of the most elementary nature and is often attainable by individual or family effort. The cost may be only the labour of the individual. There is often a case for leaving things as they are. To insist on a more sophisticated system where a lesser one is adequate is to divert State capital from ventures which may eventually be of greater benefit to the whole country.

As the size and concentration of a community grows there comes a stage where individual effort no longer suffices. It becomes impossible for the individual to protect himself from the interaction of the insanitary practice of his neighbour on himself. Community effort is now imperative in both water supply and sanitation — the lack of either is equally fatal. In this situation it is no longer: 'Can the community afford sanitation?' but 'Has the community the best sanitation it can afford?'. Public money committed for this purpose should be allocated on the basis of water supply and sanitation adequate for the situation. The criteria for selecting the particular system are efficiency at lowest cost and the expectation of continuous function, not the sophistication of the system. If a community desires more sophisticated systems it must provide the capital out of its own resources and cannot look to the State for such aid. It is only by such harsh realism that the State can be equitable to all its citizens and yet leave as much capital as possible for development. As it is the bigger, denser communities which are the most vulnerable to insanitation, this is where the major effort at solution must be directed.

Acceptance of this philosophy will assist in getting health 'purists' to rethink their approach that only the latest and most sophisticated (often confused with high standard) installations should be allowed. It may also induce State purse strings to be loosened sufficiently to allow health measures, such as adequate sanitation, to safeguard the investments in training and the acquisition of skills. A budget which withholds adequate expenditure on sanitation is no worse than a 'health' policy which, by being inflexible in its demands for specific systems, places adequate sanitation for many communities beyond available funds.

Shelter

Examining causes of death in the tropics and subtropics from exposure and from lack of water supply and sanitation respectively, it is evident that exposure is a relatively minor cause whereas insanitation is a major one. Where there is a deficiency of public capital for Housing it seems logical that the available funds should be spent on providing those components of Housing which, from a community point of view, are most likely to bring about the maximum return in improved health and those which it is not possible for the individual to supply himself. On this basis the Housing needs that must be the prime responsibility of the community are access, water supply and sanitation, drainage and a plot for a house. Only secondary is the community's responsibility for providing a house.

A means of implementing this concept is the site and service scheme: the site plus services (roads, drainage, water supply and sanitation) are provided from public funds and the 'shelter' is provided by the occupier of the site. His contribution will depend on his financial resources and his interest. It may be a crude shelter wherein he may live in physical discomfort but, from the communicable diseases point of view, in a tolerably healthy environment. As the costs of services are usually much less than the costs of a house plus services more plots can be provided for the same capital expenditure so that overcrowding is less likely and easier to control. More people will also be able to rent or own their own plot. An important point about the site and service concept is that it devolves the contribution of the occupier on that aspect – the house – most likely to evoke his self-interest and endeavour. This is never possible with services in high density areas.

Site and service schemes usually present a very untidy appearance due to the low standard of building by the occupiers. Untidiness does not make a suburb a slum; a slum is characterised by overcrowding from too high rentals making it imperative for individuals to band together in existing houses in order to afford the rental. It is ironical that the public-spirited by demanding that 'houses' must be provided may unknowingly be guilty of helping to create Housing shortages and slums.

When planning site and service schemes the end results must be clear in the mind of the planner from inception. Planning should envisage ultimately the installation of individual waterborne sanitation and water supply, adequate roads, drainage and lighting. Initially it may be necessary to accept lower standards – for example pit latrines or unpaved roads – in order to deal with an interim situation or emergency. The planning must however be such that improvements can be made as the financial situation allows without existing services being disrupted. Pit latrines, for example, should not be sited where eventually a sewer should run.

Planning on a lavish scale in terms of space or frontage may well result in substandard services which cost as much per plot as full services in a well designed low cost high density area. Service costs are linearly related to the frontage width of plots, hence deep plots with narrow frontages are indicated. The services – light, water and sewerage – should also be located on the inner boundaries of the plots on the centre line of the blocks instead of in the street.

DETERMINATION OF HOUSING NEEDS

Quantitative determination of Housing needs is extremely difficult. House-to-house surveys are normally completely misleading as house owners misinform on the number of residents for fear of penalisation under regulations limiting the number of people per house. This also applies in illegal squatter settlements around a city. In the latter case useful information is obtained from aerial photographs which not only show the exact location of each shanty but also give a measure of the number of families involved.

An effective means of assessing the magnitude of the Housing problem is to attempt a knock-out blow in a typical town, not too large for the cost may be too high. The surveys show that 4—6000 houses are required. A thousand are built and, say, four thousand site and service plots are prepared. Very likely a number of houses are not rented. This gives a sample by which the preference of the occupiers can be determined and of how much they are willing to pay in rental. By this procedure it is possible to discover the real needs of the people and their ability to pay instead of relying on postulations of local authorities as to needs and ability to pay.

In attempts to satisfy each town's needs proportionally it often seems impossible to make an indent into the problem or to define it quantitatively; the situation looks insoluble. Because each town receives only a little help the real needs are obscured — whatever is built is snapped up in competition. Concentrating on one experimental scheme also makes it possible to bring to bear a greater variety of specialist skills, allows thorough investigation and experimentation with alternative layouts, sanitation systems and construction methods which can subsequently find application elsewhere. Piecemeal schemes tend to evade the issues and to perpetuate out-of-date layouts and substandard services.

EVALUATION OF SANITATION SYSTEMS

The standard of one latrine and one water tap per family has received international recognition, but the most suitable type of latrine or system to adopt in any particular situation remains for many developing nations a perplexing question. The water supply aspect is usually accepted with little resistance, but this is not so with the provision of the latrine. It is more expensive than the water supply, its need is not so immediately apparent (it is sometimes looked upon as a luxury), and money invested in it is not directly productive of material wealth. Experience with many sanitation systems in developing countries has often also been a dismal failure so that there is justification for opposition to their installation.

Before pushing for 'improved' sanitation one must be clear on expectation of success of a particular system. For this purpose it is necessary to enquire into the background, health, education and level of technology of the people before evaluating a system in any particular situation.

Background

Most low cost high density housing areas in Africa are occupied by people who moved from agricultural areas. Their behaviour pattern and habits are those of people living in a dispersed fashion where they are totally unaware of the dangers to health when living in close proximity. In a sanitary sense they have no health protective habits for urban living. Consequently they need in general very efficient systems which will work despite their lack of understanding.

Health education

In societies with a high general standard of health education sanitation systems operate satisfactorily because of the safeguard of awareness in the users, that they know how to operate the system and to correct minor malfunctions. In societies where community health education is low this awareness is rudimentary so that there is often no effective attempt to correct even the most elementary malfunctions. In general, the lower the standard of health education the more necessary it becomes that the sanitation system will be self-operating and non-fouling.

Technology

In technologically developed communities there is a tendency to equate efficiency with technological sophistication. Whereas this is normally true, sophisticated systems developed within these communities maintain their efficiency because of an 'infrastructure' of health education and the availability of operational 'know-how'. When such a system is transplanted to a different environment it does not necessarily follow that the same high standard of performance will be achieved on an enduring basis. One example will illustrate: with the flush system toilet paper is in fact a major consideration, although because of its ready availability in richer communities it would not be considered to be a factor of any significance. However, in poor communities where corn cobs, stones, leaves, twigs, newspaper, cement bags, etc. are often the only cleaning materials readily available or within their means, the incidence of blockages increases sharply and additional supervision and expenditure are required to attain even spasmodic efficiency.

With these comments as a background one can now evaluate existing systems for the situation which pertains in low cost high density housing areas.

Pit latrines

The system does not dispose of the household waste water; in particular where piped water is provided, waste water which cannot seep away increases insanitary conditions. Prevention of access of flies to the latrine is always a problem.

The life of the latrine ranges between 2½ to 8 years with an average of, say, 5 years; this implies a minimum of 5 relocations in a housing scheme with a life of 40 years. Digging pits, erecting or re-erecting the superstructure are recurrent expenditures, the cost per latrine ranging from 20 to 40 L.sterling. Pits are difficult to excavate in hard and rocky soils and penetration may require blasting or the use of mechanical tools, greatly increasing the cost. Seepage from such pits may be poor, the pit fills with liquid reducing its useful life. Due to adverse soil conditions it may not be possible to sink pits on all plots so that otherwise good buildable ground has to be rejected.

To reduce the problems of odours it is generally accepted that pit latrines require larger plots than with other forms of sanitation to provide sufficient distance between the latrine, the household and the neighbour's household. The recognised minimum distance from a house is usually 7 m, hence there are requirements of extra land, possibly increased street frontages, with increased cost of attendant services such as piped water to be provided now or in the future. In areas with seasonally high water tables the pit may fill with seepage water up to ground level causing most insanitary conditions. In sandy and soft soil areas the sides of the pits often collapse causing the superstructure to overturn or subside into the pit.

Aqua privies

In the aqua privy the human wastes are discharged directly through a chute into a form of septic tank where the organic solids undergo anaerobic decomposition. For the proper functioning of the aqua privy a certain liquid level must be maintained by periodic addition of water through the chute; if not the water seal around the chute is broken and nuisances and health hazards arise from gases of decomposition and from fly and mosquito breeding in the tank liquor. The overflow from the tank is generally disposed of to a soakaway.

Experience in Africa is virtually general failure of this system due to the omission of users to add water to maintain the seal. Tanks eventually lose all liquid and become effectively pit latrines. If gangs are employed to add water at intervals the system works reasonably well, but failure of the soakaways due to impervious soils, porous soils becoming clogged with organic matter, or seasonal high water tables is common and continuous. Where the washing troughs discharge to the tank the seal is maintained, but the problem with the soakaway is aggravated.

Pail latrines

Odour nuisances and health hazards from fly breeding can arise in the use of the pail latrine and in the removal and disposal of excreta. The system does not dispose of waste water. Removal, cleaning of pails and excreta disposal require constant supervision of a high order. It is difficult to retain supervisory staff and labour is always a problem Labour, transportation and disposal costs can be high.

Waterborne sewerage to conventional works

In low cost Housing areas the system uses an appreciable proportion of the water supply and may be the cause of water shortages. Where health education is rudimentary the lavatory units are incorrectly used, not flushed, and blockages occur due to the use of materials other than toilet paper. In consequence costly clearance services are required to keep this system functioning.

In areas of flat topography to ensure self-cleansing velocities in the sewers deep excavations are required and often pumping must be resorted to. The cost of trenching in hard soils and with high water tables is high. To allow for transportation of solids, sewer diameters are limited to a minimum of 150 mm. In small communities the sewers therefore rarely run at capacity.

The capital outlay per unit capacity for disposal works can be very high in small communities. Mechanical plant is a necessary adjunct in conventional waste water treatment plants, making skilled supervision and a power supply a necessity for continuous operation. Replacement costs of plant are appreciable and continuous, and breakdowns and delays are common. Chemical and bacteriological tests are rarely done. In the majority of cases the efficiency of these plants is unsatisfactory as the communities have insufficient resources to employ technically trained personnel.

Communal facilities

Communal latrines are rarely, if ever, satisfactory no matter what system is provided. There is no individual responsibility. In low cost Housing areas public latrines are usually provided because they are considered the most inexpensive. The price is paid in the consequences of the resulting low standard of sanitary conditions. Once a latrine is fouled due to misuse or accident, the next user may have no choice but to foul it further. Thus a chain reaction sets in and the latrine rapidly becomes unusable. Communal latrines may be up to 100 m from some houses. Consequently at night the latrine is little used; this is particularly true if there is no street lighting. If the latrine itself is not well lit at night nobody uses it. To achieve even a minimum standard of cleanliness strict control and constant attendance of cleaners are required; each time a cubicle is used it must be inspected and cleaned if necessary. Sanitary paper must be provided otherwise blockages due to stones, sticks, leaves and grass in flush systems are frequent.

With communal standpipes and communal ablution and washing facilities water waste can be appreciable. There is no individual responsibility to see that taps are closed and water consumption *per capita* may exceed that where each family has its own individual water supply. Invariably soakaways of communal ablution facilities fail.

In assessing a sanitation system it is important not to falsely identify the situation. Usually if there are no flies, smells or unsightliness the existing system, which may on first appearance not seem to be acceptable, might prove to be quite a reasonable one. Particularly in urban areas beware of the so-called 'cheap sanitation'. Consider the problem of total costs involved, i.e. capital, operational and full maintenance costs, and when comparing systems include the costs of overcoming any deficiencies such as fly breeding in pits, recurrent digging of pits and soakaways. Ground required for effective soakaways of domestic water and relocation of these require considerable areas; bearing in mind that the sewer line can be along the back of the houses excavations for soakaways or pits may be greater than for sewers. Then there is still the cost of stone and cover materials. If soakaways can be successfully and economically excavated, the area is similarly proved for sewers. Often the high cost of sewers is due to rocky areas. If this is true for sewers it is also true for pit latrines. Essentially if the soil conditions are known the forms of sanitation suitable for these conditions are also fixed.

With adequate water supply and sanitary provision higher densities can be allowed. Health authorities intuitively recognise that there is a relationship between densities and sanitary provision. For example, they will generally permit one latrine for every 5 workers with the pail latrine system, but 8 to 12 workers with a flush system; provision of adequate water supply and sanitation allows the occupational ratio of people to land and buildings to be higher.

Reviewing the problems with existing systems, what are then the minimum requirements for a satisfactory system of sanitation in low cost high density Housing? One can put forward the following requirements: the system must be cheap, not communal, use little or none of the potable water supply for the operation, work despite misuse, require little supervision, not use soakaways, dispose of all the waste water, treat the waste water to a degree where it can be discharged with little subsequent danger to users, require no mechanical equipment.

THE SELF-TOPPING AQUA PRIVY SYSTEM

. During investigation of the existing systems for Zambia the potential of sewage stabilisation ponds as an economic and effective form of sewage purification came to notice. This led to a re-assessment of two of the systems investigated, i.e. the aqua privy and the conventional waterborne system.[2] It was decided to combine the desirable features of these two systems in conjunction with stabilisation ponds as follows:

(1) To discharge all the waste water from the household into the aqua privy tank, thereby automatically retaining the seal around the chute. By this means one of the major objections against the aqua privy was overcome, but means of disposing of the increased effluent flow had to be found.

(2) To dispense completely with soakaways by discharging the effluent from the aqua privy tank into sewers to stabilisation ponds.

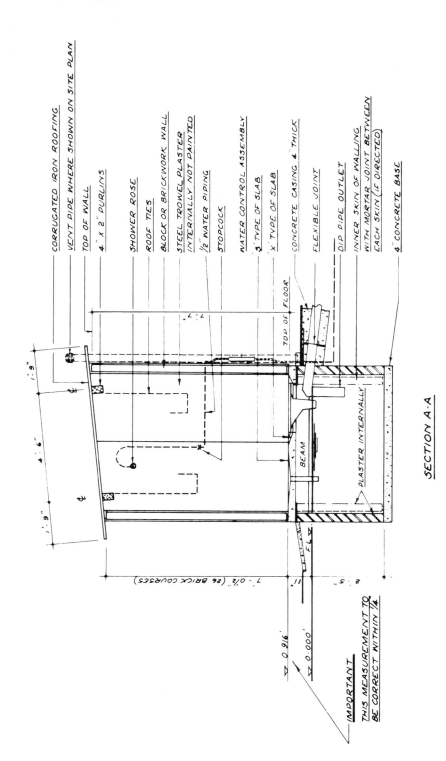

SECTION A-A

Fig. 1. Cross-section of self-topping aqua privy.

The site proposed for the first experimental system was so flat that conventional sewer grades could not be used without the necessity of pumping. As this was most undesirable it was decided:

(3) to use the aqua privy tank as a sedimentation tank to retain all inorganic solids and to pretreat the organic solids to a form more amenable to transportation in sewers with flat gradients. Grades giving a minimum velocity of flow in the sewer of 0.3 m/sec at design flow instead of the conventionally accepted 1 m/sec were adopted as an experimental measure.

The experimental system proved completely successful and led to the adoption of the self-topping aqua privy system in many communities in Zambia.

Water supply preferably, but not necessarily, piped is essential for the operation of the system. In detail the system consists of:

Sanitation block: This comprises 1 to 4 units in one block with one unit per family. The block is located astride the common boundaries or corner of the plots it serves. Each unit comprises a latrine cubicle with an aqua privy squat plate and chute, an ablution cubicle in which a shower can be installed, and a wash basin under cover of a roof. Where possible pipe water is supplied to each basin and shower. The waste water from the wash basins and ablution cubicles drains directly into a tank underneath the building, thereby maintaining the seal around the chute. (Fig. 1.)

Water-carriage system: Overflow from the tank passes into a collecting sewer. Because no sand, stones, etc. enter the sewer and the digested organic solids are finely divided, the design velocity of flow can be reduced to 0.3 m/sec and the minimum size of sewer to 100 mm.

Stabilisation ponds: The effluent from the tanks is most amenable to treatment in stabilisation ponds. The tank reduces the BOD about 50% and as the fermentation potential of the effluent is reduced no problem with rising sludge in the ponds is encountered during summer. Sludge and detritus banks do not form at the discharge point as the sand is retained in the tanks and the inlet arrangements to the pond can be next to the bank. To reduce bacterial concentrations 3 ponds in series are used with 14, 7, 7 days retention time respectively.

Discussion of the system

Individual responsibility: Each family has its own water tap, latrine, ablution cubicle and wash basin. Although the units are grouped in one building, private access is given directly from each plot to the family's unit in the block.

Simplicity of function: It requires nothing beyond the normal household activities to keep it functioning properly and the most elementary hygiene practice to keep it clean.

Complete disposal: It disposes of all liquid waste and excreta from the household.

Site development: It enables sites with very flat topography to be developed with a water-carriage system without the necessity for pumping. Sewers 100 mm diameter at slopes of 300 to 1 have been operated without blockages for a number of years. Consequently excavation for sewers need rarely exceed 1 m and the majority of sewer lines are above the water table. No system in Zambia or South Africa has yet required pumping.

Water consumption: Only waste water is used to flush and transport waste products. Piped water is not essential as the aqua privies can be connected in series receiving effluent from the communal washing point at the head of the sewer, or from hand-drawn water.

Adaptability: If the aqua privy squat plate is unacceptable a flush system discharging directly into the tank can be substituted.

OXIDATION PONDS

Treatment of effluent by oxidation ponds is virtually the only system that will work successfully on a continuing basis for small and medium sized communities. An understanding of this process is therefore crucial to the successful implementation of the aqua privy system or when treating raw sewage.[3]

About 60% of the influent BOD entering a pond settles out as a sludge, the rest enters into the liquid body of the pond. In the sludge layer anaerobic fermentation destroys some of the chemically bound energy, releases methane gas to the atmosphere, and discharges nutrients (BOD) to the supernatant pond water. On average approximately 30% of the original influent BOD is released as methane by the sludge layer. The layer therefore contributes significantly to the reduction of the total pollution load.

In the supernatant, if oxygen is present, reduction of BOD takes place by aerobic degradation. Oxygen is supplied principally by the photosynthetic activity of the algae.

The rates of reaction of the degradation process in the sludge layer and in the pond liquid are both temperature dependent. In the sludge layer the rate increases approximately sevenfold for every 10 degC rise in temperature. Below about 15°C the activity in the sludge is negligible and the pond bottom becomes merely a sludge storage area. Above 23°C fermentation is so intense that the methane gas is generated at a sufficiently high rate to lift the sludge off the bottom. The degradation rate in the pond liquid increases only about two times for every 10 degC rise in temperature and the rate is still appreciable down to low temperature values.

During winter the supernatant deals only with the non-settling influent BOD fraction; during summer with the non-settling influent fraction plus the products of fermentation of the sludge which may, in some situations, increase the BOD load in the supernatant from 5 to 8 times the winter load. Although one would expect that the effluent BOD should show a concomitant rise in summer, this does not occur for the degradation rate in the supernatant has also increased and attenuates the expected rise. However, the oxygen demand rate of the supernatant increases manifold in summer compared to winter. This demand, we have seen, must be supplied by the algae. The factors influencing algal growth are therefore important.

In warm-temperate and tropical climates radiation does not appear to be a critical factor for algal growth and oxygen production; neither does the temperature. The critical factor appears to be the mixing conditions in the pond. In ponds, turbidity of the liquid is high, and light penetrates only to about 0.5 to 1.0 m into the water. The absorbed energy causes a differential of temperature between the top and bottom layers of the pond, inducing a condition of stratification. Mixing of the pond liquid is principally due to wind action and, to a minor degree, thermal cooling at night. Due to the pond's large surface exposed to the air and the shallow depth, the pond rapidly loses and gains heat, so that with normal windy conditions the pond passes through a state of mixing and stratification every day in a fashion similar to a lake over a period of a year. Generally, pond stratification tends to be more intense during summer than winter, but this depends significantly on the wind velocities over these two seasons. During summer, if wind velocities are low, the pond can remain stratified for weeks on end. With low wind velocities in winter, mixing is often initiated during the night by thermal cooling.

Algae in ponds can be divided into two classes: motile and non-motile forms. Motile algae, of which the principal example is the *Euglena* species, possess flagella by means of which they move themselves slowly in the water. The non-motile forms, for example *Micractinium,* possess no means of locomotion and depend on mixing action to disperse themselves throughout the body of the pond.

If mixing conditions are good non-motile algae increase rapidly to dominance reaching concentrations of up to 20 million per ml. If a pond is stratified the non-motile algae below the photic zone remain permanently in darkness and die off and, generally, result in a drastic decrease in their concentration. The motile organisms, however, move in and out of the photic zone accordingly as conditions are favourable or unfavourable. Their concentration during stratification never reaches that of the non-motile organisms during mixing but may be as low as 50,000 per ml.

Dispersal of oxygen in the pond also depends on mixing of the contents. The oxygen concentration is uniformly dispersed throughout the pond depth during mixing, but during stratification oxygen is only found in the upper 0.5 m of the pond, the major part of the pond remaining anaerobic.

The situation now arises where during summer, if wind velocities are insufficient to break stratification, algal concentration is low, hence the rate of oxygen production is low and the oxygen is not dispersed through the pond. At the same time the BOD feedback from the sludge is high and the rate of oxygen depletion is high. If the reduced reoxygenation capacity is insufficient to meet the increased oxygen demand, the pond turns anaerobic. Even where the pond does not turn anaerobic sludge rising to the surface may cause odour problems.

From the foregoing it is clear that in warm-temperate climates the loading rates of ponds can be increased if during the summer period (a) the BOD feedback from the sludge layer can be decreased, and (b) the reoxygenation capacity of the supernatant can be increased. The former is attained by anaerobic pretreatment in the aqua privy or special pretreatment tanks, the latter by artificial slow stirring of the primary pond. Stirring devices, however, are only possible for relatively large installations. With anaerobic pretreatment the loading of the primary pond may increase to 7500 to 10,000 people/ha. Stirring, particularly during summer months, will also allow appreciable increase in load during this critical period.

CONCLUSION

'Considering the problems and real cost of attempting to maintain substandard sanitation, the question appears not to be can we afford to supply every house with its own water supply and water carriage system of sanitation, but rather do we want it? We are already paying for it.'[4]

REFERENCES

1. Ministry of Health, Government of India. 'Proceedings and Recommendations.' *Seminar on Financing and Management of Water and Sewage Works.* Vigyan Bhawan, New Delhi. 1964.
2. Vincent, L. J., Algie, W. E. & Marais, G. v.R. (1963). A system of sanitation for low cost high density housing. In: *Symposium on Hygiene and Sanitation in relation to Housing, CCTA/WHO, Niamey, 1961,* London Commission for Technical Co-operation in Africa, p.135.
3. Marais, G. v.R. Dynamic behaviour of oxidation ponds. *2nd Int. Symp. for Waste Treatment Lagoons,* Kansas City, Mo., USA, 1970.
4. Vincent, L. J. Improving the standard of existing housing. *Proc. Conf. on Urban Housing, Ministry of Housing and Social Development,* Lusaka, Zambia. March 1964.

THE DAN REGION LARGE SCALE OXIDATION PONDS

Y. FOLKMAN
Project Manager, Dan Region Sewage Reclamation Project,
Tahal — Water Planning for Israel, Ltd.

P. G. J. MEIRING
Consulting Civil Engineer, Pretoria.

M. KREMER
Manager of Dan Region Sewage Reclamation Project,
Mekorot Water Company Ltd., Israel.

INTRODUCTION

Oxidation ponds are looked upon by the more advanced countries as suitable for use only by small and isolated communities, or if not isolated, as temporary measures for use only until such time as a conventional mechanical type of works can be built or afforded. Oxidation ponds have also been sneered at in some quarters as a 'hole' in the ground. Climatic conditions, but also a lack of understanding, may account for this attitude.

While it must be conceded that oxidation ponds have some distinct disadvantages if compared to alternative biological sewage purification systems, they also have some important advantages which cannot be overlooked. These advantages have been pointed out by many workers and an eminent body such as the World Health Organization was quick to realize what tremendous advances to environmental health conditions they could bring about in 'developing countries'. It so happens that the majority of today's 'developing countries' lie in regions where climatic conditions are warm and temperate and favour the use of oxidation ponds.

What, then, are the disadvantages of oxidation ponds? First, the area of land required can be quite prohibitive if the loading rate for facultative ponds is to be employed. This in turn results in excessive seepage and evaporation losses and if, so as to ensure a good quality effluent, primary as well as secondary ponds are employed, the volume of the final effluent may be reduced to only 25 per cent of the inflow of raw sewage.

Short-circuiting, thermal stratification, scum formation, excessive sludge accumulation and mosquito breeding are some of the other factors which also put constraints on the use of oxidation ponds, and have virtually limited their use to all but small scale application. In many cases lagoons have been discredited because overloading, due to a lack of understanding of the underlying principles, has given rise to septic conditions and unsatisfactory effluents.

The advantages of oxidation ponds have been enumerated in the literature,[1, 2, 3] but reference is usually made to the relatively small scale application of ponds only.

A better understanding of the processes involved in the performance of ponds and pioneering work in many parts of the world have enabled lately the designer to overcome many of the shortcomings of oxidation ponds. In the instance of the Dan Region oxidation ponds are now being used to treat successfully a flow of 25,000 cubic meter of sewage per day.

PIONEERING WORK

In a facultative oxidation pond a rather delicate symbiosis exists between the bottom anaerobic layer and the overlying layer of algal laden water which is kept aerobic by photosynthetic chlorophyl activity.

This process metabolizes an end product of the anaerobiosis on the pond bottom, carbon dioxide, to synthesize new algal cells. At the same time oxygen is released to the water. However, should the anaerobic conditions on the bottom and the oxygen demand exerted thereby begin to overtax the oxygenation capacity of the algae in the overlying water, a vicious circle is triggered and the algae would start to disappear leaving the whole pond anaerobic, if not septic and malodourous.

A criterion commonly used for the design of facultative ponds in warm temperature and subtropical climates and one based on full-scale studies by workers such as Marais & Shaw,[4] stipulates a primary pond loading rate which will not exceed 120 kg BOD/day/hectare. However, by using this loading rate for the design of large installations, the area required is extensive. Also, if the primary pond effluent before discharge is required to be of a high quality, it would require further improvement in secondary and even tertiary ponds whereby the area of land required would be doubled. In this way, with an overall loading rate of 60 kg BOD/day/hectare a BOD-reduction of some 90 per cent can be achieved, but seepage and evaporation losses can reduce the outflow from these ponds very substantially.

Realizing the advantages of reduced land requirement of a completely anaerobic lagoon properly acclimatized to ensure the establishment of a healthy methane fermentation, Parker in Australia has done valuable pioneering work on which he reported as early as 1950.[5] He suggested a design loading for anaerobic primary lagoons of 900 to 1200 kg BOD/day/hectare in summer and 657 kg BOD/day/hectare in winter for the Melbourne area to bring about a BOD reduction of 60 to 70 per cent. Based on his observations he suggested a secondary (aerobic) pond area ratio of 6 to 1. If the winter design loading is accepted as the governing factor an overall loading figure of 110 kg BOD/day/hectare is arrived at which is about twice the loading a facultative system can handle under the same climatic conditions.

Unfortunately, these anaerobic ponds have some serious drawbacks: they are not consistently free of odour nuisances and require desludging every few years, which in itself can be very malodorous operation. It would also seem that upscaling of these ponds is subject to hydraulic distribution limitations as overloading of the pond area around the inlets is to be reckoned with. If septic conditions due to overloading should get the upper hand at the inlet end these conditions, which are toxic to methane fermentation and are irreversible unless the flow is stopped, will spread itself.

The drawbacks of the anaerobic-aerobic pond system were to a large extent overcome by Van Eck[6] in Durban, South Africa, in the early sixties. By employing a 15 per cent recirculation of secondary pond effluent of algae laden water to the inlet of the primary anaerobic pond, odour nuisances were overcome largely. Of particular interest is the fact that the loading rate of the relatively small primary pond was 0.410 kg BOD/day/cu.m which amounted to a surface loading rate of 8000 kg BOD/day/hectare.

However, for continuous operation and to prevent carry-over of sludge into the secondary aerobic pond Van Eck advocated the use of three primary ponds in series which could be alternated and cleaned out regularly. This system did not overcome the problem of releasing odours during cleaning operations and required great vigilance to maintain them odour-free. Scum nuisance and fly breeding were very evident at times. An overall pond loading rate of 100 kg BOD/day/hectare was achieved.

Anaerobic pretreatment of raw sewage in an enclosed structure more sophisticated than an anaerobic lagoon or even a septic tank so as to ensure better mixing of the incoming flow with the contents of the tank and also to facilitate intermittent desludging without having to switch to a second unit, has been affected successfully by Marais[7] in Zambia, admittedly only on a relatively small scale. A system such as this even if a more sophisticated digester is used employing mechanical mixers and even return of sludge for contact stabilization, deserves careful consideration when the areas required for aerobic ponds are to be limited for the sake of reducing water losses due to evaporation and seepage.

So as to ameliorate the odour nuisance created by a series of sewage lagoons 'of doubtful reputation' which was operated as an interim winter measure for more than ten years at Cape Town, South Africa (these surroundings enjoy a mediteranean climate with a winter rainfall), Abbott[8] in 1958 recycled 'oxygen rich' quarternary pond effluent to the primary septic pond in a ratio of 1 : 1 with results which he found unexpectedly beneficial. Not only did the primary recirculation pond stop creating an odour nuisance, but it turned green in spite of a surface loading of some 1200 kg BOD/day/hectare. These healthy conditions persisted throughout the recirculation pond system and an overall loading rate of the recirculation pond system in excess of 250 kg BOD/day/hectare seemed possible even in winter. A feature of these Cape Town ponds is the high algal counts which are encountered round the year, both in winter and summer and the concomitant dark olive green colour associated therewith. These ponds are completely odourless and a remarkable and hitherto unaccountable feature is the very limited sludge build-up. Abbott also reported excellent reductions in algal density if the recirculation pond system (which may consist of any number of ponds in series) is followed by secondary or maturation ponds (without recirculation) which have an area ration of 0.83 : 1 to the primary (recirculation) pond system. An overall loading rate of both primary (recirculation) and secondary (maturation) ponds of 137 kg BOD/day/hectare is thus feasible.

A system of recirculation ponds is remarkable in that although the effective loading of the first pond in this recirculation system may be extremely high, say 5000 kg BOD/day/hectare, it would not turn anaerobic provided the acceptable overall surface loading rate is not overstepped. The continuous return of oxygenated water containing dense algal concentrations which in itself affects an oxygenation capacity, takes care of this even though anaerobic conditions may beneficially occur on the pond bottom.

THE DAN REGION SEWAGE RECLAMATION PROJECT

The disposal of sewage from the Tel Aviv Metropolitan Area and surrounding towns has been affected until recently by means of sea outfalls into the Mediterranean Sea and to a lesser extent by discharge of sewage deriving from the southern areas into sand dunes to the south. Pollution of the bathing beaches has become evident, whereas the temporary disposal of sewage in the dunes gave rise to a considerable smell nuisance. These methods of sewage disposal are wasteful and cannot be afforded in a country where water is very scarce and therefore constitute an extremely valuable commodity.

Some system whereby the disposal of sewage and the reclamation of water could be integrated was therefore to be devised since it would offer the obvious solution to the disposal problem and further the cause of water optimization.

Stabilization ponds for the disposal of sewage have been used successfully for many years in Israel [9, 10] where climatic conditions are favourable and this relatively low-cost method of disposal has proved itself very reliable. Concomitant circumstances which influenced the decision to use ponds for the initial stages of the Dan Region Project at least, were the availability of a large stretch of waste land which was suitable for the construction of ponds but otherwise had little economic value, and also the urgency of the matter which allowed no time for the planning and construction of a conventional sewage purification plant. An especially fortunate feature of the area which was available is the existence on part of it of a deposit of clay some five to eight meters below the surface which would effectively seal the pond bottoms and reduce seepage losses.

It was considered possible to produce an effluent compatible with quality requirements which would permit the spreading of the final effluent on nearby spreading basins for recharge of existing underground waterbearing formations. These formations would then serve as a source of water which, after extraction and chlorination, would be delivered to the National Water System and conveyed to the arid southern parts of the country. [11, 12]

It was the initial intention to operate the ponds built under the first stage of the program as a series of anaerobic — aerobic ponds by subjecting the primary anaerobic pond to a loading of 900 kg BOD/day/hectare. This approach did not prove to be entirely successful and inspite of assiduous efforts by the operators to control the situation, complaints of odour nuisances came from nearby towns. The inflow was thus stopped and the ponds adapted so as to operate as recirculated facultative ponds as described by Abbott.

The ponds A,B,C,D,E,F built during 1969 as an anaerobic — facultative — aerobic pond system and converted in 1970 to recirculated aerobic ponds are shown in Figure 1. The ponds have levelled bottoms and an average depth of 2.5 meters. The embankments between ponds are of compacted sand, they have a slope of 1 : 3 and are covered with mortar to protect them from erosion and an ingress of vegetation.

Ponds A and B are used in parallel as primary ponds and ponds D, E and C (in that order) are subsequent ponds in the same recirculation system. Effluent from pond C is recirculated to the inlet channel 300 meters upstream of ponds A and B at a recirculation ratio of 1.5 : 1. Good mixing of the recycled effluent and the incoming sewage is thereby achieved prior to their entering of ponds A and B. The ponds are all of the same size each covering an area of eight hectares. Multiple inlets and outlets and baffles where necessary, prevent short circuiting and stagnant corners. Pond F serves as a maturation pond but later on as part of the eventual water reclamation project, it will receive lime-treated effluent and be linked in series with pond 12.

Satisfactory performance of the ponds built during the first stage induced the planners also to select recirculated aerobic ponds for the treatment of the increase in flow expected during the second stage of sewerage development in the Dan Region. These ponds are presently (1971) under construction and are indicated as ponds 1 – 8 in Figure 1. Ponds 9 – 12 are secondary ponds which will eventually be used as ammonia stripping

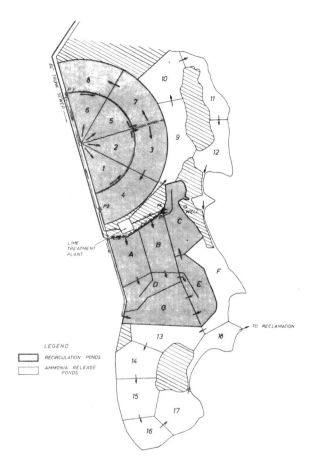

Fig. 1. Existing and second stage ponds

ponds when the effluent from the recirculation pond systems will be limed in a physical chemical treatment plant, for the removal of algae and some of the remaining chemical oxygen demand and phosphates.

The all-inclusive cost of constructing the second-stage pond of the Dan Region Project amounts to $48,500 per hectare of pond area. At a loading rate of 230 kg BOD/day/hectare this represents a treatment cost of 2.6 US cents per m^3. This cost is relatively high as far as the normal cost of treatment of sewage in ponds goes, but this is due to the fact that provision has been made for the lining of these ponds with a layer of clay brought in from elsewhere, and to the fact that the sand dune topography of the available land necessitated much more earth moving than is normal for the construction of ponds. However, this treatment cost compares favourably with all conventional treatment methods for the production of an equivalent quality of effluent, and is therefore fully justified.

STAGES OF OPERATION

The period of operation of the system of recirculated facultative ponds which began in June 1970 can be divided into three stages: the stages of filling, of gradually increasing the organic load, and of operation at a full load. The main difficulty was encountered at the stage of filling the ponds: there was as yet no effluent for recirculation, and on the other hand a period of acclimatization was required for the development of methane fermentation bacteria on the pond bottoms; it was thus necessary to begin at very low organic loads. As a result, and because of the permeability of the soil, there was a danger that all influent might be lost to seepage and that it would thus be impossible to fill the ponds.

In order to overcome the danger, effluent of high algae concentration which still remained in pond C from the period of operation as anaerobic-aerobic ponds was pumped together with the raw sewage into ponds A and B. In addition, clean water was brought in from a well in the pond area and from the National Water System. Influent thus exceeded seepage, and the ponds were quickly filled.

Ponds A, D and B were the first to be filled. They were then operated as a small recirculation system in which effluent was pumped out of pond B and made to enter pond A together with the sewage. Surplus effluent was used for filling pond C and then pond E. By December 1970, all five ponds had been filled, and operation of the big recirculation system as planned could be started. In order to avoid disturbance of the biological acclimatization of the ponds in winter, operation began at low organic loads of about 130 kg BOD/day/hectare and a sewage flow of 15,000 cu.m/day; the loads were then gradually raised and sewage inflow increased. At the same time, as the perched aquifer underlying the ponds became saturated, seepage decreased. The increased sewage inflow and the decrease in seepage made it possible to discontinue withdrawing water from the National Water System and also gradually to cut pumpage from the local well. By mid-February 1971 the organic load had been raised to 185 kg BOD/day/hectare at a sewage flow rate of 22,000 cu.m/day, and the ponds were operating with the full quantity of sewage as planned, without causing nuisances. In spring and summer, the sewage flow was gradually increased to 25,000 cu.m/day at an organic load of 230 kg BOD/day/hectare.

At this stage the seepage rate was of the order of 7 cm per day so that seepage counterbalanced the influent of raw sewage, but it would appear that a progressive sealing of the pond bottoms are taking place and at the time of writing (Oct. 1971) seepage has decreased to 3.5 cm per day and there is a gradual increase in the volume of effluent leaving pond C.

POND PERFORMANCE

Monitoring of pond performance has been conducted ever since the start-up. The main objectives of this follow-up were:

— to examine the parameters which are indicative of conditions of satisfactory (aerobic) performance

— to study treatment efficiency with regard to the removal of COD, BOD and nutrients.

Dissolved oxygen concentrations, pH, temperature at different depths and locations in each pond were examined at different times of the day, and algae identified and counted. A thorough examination of sewage and effluent from each pond was made once a week.

Fairly strong winds — with mean velocities varying between 10.8 km per hour in summer and 17.6 km per hour in winter blow over the ponds. Wind velocities sometimes reach 70 km per hour. Average daily temperatures range from 10°C in winter to 27°C in summer. The winds and the large pond areas appear to induce a marked degree of mixing of the pond contents. Stratification has never been observed and this phenomenon is apparently responsible for the predominance of non-flaellate algae, mainly of the type chlorella. The chlorella ranged in numbers from 500,000 to over 2,000,000 per cubic centimetre.

The high algae concentration causes a high dissolved oxygen concentration and a rise in pH in the early afternoon, Table 1.

TABLE 1

Number of Algae and Concentration of Dissolved Oxygen
(Average Values for July 1971)

Pond	Number of chlorella	D.O. concentration mg/l at top layer		
	per cu.cm	at 8.00 h	at 14.00 h	at 02.00 h
A – B	1,600,000	0.9	0.7	0.4
D	2,200,000	1.0	8.0	0.5
E	1,600,000	0.8	8.2	0.5
C	900,000	3.4	11.2	0.5

The table shows that whereas the dissolved oxygen concentration in the secondary ponds is above saturation at midday, it is low at night and in the early morning. In the primary ponds (A-B) on the other hand, the oxygen depletion rate is so high that even the high algae densities do not cause the dissolved oxygen concentration in the early afternoon to rise to that of the secondary ponds. This phenomenon occured most of the time, and quite frequently the dissolved oxygen concentration in the primary ponds was below 0.5 mg/l during daylight hours.

The high dissolved oxygen concentrations first occur at the water surface, but these are normally rapidly extended to the bottom as a result of the slow mixing caused by the winds. Under these circumstances only a slight difference in temperature was observed in the various layers.

There were also the rare occasions when stratification did occur and when an inversion during the night was observed, see July 6, 1971, Fig. 2.

At certain times algae predators developed in the secondary ponds, and algae virtually disappeared from those ponds. However, despite the recirculation, the algae predators did

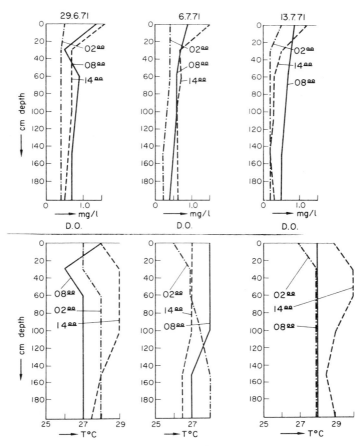

Fig. 2. Distribution of D.O. and Temperature at different hours on three summer days in pond
A – B

not develop in the primary ponds, presumably because of the absence of dissolved oxygen
in the early morning. Raw sewage was let at times into the secondary ponds to deplenish
the dissolved oxygen, and this helped to destroy the algae predators. Another
phenomenon worth mentioning was the formation of scum, consisting mainly of algae
and sludge in summer. This scum is carried by the wind to the corners of the primary
ponds where it accumulates in large quantities and create some odours.

The sewage being treated in the ponds is relatively strong having an average BOD of
340 mg/l, a COD of 760 mg/l and a total nitrogen concentration of approximately 98
mg/l in summer. After admixture with the recirculant from pond C the incoming sewage
enters the primary ponds A and B. The BOD of the effluent from these ponds is reduced
to approximately 40 mg/l and the COD and total nitrogen values also show substantial
reductions.

During the monitoring period the average quantity of influent and effluent from each
pond was calculated, account being taken of seepage and evaporation losses. The drop in
total and in dissolved BOD and COD (BOD_f and COD_f), total Kjeldahl nitrogen (N_k) and
dissolved nitrogen (N_{kf}),ammonia, and pH as a function of detention time in the ponds in
summer and winter, are shown in Figures 3 and 4. These values are weighted averages
computed for a period of over 30 days per season.

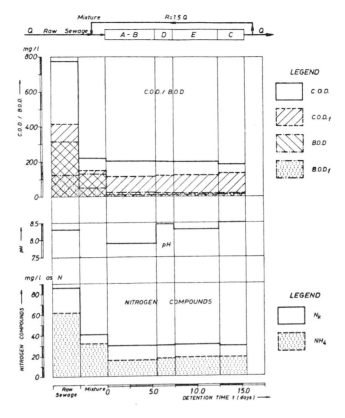

Fig. 3. Treatment efficiency of facultative recirculation ponds (winter 1971)

DISCUSSION

The overall purification achieved for winter and summer through the entire recirculation pond system is BOD 90 and 89%, COD 60 and 65% Kjeldahl-N 65 and 65% and ammonia 70 and 77% respectively. These figures are for unfiltered samples which still contain the biomass consisting of algae and dispersed bacteria. It is also for this reason that an actual increase in BOD and COD has been recorded at times in successive ponds in summer.

The excellent decrease of total nitrogen which is very high to start with, took place in parallel to the decrease of ammonia which points to the fact that the decrease of nitrogen is caused mainly by the release of ammonia to the atmosphere due to the elevated pH of approximately 8.5. At this pH about 20 per cent of the ammonia is found as dissolved gas tending to be released to the atmosphere. The extent to which this will happen is determined by detention time, temperature and wind velocity.

From Figures 3 and 4 one may conclude incorrectly that the system could as well do without the subsequent ponds following ponds A and B since hardly any further purification is achieved therein. In this regard, the following figures extracted from a paper by one of the authors[13] on the performance of these same ponds are of particular interest.

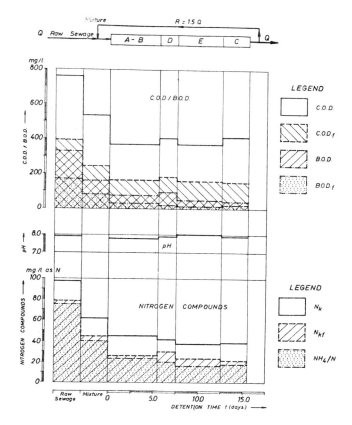

Fig. 4. Treatment efficiency of facultative recirculation ponds summer 1971 (June—July)

TABLE 2

Pond Performance during January — April 1971

	Sewage	Effluent Pond A & B	Effluent Pond D	Effluent Pond E	Effluent Pond C
pH	8.3	7.8	8.0	8.2	8.3
Volatile Acids mg/l as Acetic Acid	130	70	75	65	20
Dissolved Oxygen morning (mg/l)	0	2	5	6	6
Dissolved Oxygen evening (mg/l)	0	5	8	15	20
Fats (mg/l)	180	10	8	7	4
Algae — Total Count	0	2000	2500	3000	3700

From these figures it appears that the subsequent ponds D, E and C play an important role in the pond ecology even if only to allow time for algae to grow so as to provide an algal rich recirculant for admixture with the raw sewage inflow. It must, however, be conceded that the figures do not necessarily indicate that the loading of the ponds cannot be increased. Indeed, this will be done in future but only gradually so as not to turn the ponds anaerobic during the critical season of spring.

FINANCIAL CONSIDERATIONS

It has been pointed out earlier on that even at a cost of 2.6 US cents per m^3 the treatment of sewage in large scale oxidation ponds is economically justified. The lining of the ponds which have been effected as an additional precautionary measure is in turn justified by the protection it will afford to the quality of the underground water.

Should it be decided to build a conventional sewage purification plant during the later stages of the project the ponds will become an extremely valuable asset, as it can be shown that if the ponds are then employed to receive all peak flows and leave the conventional plant to treat a constant flow, a very substantial saving could be brought about which more than offsets the cost of the ponds.

In a similar fashion could the ponds be used to receive plug loads of toxic or inhibiting substances which may be discharged inadvertently to the sewers and thereby to safeguard the conventional plant being operated in parallel to the ponds.

CONCLUSION

The large scale oxidation ponds built to treat a flow of 22,000 cu. meter per day of raw sewage is giving very satisfactory performance with excellent reductions of all indicators of pollution. A 65 and 70 per cent reduction of Kjeldahl and total nitrogen respectively in winter (and better in summer) is truly remarkable and very gratifying since the total nitrogen in the Tel Aviv sewage is uncommonly high, 90 mg/l. To remove this amount of nitrogen by any other biological means would be very costly.

The ponds built during the first stage of the project situated above a natural layer of clay but it is still too early to conclude whether effective sealing of the pond bottoms will be achieved in course of time. The second-stage ponds are being lined artificially with a layer of clay as a precautionary measure and this has increased the cost of treatment in these ponds to 2.6 US cents per cu.meter of sewage. This cost is all-inclusive, including also the cost of pumping, maintenance, and operation, and compares favourably with the cost of treatment to the same standard in any other type of treatment plant.

Operation of these ponds is very simple and straightforward and the performance of the ponds have been very steady and reliable. No upsets due to biological fluctuations, such as algal blooms or surges in predators and the resultant virtual disappearance of algae in the secondary ponds (of the recirculation pond system), have occurred.

Stratification of the pond contents have been observed only on rare occasions. Windy conditions throughout the year account for this but as a guarantee against possible stratification which may hamper the growth of non-flagellate algae, mechanical stirrers, as suggested by Marais,[14] have been installed in some of the ponds.

The ponds are at present operating at their overall design load of 230 kg

BOD/day/hectare, but it seems most likely that the loading can be increased quite substantially. This will be done very slowly.

An interesting feature of these ponds is that the dissolved oxygen level in the very heavily loaded primary pond may fall to virtually zero for the greatest part of the night, and rise to only 0.5 mg/l during the afternoon without any odours emanating from the pond or other signs of overloading. The performance of the primary pond is completely dependent on the high rate of recirculation (1.5 : 1) of effluent from the last pond in the recirculation pond system and the secondary ponds perform an indispensable function in the pond ecology.

The initial indications are that the build-up of sludge on the bottom of the primary ponds is very slow, much more so that what is normal in a facultative pond without high rate recirculation. This aspect warrants further investigation.

It is concluded that the large scale oxidation ponds of the Dan Region Sewage Reclamation Project operated as recirculation ponds are a success.

REFERENCES

1. CPHERI *Design Construction and Operation of Waste Stabilization Ponds in India*, Nagpur, India 1970.
2. Oswald W. J., Golueke C. G., Cooper R. C., Gee H. K., and Bronson J. C. Water reclamation, algal production and methane fermentation in waste ponds. Intern. Conf. of Wat. Pollut. Res. 1960 Paper 25 Part II.
3. Meiring P. G. J., van Eck H., Drews R. J., and Stander G. J.: *A guide to the use of pond systems in South Africa for the purification of raw and partially treated sewage.* CSIR Spec. WAT 34, Pretoria 1968.
4. Marais G. van R., Shaw V. A. A rational theory for the design of sewage stabilization ponds in Central and Southern Africa *Civ. Engr. S. Afr.* Vol. 3, No. 11, Nov. 1961.
5. Parker C. D., Jones H. L., and Greene N. C. Performance of large sewage lagoons at Melbourne, Australia. *Sew. and Ind. Wastes.* Vol. 31 No. 2, Feb. 1959.
6. Van Eck H. and Simpson D. E. The anaerobic digestion pond system *J. Proc. Inst. Sew. Purif.* Part 3, 1966.
7. Marais G. van R. Overloading of Oxidation Ponds. Paper read at Annual Conf. of the Inst. of Municipal Engrs. of South Africa. Salisbury 1970.
8. Abbott A. L. Oxidation Ponds. *Proc. Diamond Jubilee Conf.* 1963 S.A. Inst. Civ. Engrs.
9. Watson J. L. A. Oxidation ponds and use of effluent in Israel, *Proc. Inst. C.E.* Vol. 22, May 1962.
10. Sless J. B., Samsonov E. B., and Wachs A. M. *Performance of Waste Water Stabilization Ponds at Tirat Hacarmel under various loadings.* Technion, Israel Institute of Technology, Dec. 1964.
11. Wachs A. M., Avnimelech Y., and Sandbank E. Effect of the irrigation with stabilization pond effluents on the concentration of nitrates in underground water. *Annual Report* Sanitary Engineering Labs. Technion, Haifa 1970.
12. Folkman Y. and Wachs A. M. Filtration of Chlorella through dune-sand *ASCE SA 3* June 1970.
13. Kremer M. *Dan Region Sewage Reclamation Project.* IAWPR Workshop, Vienna 1971.
14. Marais G. van R. *Oxidation Ponds.* Proc. Convent: Water for the future. Pretoria 1970

Discussion *by* Leonard B. Dworsky,
Director of the Water Resources and
Marine Sciences Center,
and Professor of Civil Engineering,
Cornell University, Ithaca, New York.

I am impressed with the process by which uncertainty is managed in a program such as the Dan. Pioneers in lagoons — anaerobic, aerobic and faculative, — earlier explored various features of the technology and managed to convey their positive and negative findings. The ability to assemble and assess technological intelligence, to accept and discard, to innovate and integrate, to determine acceptable risks and define with reasonable clarity the uncertainties, and be willing to risk millions of dollars, and reputations, and government policies is a challenge that does not come too frequently. The Dan project will stand as a major pioneer in water reclamation, conservation and use in vast parts of the world.

We should recognise the benefits of an organization structure that can, with limited resources, plan and assess needs, establish priorities and schedules, and have responsibility for implementation policies, subject only to review by the most senior executive and legislative bodies for conformance to public policy.

The Dan project provides an excellent illustration of the need to exercise great caution in organizing to meet environmental problems. In Israel, for example, water resources management already is consolidated in the Ministry of Agriculture. Recent laws have also provided water pollution responsibility to the same agency. Chemical contamination of water by nitrates, phosphates, pesticides and other toxic materials are deep concerns of the Agricultural Ministry and they are in a position to affect the contribution by the agricultural sector of the economy as well as by municipal and industrial waste water.

Water renovation, recycling and appropriate use, too, are within the purview of the Agricultural Ministry. Thus, by these illustrations and other actions, an interesting set of questions are posed relating to organizational alternatives.

In the U.S.A. and in many states, E.P.A.s have been formed to manage environmental issues. Perhaps, since we in the United States have so much to do, and have larger resources to use, an advocacy agency for the environment may be the useful way for us to proceed.

In developing countries, however, with limited resources, the trade-offs become more important and the priorities, schedules and program strategies take on a new light. In my mind, the presence of an advocacy agency becomes less important than an agency that can effectively design necessary trade-offs and implement the decisions made.

These are a few of the implications I see in the Dan project. It will be very important to watch this development as it proceeds for it has large meaning to arid and semi-arid lands everywhere as well as to locations seeking means to practically eliminate pollutant discharges to water resources.

LOW COST METHODS OF WASTEWATER TREATMENT

C. D. PARKER

Director, Water Science Laboratories, Melbourne, Australia

INTRODUCTION

Ever increasing demands for control of water pollution, in most instances involving construction of facilities for treatment, create massive requirements for capital expenditure. The lower the unit cost to cope with each situation, the more rapidly will the overall problem be brought under control. Recognition that major pollution situations exist in the less developed areas where capital resources are limited and particularly in association with the processing of agricultural products where profit margins are small, further emphasizes the obligation of scientist and engineer to devise and provide satisfactory methods of control at the lowest possible cost. While it is expected that discussion will involve those processes (irrigation, lagoons, oxidation ditch) which by convention have been considered 'low cost', in comparison with the established mechanical processes of trickling filter, activated sludge and sludge digestion, it is well to recognize there may well be scope for increasing the efficiency of the latter processes by modification of design. The use of alternative materials of construction may also achieve lower costs. Where skilled operation backed by adequate laboratory facilities, is available, maximum plant performance can be achieved. Frequently design engineers need to overdesign facilities to ensure a satisfactory performance by unskilled operators. One means of reducing cost is to have available properly trained operators backed by proper laboratory facilities.

IRRIGATION

The earliest means of disposal and treatment of sewage was the 'sewage farm' where surface irrigation of raw or settled sewage over land achieved disposal, and under proper conditions a high degree of purification, albeit with some level of odour and possible public health risk. With increasing urbanization and the large areas of land required, this method has largely given way to the more land intensive mechanical processes. However, recognition on the one hand of the need to conserve water resources including the nutrients available in sewage, and concern for the eutrophic effect of nitrogen and phosphate in the effluent from conventional mechanical processes, has revived interest in the transformation in water quality which can be achieved by passage through or over soil. A further stimulus to the study of soil passage as a process for purification is the unavoidable reliance on soil disposal of sewage from the increasing numbers of large urban populations located in unsewered areas.

Apart from the application of soil processes for purification, the need for reuse of wastewater has established irrigation of agricultural crops as one of the simplest and most common applications. This has stimulated interest in the evaluation of plant fertilizing value of wastewater and the public health risks when used to irrigate food crops.

Probably the largest existing irrigation scheme for the treatment and disposal of sewage, is at Werribee in Victoria, Australia where substantially the whole of the sewage from the City of Melbourne (population 2,300,000) is purified on an area of 27,000 acres. During summer raw sewage is surface irrigated for one day over prepared pasture areas on an 18-day cycle. The land required is 250 acres per 1.0 mgd of flow. After irrigation the pasture is grazed by 17,000 head of cattle for over two weeks prior to the next irrigation. The area is drained by (6–8 ft) deep cut-off drains spaced at half mile intervals through the irrigated areas. An analysis of drainage water compared with the irrigated raw sewage, is shown in Table 1.

TABLE 1.

Composition of Drainage from Sewage Irrigated Pastures

	Raw Sewage	Drainage
B.O.D. ppm.	600	4.0
Org.-N. ppm.	35	2.5
Ammonia-N. ppm.	40	1.0
Nitrate ppm.	0	2.5
Phosphate ppm. (PO$_4$)	25	4.5
Chloride ppm.	600	1200
A.B.S. ppm.	9.5	0.1

During the wet winter period, settled sewage is purified by continuous surface passage over a grassed surface with continuous outflow at the lower end of accurately graded irrigation bays. The area required is 35–40 acres per mgd Treatment achieves a reduction of BOD from 600 to 15–20 ppm but with only moderate nitrogen, phosphate and detergent removal. The effectiveness of this process is dependent on the development of a prolific growth of a water tolerant vegetation. A similar process has been developed by Glide[1] for the purification of cannery wastes. In this case the wastewater is distributed over sloping grassed areas by sprays and the runoff collected in trenches. The area required is 100 acres per mgd in summer and 200 acres per mgd in winter. The process produces an effluent BOD of 9 ppm with 40% removal of phosphate and 85% removal of nitrogen. Another application of spray irrigation is for the purification and disposal of milk processing wastes. Spray irrigation of high strength wastes from cheese and casein manufacture is extensively practised in New Zealand,[2] although in most cases it is expected that the wastewater will be absorbed on the land and there will be no surface runoff. Even very high strength wastes such as cheese and casein whey if moderately diluted can safely be irrigated over pasture.

Raw sewage and sewage effluents are extensively used in South-East Asia for the irrigation of food crops; this represents a substantial public health risk if the crops are eaten raw. In Israel[3] it is understood primary and secondary treatment by lagoons is required as a prerequisite to the use of wastewater for irrigation. In India[4] extensive studies have been made of the growth potential for the irrigation of variety of crops by lagoon treated sewage. In Australia raw sewage has been extensively used for the irrigation of grazed pasture, and investigations have shown the health risks of irrigation of unchlorinated effluent onto a variety of food crops to be slight. Irrigation has several different applications with regard to treatment of wastewater. Raw or primary treated sewage or organic type food wastes may be purified with regard to BOD by irrigation over pastures with substantial removal of nutrients, detergents and bacterial count. Similar removals may be achieved by soaking of septic tank effluents through soil in absorption or percolation trenches. Studies made in Queensland[5] have shown that with porous sandy soil, most organic constituents of septic tank effluent may be removed by travel through soil for distances less than 50 feet. Water quality transformations in soil are relevant to ground water studies, wastewater reclamation and specific use to achieve purification.

Ability to achieve purification by soil passage is undoubted. However, there is an urgent need to quantitatively document the relationships between removal and surface area loading, rate of travel in soil, and soil composition with regard to clay type, organic content and mechanical analysis. There is considerable scope for adaptation of these mechanisms either by using the soil profile *in situ* or developing artificial beds of soil particles for particular purposes.

LAGOONS OR STABILIZATION PONDS

Passage of raw or partially treated sewage through ponds confined by earth banks has been demonstrated in California and Australia to be an effective and inexpensive means of purification of sewage for over 30 years.[6,7] Within the last decade these early studies have been developed as a low cost form of treatment in many parts of the world, i.e. South Africa, Israel, India, South America, Rhodesia and other parts of Africa. The obvious cost advantages where adequate areas of land are available has stimulated widespread study of the mechanics of purification processes, various types of application, and the quantitative effect of various factors on performance.

Discussion of the applications of lagoons can be clarified by recognition of the different applications of the general principles. The early concept was to use confinement of wastewater in ponds to exploit purification by the natural development of anaerobic fermentation and the aerobic environment which could be established by the massive development of algal photosynthesis in the semipurified water. Systems were variously described as anaerobic lagoons, aerobic lagoons, facultative lagoons, stabilization ponds, oxidation ponds and maturation ponds. More recently to minimize the land area involved while taking advantage of the low cost of earth structures, there has been developed the aerated lagoon where some of the above natural processes have been assisted by mechanical means of aeration.

Natural Lagoons

By appropriate design of layout and loading, lagoons may achieve in whole or part the functions of primary treatment for solids removal and stabilization, secondary treatment for BOD removal to acceptable levels for discharge to streams, reduction of coliform count and substantial reduction of nitrogen and phosphate content. Such ponds may be used to treat raw sewage, settled sewage or the effluent from conventional secondary processes.

Anaerobic Lagoons

By adjusting the pond size and the load of raw wastes onto first stage ponds, very high BOD and suspended solids loadings and removal may be achieved and strictly anaerobic conditions exist in such ponds. As shown by Parker[7,8] the presence of adequate quantities of actively fermenting sludge solids in the pond are essential for maximum digestion of solids and removal of soluble BOD from the waste. Because of the strictly anaerobic condition of such ponds very little algal growth can occur and the usual presence of some sulphide is responsible for a moderate level of odour from such installations. The commencement of operation of such ponds can be associated with objectionable odour problems. If the full design load is applied initially without the presence of the essential actively digesting solids, very high levels of sulphide can be generated with massive odour nuisance. To prevent these conditions it is essential that the initial load during filling shall be kept down to a level to ensure the rapid development of massive algal growth (60–100 lbs BOD/ac/d). Development of full performance can take 12 months. It was realized that by the development of actively fermenting sludge deposits in such ponds BOD removals considerably in excess of that associated with the removal of suspended solids can be achieved. The process involved is essentially that of methane fermentation normally associated with sludge digestion. By adequate detention of the sewage in contact with this digesting sludge a considerable removal of soluble BOD is achieved (70–85%). One interesting aspect of the gas produced is the presence, in addition to the expected methane and CO_2, of significant quantities of nitrogen, for which no satisfactory microbiological explanation has been advanced.

One variable of considerable importance in design and performance is the depth of anaerobic ponds. The first installations were built 4 ft deep, more recently they have been built 6–10 ft with the claim that deeper ponds take less area, promote more effective anaerobic action, conserve heat in cold climates and permit greater accumulation of sludge solids. Some of these claims are incompatible. It should also be pointed out that where land is readily available at low cost (less than $1000/acre) the increased cost of building high earth banks offsets the reduced area required.

Temperature has a considerable influence on the rate of methane fermentation, much higher performance of anaerobic ponds is possible under tropical or semi-tropical conditions. Attempts have been made to predict the relationship between performance and temperature. Australian experience[9] suggests a permissible load of 1000 lbs/ac/d at 23°C and 400 lbs/ac/d at 10°C for BOD removal of 70–80%. Arceivala[4] reports for Indian conditions design loads up to 3000 lbs/ac/d. There is a need for a detailed evaluation of the effect of temperature on the performance of this type of pond. It is necessary to separate out removal achieved by surface aeration, solids deposition and methane fermentation and to ensure that conditions are such that algal photosynthesis is not involved.

The BOD/nutrient ratio is important in aerobic processes but has received little attention with regard to lagoons. With domestic sewage this is probably of no importance but where ponds are used to treat industrial wastes, nutrient content is important. Cannery wastes can be treated in anaerobic-aerobic lagoon systems (Parker [10]) provided their high carbonaceous content is supplemented with nutrients from sewage effluent or by chemical addition. Recent work suggests that the optimum BOD/nitrogen value is about 100/2.5 (Parker, unpublished).

Facultative Lagoons

On one hand it has been shown that high BOD removals can be achieved by strictly anaerobic type ponds but with some level of odour. On the other hand by using very shallow ponds with appropriate loading intensely aerobic conditions associated with dense algal growth, can be developed. Raw sewage or industrial waste can be treated in a first stage pond at relatively high rates and with minimal odour level provided the loading is such as to permit algal development in the upper layers while methane fermentation proceeds in the bottom sludge solids.

Studies have shown that provided the contents of the pond have a BOD below certain defined levels, algal proliferation will occur. The maximum permissible loading to achieve these conditions depends on the extent of solar energy, temperature, water depth, wind and the nature of the waste. Of these there is considerable agreement that temperature is the most important. Gloyna[11] has propounded a formula for the calculation of required area related to the BOD load per day and a temperature factor. Temperature also has an influence on the type of algae which develop. Higher temperatures encourage blue-green forms. Water depth influences performance in several ways. As light penetration is limited to the top 18–24″ increased depth has no value with regard to photosynthetic production of oxygen. Increased depth gives greater scope for vertical stratification and high activity in the anaerobic bottom sludge layer. Mixing is important to prevent localized anaerobic surface areas and to take advantage of the oxygen gradient from surface to bottom. Wind is responsible for considerable pond mixing. A depth of 4–5 ft in many cases is adequate to preserve the separation of surface photosynthesis and bottom sludge anaerobiosis with sufficient surface mixing to prevent local concentrations.

The composition of the waste itself has some bearing on performance. Toxic constituents obviously inhibit activity. Where these wastes are deficient in heavy metals, nitrogen or phosphate they need to be added. Hardness of the wastewater has a bearing on the algal groups which develop. Detailed study of cannery wastes has shown that the composition of the pond at which algal growth will proliferate varies with the nature of the waste. It has been shown (Parker[10]) that for water temperatures of 20°C facultative conditions can be maintained treating fruit cannery wastes (peaches and pears) at a BOD loading of 600 lbs/ac/d and pond contents BOD of 400 ppm whereas vegetable cannery waste (tomato) require a loading of 400 lbs/ac/d and pond contents BOD of 80–100 ppm. For domestic sewage pond contents should not exceed a BOD of 100–130 ppm.

The design of pond inlets and outlets, the shape of the pond and pond layout for multi-cell installations also influence performance. First stage ponds treating raw wastes should be square if possible or near to it; the secondary stage units should consist of three or four ponds in series. As shown (Parker[12]) for a given detention time short circuiting is minimized and maximum BOD and coliform reduction is achieved. If the area of first stage pond is such as to require subdivision into multi-cells these should be operated in paralled not series, to minimize local overloading. Sludge accumulation can be largely limited to these first stage units.

Aerobic Lagoons

Considerable attention has been given to the possibility of harvesting algal cells from lagoon effluent for use as animal food. Maximum cell yield is important under these conditions and attention has been given to shallow aerobic ponds to maximize cell growth. More detailed consideration of this subject will no doubt be presented by Professor Oswald.

Tertiary & Maturation Lagoons

Ponds have also been used to improve the quality of effluents from conventional secondary treatment processes (maturation ponds). The principle purpose is to achieve reduction in faecal coliform count, a detention of 7—10 days is suggested. In Southern Australia it is conventional to allow 30 days detention after satisfactory removal of BOD by primary and secondary lagoons to ensure satisfactory coliform reduction. Some attention has been given to the removal of nitrogen and phosphate in ponds. In general, ponds designed to achieve adequate BOD reduction also achieve 50—60% removal of nitrogen and 30—40% removal of phosphate. Prolonged detention can increase this to 80—90% removal of nitrogen and 60—70% removal of phosphate.

Public Health Aspects of Lagoon Treatment

The substantial reduction of coliform count that is associated with BOD removal in lagoons was early recognized and the rate of die away has been extensively studied. Marais[13] has presented a theoretical treatment of the relationship between pond operation and coliform die away. It appears that coliforms, faecal streptococci, and bacterial pathogens, die off at approximately the same rate, that reduction in count is much less in anaerobic than facultative or aerobic type ponds; the rate increases with temperature; and that complete reduction in bacterial content takes much longer than BOD removal. As shown by Parker[12] and later confirmed by several authors, bacterial reduction is much more effective in multi-cell series ponds. The reason for bacterial reduction has variously been ascribed to algal toxins, high pH, destruction of bacterial nutrients, sunlight, etc... Reduction of virus content through ponds has received considerable attention particularly with regard to the reuse of pond effluent at Windhoek and Santee. Again substantial reduction is achieved but complete sanitary disinfection cannot be assured. For complete disinfection chlorination is indicated, but treatment of algal laden effluents high in ammonia nitrogen raises problems. Ideally one would hope to achieve bacterial and viral disinfection with minimum kill of algal cells. Hom[14] has reported evidence to suggest that by appropriate dosage and contact time, effective bacterial kill can be achieved with minimal algal destruction. It would be desirable to extend this work to cover viral aspects as the chlorine was presumably present in the combined form. Studies in India[4] indicate that lagoon treatment effectively removes parasitic cysts and ova from sewage. Provided ponds are properly maintained and vegetation is prevented from spreading across the surface, mosquito breeding does not occur. However, there are several instances recorded where prolonged retention in ponds has been responsible for the development of prolific midge populations.

Lagoon Treatment & Water Reclamation

Lagoons are a very inexpensive means of removing BOD providing sufficient land is available. They can also achieve a very substantial reduction in bacterial, viral and parasitic pathogens. The effluents are in general turbid due to the presence of algal cells. Where water is to be reclaimed for agricultural use lagoon treatment provides a public health safety margin and permits the water to be stored for use as required. Reuse for, industrial or domestic purposes requires removal of algal cells and stabilization against algal regrowth by nutrient removal. Experiemental work at Lancaster in California[15] and in S.W. Africa[16] has demonstrated the feasibility of these processes. Where prolonged retention at high temperatures is required for purification, water loss by evaporation may be considerable.

Industrial Waste Treatment by Lagoons

Most studies of lagoon treatment have been in relation to the treatment of domestic sewage. Organic wastes with balanced nutrient content, such as meat packing and abattoirs, have been shown to be readily purified in the same way. By a proper understanding of the nutrient requirements and physiology of the bacteria and algal forms on which purification depends, a wide variety of other organic wastes can be similarly treated. By addition of suitable algal nutrients phenolic oil refinery wastes can be purified. By admixture of fruit cannery wastes with milk, abattoirs or domestic wastes at Shepparton, Australia a BOD load of 70,000 lbs/day can be completely purified by lagoons. Strawboard, paper pulp and other cellulosic wastes represent considerable pollution loads from industries with small profit margins. By appropriate concern for nutrient addition it has been shown that these wastes can be purified in lagoons at low cost. Treatment by conventional methods of activated sludge would be prohibitive in cost. Wastes from milk processing to casein and cheese can be purified at comparatively high rates of application to multi-stage lagoon installations. Farm wastes from cattle and hogs can be stabilized in anaerobic-aerobic systems.

Aerated Lagoons

Provision of external means of aeration in pond structures has been developed with various objectives and in different ways. Initially the purpose was to supplement the oxygen provided by photosynthesis by mechanical aeration, such installations have been described by Benjes[17] as aerated oxidation ponds. In other types of installation most of the contents of an earthen pond are circulated and aerated but some solids deposition and accumulation occurs in stagnant areas. The primary purpose is to achieve purification by oxygen transfer from the aerator. These have been described as facultative aerated lagoons. The third form is one in which complete mixing of the basin is achieved; it is essentially an activated sludge aeration basin without sludge return. Attempts have been made by McKinney, Eckenfelder and Monod to develop mathematical analysis for such completely mixed systems. Other forms of aerated basin are less well documented. While the introduction of mechanical or diffused aeration into lagoons undoubtedly results in increased BOD removal per unit area of land there is definite need for critical evaluation of costs of such achievements compared with other means of achieving the same result e.g. by conventional activated sludge or by the provision of additional areas of natural lagoon.

PASVEER OR OXIDATION DITCH PROCESS

This process developed in the Netherlands by Pasveer is essentially a completely mixed activated sludge process using in the simplest form the minimum of expensive structures and mechanical equipment. The continuous channel can be built of earth usually with a water depth of 3 ft. and the horizontal cage type rotor is a very simple aeration device of at least equal efficiency with vertical type mechanical aerators or diffused air systems. Capital cost is very small and for equal BOD removal may be comparable with that of facultative type lagoons. Because of the immediate complete mixing achieved, it is very stable against shock loads. It has been used to treat a variety of industrial wastes, distillery, milk, cannery, etc. . . . By using the process following an initial 70–80% BOD reduction in anaerobic type lagoons, a very economical process can be developed to treat seasonal wastes such as those from fruit and vegetable canning Parker[10] and excess sludge can be wasted to the anaerobic lagoon.

COST CONSIDERATIONS

Because of marked variations in cost structures, labour cost, mechanical equipment, land, and power, from country to country, it is impossible to make any generalized statement on costs. Each situation needs to be evaluated with regard to capital and maintenance cost of alternative means of treatment and with consideration of effluent standards to be met. However, in general it can be said that where natural bacteriological and algal activity can be utilized by holding wastewater in simple earth structures or passing it through natural soil profile, effective processes for removal of BOD, bacterial, viral and parasitic ova; nutrients and detergents can be utilized at low cost.

SUMMARY

Irrigation with suitable soil drainage is an effective means of treatment and dosage of raw wastes but is subject to some odour and public health concern and is ineffective in wet weather. Processes of BOD, nutrient and bacteriological reduction can be achieved. Further study of these processes is necessary.

Natural lagoons may achieve BOD, nutrient, bacteriological, viral and parasite ova reduction and this is most effectively achieved with the minimum of odour and other nuisance by provision of facultative type lagoons loaded appropriate to local temperatures operated in parallel and followed by at least 3–4 cells in series to achieve the required bacteriological reduction.

Aerated lagoons are a means of achieving increased BOD removal with reduced land area but various forms need critical cost evaluation in comparison with other processes.

The Pasveer oxidation ditch is an effective low-cost means of applying the activated sludge process where moderate areas of land are available and structural appearance is not important.

Most of these low-cost treatments by proper consideration of microbiological nutrition and physiology can be adapted to the treatment of a wide variety of organic industrial wastewater.

REFERENCES

1. Gilde L. C. (1970). Proc. Nat. Symp. Food Process Wastes. Portland U.S.A. p. 311.
2. McDowell F. H. and Thomas R. H. Disposal of Dairy Wastes by Spray Irrigation on Pasture Land. Dairy Wastes Committee. Pollution Advisory Council. Public. No. 8. Wellington N.Z. March 1961.
3. Shuval H. I. (1963). Proc. 17th Ind. Wastes Conf. Purdue Univ. p. 652.
4. Arceivala S. J. (1970). Waste Stabilization Ponds. Cent. Pub. Hlth. Eng. Res. Inst. India.
5. Parker C. D. (1970). Unpublished.
6. Cadwell D. H. (1946). S. W. J. Vol. 18, p. 433.
7. Parker C. D. (1950). S. I. Wastes. Vol. 22, p. 760.
8. Parker C. D. (1968). J. Wat. Poll. Cont. Fed. Vol. 40, p. 192.
9. Parker C. D. (1959). S. I. Wastes. Vol. 31, p. 133.
10. Parker C, D. (1966). Proc. 21st. Ind. Wastes Conf. Purdue Univ. p. 284.
11. Gloyna E. F. (1971). Waste Stabilization Ponds. Wld. Hlth. Org. Monograph No. 60.
12. Parker C. D. (1962). J. Water Poll. Cont. Fed. Vol. 34, p. 149.
13. Marais G. R. (1970). 2nd. Int. Symp. for Waste Treatment Lagoons. p. 15.
14. Hom L. W. (1970). 2nd Int. Symp. for Waste Treatment Lagoons. p. 151.
15. Dryden F. D. and Stern G. (1968). Env. Sc. and Tech. p. 268. Lancaster.
16. Cillie G. G. et al. Proc. 3rd Int. Conf. I. A. W. P. R. Munich 1966. Vol. 2, p. 1.
17. Benjes H. (1970). 2nd. Int. Symp. for Waste Treatment Lagoons. p. 210.

Discussion *by* F. Engelbart, Wasserwirtschaftsamt
Hannover, Germany
and
W. Englebart,
Institut f. Gärungsgewerbe u. Biotechnologie,
Berlin, Germany

"Low cost" must not mean "low efficiency". – Equal efficiency is the basis for all comparisons of methods.

The result of treatment is quite differently influenced by sunlight radiation, temperature and other meteorological conditions. Comparison is to be based on definite assumptions.

Costs for the land required, for construction and equipment, and for maintenance vary considerably either in relation to each other or between different countries. Therefore, they should be specified so that comparisons with other countries are possible.

Material costs for maintenance mostly result from the demand of energy.

These costs differ between countries and even between regions. But this part of total costs should not be given too much weight.

Costs for wastewater treatment with high requirements of personal service will increase more and more, for wages tend to increase faster in relation to other costs.

Removal of sludge may be a problem and costs for this have to be taken into account.

In view of these varying relations only general conclusions can be drawn:

a) Irrigation of sewage, being an effective and economical treatment in arid zones and in countries having low wages, is the most expensive process in Germany even when profitable agriculture is taken into account. But in the case of very highly loaded organic wastes, irrigation is still recommended in densely populated countries like Germany.

b) Shallow oxidation ponds, having good results by algal oxygenation in warm and sunny zones the whole year, fail in regions with cold and dark winters and are not competitive where land is expensive.

c) Use of highly loaded facultative and of anaerobic lagoons is restricted by the problems of odour.

d) Of the methods mentioned by the author only the oxidation ditch process (Pasveer) can be used everywhere. This process is an activated sludge treatment with total circulation in the ditch, with simple construction and equipment. If the load isn't too high, treatment can be done without continuous recirculation of sludge.

A broad spectrum of applications, a minimum of personal service and fool proof function are the desired advantages of a new method, which we would like to be discussed here additionally:

The starting point for the development of the "simultaneous treatment" was experience with several methods of wastewater treatment but also accurate observations of natural ponds and flowing waters.

These have resulted in the development of new devices, viz: –

The "Rotating Pipe Filter" which can be described as a combination of an effective screw pump (normally with horizontal flow) and a biodisc. It consists of many layers of plastic tubes which are wound up spirally on a cylindric drum. A biological slime grows on the wall of the tubes. When rotating the tubes move alternately through water and air so that the fixed microbes come into contact with sewage and oxygen. A "Rotating Pipe Filter" of 2,5 m diameter and 10,8 m length, consisting of 50-mm-tubes with a spiral gradient of 0,85 m offers a biological slime of more than 4000 m^2. At 3.6 rpm it pumps about 650 m^3/h of water and 175 kg/h of oxygen. BOD-digestion rate will be $100 - 400$ kg/d in relation to the load. Energy demand will be $0.3 - 0.1$ kWh/kg BOD_5-digestion. – Pumping horizontally, energy is needed only to balance friction of the water passing the tube walls at an average speed of 32 cm/s. This makes it possible to work with semiclosed circuits. "Rotating Pipe Filters" are used most advantageously for the treatment of relatively high loaded wastes.

The "Line Aerator", being a narrow shaft transversally through a basin or lagoon, acts as a giant pump by supplying fine bubbles of air to the bottom. Both at the bottom and at the surface, guiding devices ensure that the vertical stream is transformed into a horizontal parallel flow in direction to the shaft at the bottom and away from the shaft at the surface, resulting in a quick and total circulation of the whole volume. The upper guiding device, consisting of small tubes, offers a large surface ($400 - 1000$ m^2/m of "Line Aerator") for a biological slime, which is overflowed by the mixture of water and bubbles at an average speed of $40 - 50$ cm/s. Efficiency of pumping is $2 - 4000$ m^3/kWh. Each meter of "Line Aerator" gives a total circulation of up to a volume of 300 m^3 with an energy demand of 0.25 kW. Oxygen efficiency will be more than doubled in relation to normal aeration with fine bubbles.

These characteristics of the "Line Aerator" make it possible to use large volumed basins or lagoons resulting in a big buffering capacity of the system. They also make it possible to use a stepwise treatment resulting in a reduced land requirement and a better economy.

A big part of the bottom of lagoons may serve as a store for the sludge. Water containing oxygen and flowing directly above the bottom, suppresses odours resulting from anaerobic processes in the sludge. Thickening of the sludge can be increased by drainage of the bottom. According to the surface loading the stabilized sludge must be removed after 1 – 10 years.

"Simultaneous treatment" reduces land requirement and it is practicable everywhere, even under an ice cover. Its quick construction, simple equipment and outstanding economy should not brand this new treatment as an inadequate method only fit for poor communities or industries or developing countries; it can be used even in highly industrialised areas.

F. Josa, Spain.

Two low-cost treatment methods not mentioned were treatment on grassland and running the sewage through channels in which plants were allowed to grow.

Bernard, France.

How frequently and by what means was sludge removed from anaerobic ponds treating raw sludge, and what were the costs of operation? What was the solids content of sludges from ponds treating milk products effluents?

Reply

Frequency depends on solids loading expressed as lbs./ac./d. Where loading has been 250 lbs./ac./d. there has been no need for solids removal after 10 years. Solids accumulation has been 6-9 inches in that time. Have no information re cost. The solids content of such sludges is 10-15%.

R.J. Davis, Israel.

R.J. Davis observed that aerobic conditions were necessary to avoid odour problems. In this connection screening and a brief period of settlement for only a few minutes was effective. By early removal of suspended acid-colloidal matter oxidation ponds operated more satisfactorily, producing an effluent of lower total dissolved solids content.

J. Galt, Israel

Tertiary and maturation lagoons have been used to improve the quality of effluents from conventional secondary treatment processes (maturation ponds). The principle purpose is to achieve reduction in faecal coliform count in 7-10 days. What are the comparable results in the plants referred to by the author?

Comparable results are taken from reference 12 J.W.P.C.F. Vol 34 (1962) from Tables 3 and 5.

TABLE 3. BACTERIAL VALUES FOR AEROBIC MULTI-UNIT STABILIZATION PONDS (Summer Temp. 70°F.)

Item	Raw Sewage	Pond 1	Pond 2	Pond 3	Pond 4	Pond 5	Pond 6	Pond 7	Pond 8
Detention Time (days)	–	3.8	8.0	13	18	23	28.5	33.5	38.5
E. Coli (No./ml)	6.0×10^5	4.0×10^5	4.0×10^4	2.1×10^4	6.6×10^3	249	15	1.1	0.13
Reduction in pond (%)	–	33	90	50	66	96	94	94	88
Reduction Overall (%)	–	33	93	97	99	99.96	99.99	99.99	99.99

TABLE 5. BACTERIAL VALUES FOR AEROBIC MULTI-UNIT
STABILIZATION PONDS AFTER HIGH-RATE TRICKLING FILTERS

SUMMER (TEMP.70°F.)					
	Bacterial Count (No./ml)			Algal Count	
Item	Plate Count at 22° C.	E. Coli	S. Faecalis	No./ ml	Areal Std. units/ml.
Secondary effluent	5.9×10^5	2.7×10^4	6.0×10^3	—	—
Pond 1	7.5×10^5	7.0×10^3	400	5.5×10^5	5.6×10^5
Pond 3	3.8×10^4	42	2.3	5.6×10^5	4.0×10^5
Overall (%) reduction	94	99.8	99.5	—	—

COMPLETE WASTE TREATMENT IN PONDS*

WILLIAM J. OSWALD
Professor of Sanitary Engineering and Public Health,
University of California, Berkeley, California 94720, USA.

INTRODUCTION

Although it is often presumed that the last three decades of the Twentieth Century will be an era of increasingly intensive mechanical, physicochemical, and biological methods of waste treatment, it may also be one in which earthwork reactors or ponds will be applied to an increasing extent. There is a growing realization on the part of those interested in the cost effectiveness of waste disposal processes that many of the unit processes for waste treatment now accomplished in masonry, concrete, or prefabricated reactors may be accomplished equally as well in ponds and usually at a substantially lower cost. One major reason for this is that the cost of a pond per unit of effective volume is rarely more than 1/10 and often as little as 1/100 that of a concrete reactor of equivalent volume and performance. It is the low unit cost of ponds which permits their use in many situations where no other waste disposal is economically feasible. Now, other properties of ponds are becoming important because of concern with ecology and environmental impact. In addition to low unit costs, ponds have the potential for multiple benefits not shared by conventional reactors. For example, they may provide opportunities for both treatment and disposal of wastewater in the same reactor. When properly designed and operated, they may qualify as open space, provide opportunities for reclamation of water and nutrients, enhance the propagation of wildlife, and contribute to the beautification of an area — all at a capital investment little more than required for waste impoundment alone.

In addition to their use as treatment systems, ponds are often used as buffer systems to increase the reliability·of, and to render fail-safe other forms of waste treatment. For example, in one of the most advanced waste treatment systems in the world (South Tahoe Public Utility District), a pond (Indian Creek Reservoir) is used to retain plant effluent and to act as a buffer and back-up treatment system should a failure occur in the complex biological, chemical and physical processes within the plant itself.[1] It is interesting to note that in a more clement environment than that in the Sierra Nevada Mountains, a properly designed pond system as large as Indian Creek Reservoir would be adequate to treat the 7.5 mgd of wastes without the preceding advanced waste treatment. However, under the climatic conditions at Lake Tahoe, this would have been impossible.

While ponds of the terminal type (i.e. having no overflow) have been used in all parts of the world, they function mainly during warm weather and are biologically dormant during the winter. Thus, if ponds are to be an active part of a continuous flow-through, complete waste treatment process which dependably produces an effluent of tertiary quality, their use is largely restricted to areas where the visible solar energy input is above 100 gm calories per cm^2 per day 90% of the time, and where freezing conditions do not persist at any time.[2] Surprisingly such conditions are met or exceeded over 60% of the

* For presentation before Workshop Panel 3 'Low Cost Waste Treatment Systems', 6th International Conference on Water Pollution Research, Jerusalem, Israel, June 19, 1972.

Earth's habitable land surface, including most of Africa, Australia, South America, half of Eurasia, and 40% of North America. Conditions favorable to flow-through ponding do not exist in those countries, such as England and Germany, where activated sludge and trickling filtration were developed and are widely and properly applied. The fish ponds of Munich and the waste ponds in Norway and Sweden are, of course, exceptions.

Restrictions on ponding are primarily due to the influence of temperature and light on algae and the effect of temperature on methane bacteria.[3] Regardless of where applied, during the warm season of the year algae develop in ponds and produce a green pigmentation consisting of the whole cells of microalgae. Some regulating agencies view green effluents with alarm and have placed microalgae in the same category as sewage suspended solids — an absurdity which may lead to unnecessary expenditures in many cases. Green algae are particularly innocuous since their cells do not decompose easily (except perhaps in a COD test), and will usually remain suspended in streams with velocities as low as 5 cm per second. Suspended algae provide a base for the aquatic food chain and become a nuisance only when their concentration greatly exceeds the food requirements of the stream biota. Thus, regulations related to algae in streams should be based on velocity, biotic conditions, and beneficial uses of each stream rather than on any arbitrary universal standards for suspended solids or chlorophyll.

The microalgae during growth contribute greatly to the high quality of pond effluents. They produce excess oxygen for waste oxidation and produce high protein, high energy cell material which may serve as food for plankton, fish, and higher organisms. The growth of algae in a waste will raise its temperature and pH, which in turn, as pointed out by Parhad[4] contributes to the accelerated death of any coliform and presumably pathogenic bacteria contained in the waste. Similarly, a rise in pH may cause precipitation of polyvalent cations and anions such as calcium, magnesium iron, and orthophosphate. This precipitation is often accompanied by natural separation (autoflocculation) of the algae and a natural softening of the water.[5] Sobsey[6] has found that a combination of algae and bacteria is essential to obtain the unique inactivation of poliovirus that has been found in ponds. Algae may incorporate substantial quantities of plant nutrients such as ammonia and nitrate nitrogen, phosphate, iron, and trace minerals in their tissues; and when separated, provide a degree of tertiary treatment. Controlled algal growth and tertiary treatment is particularly desirable in the arid zones of the world, where impoundments of reclaimed water may be subject to uncontrolled proliferation of submerged aquatic plants or undesirable blue-green algae with attendant problems of toxicity to wildlife, of mosquito and fly breeding, as well as of eutrophication.

As I have pointed out recently,[7] ponds which are especially designed to enhance the growth of algae are being increased in size. In the early 1940s, controlled algal cultures 10^1 to 10^2 liters in volume were normally used; and in 1950, cultures 10^4 liters in volume were studied. By 1960 several 10^6-liter cultures had been constructed for algal production, and in 1970 a 10^9-liter algal growth system was put into successful operation to oxygenate the wastes of the City of Modesto in California. Under design or consideration are two 10^{10}-liter cultures, one for treatment of drainage water in California Water Plan,[8] and one to produce Spirulina in Mexico.[9] Ponds of the 10^{10}-liter size have the potential to yield prodigious benefits. For example, a 10^{10}-liter algal culture is sufficiently large to oxidize and assimilate all of the organic wastes from a city of 1,000,000 persons, and at the same time could be used to cool a power plant operated to provide all of the electrical power for that same community.[10] Plants 10^{12} liters in total volume would be adequate to treat all domestic wastewater in the United States.

While for climatological and land restrictions one may not expect all waste treatment to involve immediate impoundment, the current trend in the United States is away from the direct discharge of any wastes, treated or not, to normally pristine waters. Such a policy must lead inevitably to land disposal or impoundment. In view of this national trend and the rapid increase in use and size of controlled algal cultures, it is especially pertinent at this time to review in some detail the relationships between microalgae production and waste treatment in ponds.

As noted previously the beneficial uses of algae include but are not limited to:

1. Production of oxygen under conditions whereby it can be used by bacteria to oxidize dissolved organic wastes and produce an effluent high in dissolved oxygen, and low in soluble BOD (photosynthetic oxygenation).

2. Production of algae that may be used for fertilizer or livestock feed – thus recycling nutrients (microalgae production and use).

3. Removal of nutrients by algae stripping (tertiary treatment).

Inasmuch as both photosynthetic oxygenation and microalgae production and use have been described and discussed previously,[11,12,13] I wish to restrict the balance of this discussion to a consideration of the mechanisms by which complete treatment of wastes, including both secondary and tertiary treatment, can be accomplished in ponds.

To draw conclusions regarding the application of ponding systems to accomplishment of complete treatment of wastes, one initially requires detailed information on the nutritional requirements of algae, the amounts of required nutrients available in wastewater, and the kinetics of uptake of nutrients from wastewater by algae.

Concerning the nutritional requirements of algae in general, Jewell and McCarty[14] and Foree and McCarty[15] made an extensive review of the literature to determine the reported composition of algae in terms of the mean composition of actively growing cells together with minimum and maximum values. They found the reported mean, minimum and maximum compositions respectively of the various major elements expressed in percentage of ash-free dry weight to be as follows: carbon, 53, 42.9, 70.2; hydrogen, 8, 6.0, 10.5; oxygen, 31, 17.8, 34.0; nitrogen, 8, 0.6, 16.0; and phosphorus, 2.0, 0.16, 5.0. They pointed out the importance of growth conditions in the establishment of these values.

Because oxygen and hydrogen are always present in excess in aquatic systems, a discussion of kinetics of the major elements need only involve carbon, nitrogen, and phosphorus.

PHOSPHORUS AS A NUTRIENT

Inasmuch as phosphorus is of greatest current interest, it will be discussed first. In 1951 the author and his colleagues[16] grew continuous cultures of *Euglena gracilis* in domestic sewage under environmental conditions defined as follows: temperature 25°C; light continuous at 27 gm cal per liter per min.; detention period variable; phosphate 8 mg per liter (mg/l). It was found that the suspended solids in the culture, comprised mainly of *Euglena* cells, contained 2.3% P at $\theta = 2$ days and 0.8% P at $\theta = 14$ and 20 days. In the case of the 14- and 20-day cultures, nitrogen was found to be limiting to growth, but in no case was phosphate limiting.

Zabat[17] studied the kinetics of phosphate metabolism of *Chlorella pyrenoidosa* (Emerson), utilizing continuous cultures under similar environmental conditions, but with phosphate controlled to be limiting. He found under carefully controlled conditions that a 25°C maximum specific growth rate for *Chlorella* of about 1.5 day^{-1} occurred when the background (S$_1$) phosphate value was 0.05 mg/l. Under these conditions his cells contained about 0.6% P. At one half the maximum specific growth rate, the background phosphate value was 0.020 mg/l and the cells contained 0.4% P. At the lowest growth rate explored, 0.25 day^{-1}, the phosphorus content of the cells was 0.28%, and the background P was 0.0037 mg/l. Porcella[18] working with continuous cultures of *Selenastrum capricornutum* in PAAP medium[19] found that under P-limiting conditions, the cells contained 0.41 to 0.34% P. Toerien *et al.*,[20] also working with continuous cultures of *S. capricornutum* in PAAP medium with phosphate limiting, observed a maximum specific growth rate of 1.85 days^{-1} at P values of 0.030 to 0.050 mg/l, and for 1/2 this rate a background P of 0.0037 and 0.0057 mg/l. They obtained under severe phosphate limiting conditions, the remarkable yield of 850 mg of cells per mg of phosphate. At this yield, the cells had a phosphate content of only 0.125% P or less.

In outdoor pilot plant experiments with domestic sewage containing about 10 mg/l of P, Oswald and Gotaas[21] found the mean percentage of P in algae solids to vary as a function of pond depth as follows: 5 cm, 1.1%; 15 cm, 0.95%; 30 cm, 0.7%; and 45 cm, 0.4%. Corresponding mean pH values were 10, 9, 7.5 and 7.1. Based on these pH values, some phosphate precipitation probably occurred at the high pH values, but no precipitation would have occurred at pH 7.5 and 7.1. On the other hand, at 30 cm and 45 cm, less than 1 mg/l of P was incorporated in algal cells, indicating that while P was in great excess, the cells had only about 0.5% P. Such data evidence an absence of luxury uptake in the deeper cultures. Thus, luxury uptake may depend on excess light.

Hintz, *et al.*[13] studying the nutritional value of algae grown in a 20-cm deep outdoor culture, separated by centrifugation, and drum dried, reported phosphorus in 10 samples to average 2.2% of the dry cell material. However, it is believed that in this case the algae may have contained an extracellular calcium phosphate precipitate as well.

Based on the available evidence, it appears that under the extremes of ordinary growth conditions, algae may contain from 0.25 to 1.25% P. The former occurs under severe P-limiting conditions, and the latter under conditions where P is present in large excess and nutrients are abundant, and the cells are exposed to an abundance of light, and the temperature is optimum.

NITROGEN AS A NUTRIENT

Nitrogen also is a nutrient of major interest. It may be utilized by algae in either the ammonia (or ammonium) or nitrate form. Ammonia is the form normally used by algae in waste ponds because nitrate only appears in such ponds in minute concentrations, unless the pond is preceded by an activated sludge or trickling filter plant.

Considering ammonia first, in the author's[22] studies of continuous combined algal-bacterial cultures with *C. pyrenoidosa* (Emerson), ammonia was the source of nitrogen and was usually present in excess. Under these conditions, the algae were found to contain 9.0 ± 1.0% N. Shelef *et al.*[23] studied ammonia uptake by *C. pyrenoidosa* in continuous cultures under nitrogen-limiting conditions. He found that at 19°C the growth rate was about 1.5 day^{-1}, and no increase in growth rate occurred when the background ammonia concentration was 1.5 mg/l. A growth rate of 1 day^{-1} occurred when the ammonia concentration was 0.65 mg/l. He found the maximum growth rate at 28.5°C to be 2.2 day^{-1} when the ammonia concentration was 1.75 mg/l. In pilot plant work with ammonia as a major N source, N was rarely limiting and the algal cells were found to contain 10% N ± 2.

With regard to nitrate, Shelef[24] studied the nitrate requirements of *C. pyrenoidosa* (Emerson) under continuous culture conditions at two temperatures — 19°C and 28.5°C. At 19°C he found the maximum growth rate to be about 1.45 per day; and at 28°C, about 2.22 per day. With nitrate-nitrogen limiting, the 1/2-maximum growth rates occurred when the nitrogen concentrations were about 0.9 mg/l at 19°C, and 1.2 mg/l at 28°C. At 28°C the mean yield was 12.0 mg of cells per mg of N; and at 19°C, about 14 mg of cells per mg of N, indicating 8.3 and 7.15% N respectively in cells. In optically dense cultures with nitrogen often severely limiting, Shelef[25] found yield coefficients for nitrogen as high as 17.75, indicating that the cells had nitrogen contents of only 5.65%. Levels of nitrogen of 16% occasionally reported in the literature are unlikely, since this is the level of nitrogen in pure protein (protein = 6.25 N). Inasmuch as algae must have cell walls and some lipids and carbohydrates, nitrogen levels reported to approach 16% must be regarded as questionable.

Under nitrogen-limiting conditions, again with PAAP medium Porcella *et al.*[18] found yields of 13 to 18 mg of N per mg of cells indicating nitrogen fractions from 5.5% to 7.7%. According to Shelef,[23] nitrate nitrogen is not limiting to *C. pyrenoidosa* at ordinary temperatures, if the background nitrogen concentration is above 5 mg/l; but as background nitrogen declines from 5 mg/l to 0.3 mg/l, there is a rapid decrease in growth rate and growth ceases for all practical purposes and at concentrations below 0.3 mg/l.

CARBON AND ALGAL GROWTH

With regard to carbon, little information is available on carbon kinetics under carbon-limiting conditions. Goldman, *et al.*[26] has recently published a comprehensive review of the literature on the effect of carbon on algal growth, and he is currently undertaking a detailed investigation of the kinetics of carbon utilization by *S. capricornutum* and other algae in continuous cultures under conditions similar to those applied by Shelef[24] and Zabat.[17] His study will indeed be valuable when it becomes available.

Studies of carbon in continuous culture outdoor ponds were made by Oswald and Gotaas.[21] They found cell carbon to vary from 43.5 to 60% when grown in sewage having organic plus inorganic carbon averaging 157 mg/l and ranging from 113 to 211 mg/l C. Uptake of carbon by the algae varied from 10% to 88% of the available carbon. The highest values were obtained when the pond was operated at a depth of 5 cm and the lowest when operated at a depth of 45 cm. When the carbon demand of the algae reached 88% of the available carbon, the carbon content of the cells was about 50%. Under these conditions, the daytime pH of the system was found to be 10.5; but in view of the continued high carbon content of the cells, it did not appear that carbon was the major limiting factor in these outdoor experiments.

UPTAKE OF C, N AND P BY ALGAE

In studies of the growth of algae on sewage wastes, specific characterization of sewage with regard to C, N, and P are not possible because of the well-known variability of sewage. It is, therefore, only possible to pursue a discussion of the uptake of elements by algae from sewage by making certain assumptions concerning the composition of sewage and the growth of algae.

If no substances other than household wastes are discharged in sewage and the waste volume is about 120 gal/capita/day (450 l/cap/day), sewage may contain on the average about 125 mg/l available carbon, 40 mg/l of N, and 11 mg/l of P. Based on preceding conclusions regarding the composition of algae, and assuming that environmental conditions (i.e. temperature and light) were such that each of these elements could be fully assimilated by algae growing in sewage, one may characterize sewage in terms of its input nutrient level as having an algal growth potential (AGP)* as follows:

Carbon: 125 mg/l C converted to algae containing 40 to 60% carbon – potential yield, from 208 to 312 mg/l;

Nitrogen: 40 mg/l N converted to algae containing 5.5 to 12.5% nitrogen – potential yield, from 320 to 730 mg/l;

Phosphorus: 1 mg/l of P converted to algae containing 0.5 to 1.5% phosphorus – potential yield, from 730 to 2200 mg/l of algae.

Although ranges are given above for the AGP of the various nutrients, in the interest of simplifying the ensuing discussions and sample calculations, only a mean value of the AGPs will be used which is as follows: C-AGP, 250 mg/l, 50% C; N-AGP, 500 mg/l, 8% N; and P-AGP, 1220 mg/l, 0.9% P. It should be emphasized that these are only representative values within the normal ranges discussed above. By comparing these AGP values, it is evident that unmodified domestic sewage is a nutritionally unbalanced medium lacking sufficient carbon to permit direct assimilative removal of nitrogen or phosphorus by algae, and lacking sufficient nitrogen or carbon to permit direct assimilative removal of phosphorus by algae. The design of algal assimilative systems for nutrient removal by algae must, therefore, involve consideration and modification of these nutritional deficiencies. Two types of modifications are available for consideration – either a supplementation of the amount of carbon and nitrogen in proportion to the amount of phosphorus to be assimilated, or removal of the amount of phosphorus and nitrogen in proportion to the amount of carbon to be assimilated. Obviously, various combinations of these two alternatives are also available.

In consideration of both alternatives, it is worthwhile to examine the free energy relationships involved. According to the author's studies,[22] the energy content of algae is normally about 6 gm calories per mg of algae (ash-free dry weight basis). Inasmuch as algae normally assimilate the oxidized forms of their essential nutrients, they must fix practically the entire 6 calories for each mg of cell material synthesized. An exception may be that those algae utilizing ammonia nitrogen may receive about 0.5 cal per mg of cell material through direct assimilation of ammonia, but this potential source of energy is sufficiently small to be neglected in the ensuing general discussion.

In the example, to assimilate 125 mg/l of available carbon, 250 mg/l of algae must be synthesized; and at 6 cal per mg, 1500 calories must be fixed. (Coincidentally, the synthesis of 250 mg/l of algae will be accompanied by the release of 400 mg/l of molecular oxygen, easily satisfying the ultimate BOD of the waste which normally is found to be about three times the available organic carbon.)

Fixation of light energy by algae normally is at less than 10% efficiency;   and for this discussion, an efficiency of 4% may be assumed. At 4% efficiency, to fix 1500 gm calories in algal cell material, 37,500 calories must be available. Assuming a visible solar energy

---

* The quantity of algae that will grow when no factor other than a specific nutrient limits growth.

flux of 200 cal per cm² day, one needs to expose to solar energy 187 cm²/l of algal culture for one day, or 18.7 cm²/l for 10 days. The corresponding depths would be 5.33 cm and 53.3 cm. The interrelation of the above factors were generalized by Oswald and Gotaas[11] in equation 1 (a):

$$\frac{d}{\theta} = \frac{1000\, eS_s}{h\, X} \qquad\qquad 1\,(a)$$

in which d is the culture depth in cm, θ the detention period in days, h the heat of combustion of algae in gm calorie per mg, S_s the solar energy flux in calories cm² day⁻¹, e the efficiency of light energy conversion, and X the required algal cell concentration of AGP in mg/l. Substitution of the values indicated above, e = 0.04, S_s = 200, h = 6, and X = 250 yields the value 5.33 cm per day. Thus in theory any combination of d/θ which yields the ratio 5.33 cm per day will provide sufficient energy at an efficiency of 4.0% to produce 250 mg of algae/l of waste passed through a pond.

Similar calculations substituting the assumed N-AGP and P-AGP for X give d/θ values of 2.65 cm per day and 1.07 cm per day respectively. The term d/θ may be regarded as a time energy restriction on the system; thus, d/θ = Ke and

$$\theta = \frac{d}{Ke} \qquad\qquad 1\,(b)$$

To use equation 1 (b) to determine θ, some independently valid basis must be used to select d. An experimentally determined restriction on depth is that the permissible depth of a pond to grow a specific concentration of algae cannot exceed three-fold the depth predicted for light penetration into a culture having a concentration X_d. The restriction is that if d is made greater than predicted, the experimentally determined value of X will be less than predicted, and less than the AGP. This restriction is generalized in an approximate depth prediction equation derived by the author[11] from the well-known Beer-Lambert Law and the experimentally determined facts.

$$d_L = 3\,\frac{Ln\,I_o - Ln\,I_d}{X_d\,\alpha} \qquad\qquad 2\,(a)$$

in which d_L is the required design depth, I_o is the incident light intensity at the culture surface, I_d is the compensation light intensity normally regarded to be 2.7 ft-C in approximate calculation of this type, X_d the assumed concentration of algae equal to the AGP in mg/l, and α a light adsorption coefficient in cm² per mg, and Ln refers to natural logarithms. If I_o and I_d are expressed in ft-candles, α has a mean value of about 1.5 x 10⁻³ cm² per mg. To further simplify equation 2, normally assumed values may be substituted. Thus, under outdoor conditions, the surface light intensity I_o averages about 8000 ft-C, and if Ln I_d is assumed to be 1.0 and α 1.5 x 10⁻³, equation 2 (a) reduces to

$$d_L \simeq \frac{16,000}{X} \qquad\qquad 2\,(b)$$

Substituting the assumed AGP's for X in 2 (b), one obtains:

for carbon $\qquad\qquad d_L = \dfrac{16,000}{250} = 64\ cm$

for nitrogen d_L $= \dfrac{16,000}{500} = 32$ cm

for phosphorus d_L $= \dfrac{16,000}{1,220} = 13.1$ cm

If these values of d_L are then substituted for d in equation 1 (b) and solved for θ, one obtains:

for carbon θ $= \dfrac{64}{5.33} \cong 12$ days

for nitrogen θ $= \dfrac{32}{2.65} \cong 12$ days

for phosphorus θ $= \dfrac{13.1}{1.06} \cong 12$ days

DESIGN CONSIDERATIONS

The design conditions under the various assumptions are therefore as follows: For complete carbon assimilation, d = 64 cm, θ = 12 days. For complete nitrogen assimilation, d = 32 cm, θ = 12 days. For complete phosphate assimilation, d = 13.1 cm θ = 12 days. Inspection of these values indicates that major practical problems would exist if the design of a pond were based on phosphorus or nitrogen considerations solely. For example, in the case of the phosphate assimilation criteria, waste application rates of only 1 cm per day would usually be exceed by the rate of evaporation. This would also be practically true in the case of nitrogen. Another objection to this technique is that large amounts of carbon and nitrogen would be required as supplementary nutrients to permit complete phosphate assimilation; and the application of such supplements would be costly and difficult. Thus, the application of the assimilation technique of tertiary treatment in ponds does not seem to be practical, except perhaps in the case of a system to assimilate carbon.

One is then left with the alternative of nutrient subtraction. This is done biologically with relative ease with carbon, and chemically with phosphate, but it becomes a difficult and expensive task with nitrogen, particularly when preceded by aerobic biological systems. Fortunately, an extremely inexpensive technique for nitrogen subtraction has recently come to attention. This is the deep facultative pond in which by some mechanism not yet fully explored, more than 60% of the influent nitrogen is normally converted to nitrogen gas. By using such a pond as a primary unit, it is possible to subtract sufficient nitrogen from a waste to permit the balance of the nitrogen to be assimilated by algae to a high degree in a pond designed according to Equation 1.

This design concept was applied at Saint Helena, California, in the construction and operation of a pond-system which involves a 10'-deep, 20-day detention primary pond followed by a 3'-deep, 10-day detention algal growth pond which is mixed at a velocity of 1/2 foot per second. This system has been described fully in a recent paper.[27]

The Saint Helena system was studied in detail by Meron[28] who found that without chemical separation of algae, the overall removal of BOD was 97%; COD, 93%; carbon,

78%; nitrogen, 92%; and phosphate, 64%. In studies subsequent to those of Meron, lime (CaO) at about 150 mg/l was added to the algal growth pond effluent to effect algae coagulation and removal. The remaining clear supernatant had a BOD of 0.5 mg/l, total N of 1.4 mg/l, and total P of 0.59 mg/l. These values represented removals of BOD, 99.8%; total nitrogen, 93.3%; and phosphorus, 92%. It thus appears that in the application of ponds to complete waste treatment, nutrient substraction combined with assimilation is a far more economical and simple procedure than is either nutrient assimilation or nutrient subtraction alone.

If systems of the Saint Helena type can be made to function as well at other localities as at Saint Helena, it should be possible to effect major savings in complete waste disposal, and at the same time have systems which are more fail-safe and compatible with the principles of conservation and ecology than are current mechanical plants. The Saint Helena results indicate that where climate permits, complete tertiary treatment of sewage need not always involve the complex, costly, and delicately balanced advanced waste treatment systems of the type studied at the South Tahoe Public Utilities District (STPUD) and Indian Creek Reservoir.[1] Judging from the quality of water reported in Indian Creek Reservoir, the product water of the Saint Helena system following lime treatment for algae removal would be almost as low in phosphate and lower in BOD and nitrogen than that in Indian Creek Reservoir.

In any consideration of alternatives, costs become especially significant. A comparison of costs of the STPUD and the Saint Helena systems can only be approximate because STPUD costs are based on 7.5 mgd design flow, while current flow is about 2.5 mgd; whereas Saint Helena's flow is about 0.35 mgd and the design flow 0.5 mgd. Based on the curves published by Robert Smith,[29] the unit cost of construction and operations and maintenance for an 0.5 mgd plant should be approximately twice those for a 7.5 mgd plant. Reported costs for the STPUD plant are approximately $405 per million gallons, excluding the Indian Creek Reservoir project. The cost of the Indian Creek Reservoir project is reported to be $217 per million gallons, but much of this cost involves the pipeline and pumping stations over Luther Pass. The added cost due only to storage is estimated to be $50 per million gallons for a total of $455 per million gallons. Because of unit cost differences, such a system at Saint Helena would cost about $900 per million gallons; whereas current costs at Saint Helena are estimated to be $380 per million gallons. If lime treatment for phosphate removal were employed at Saint Helena, the added cost would be about $120 per million gallons. Thus, a total cost of $500 per million gallons is likely at Saint Helena. From this it may be concluded that the process at Saint Helena costs only about 55% of that at South Tahoe; or conversely that a Saint Helena-type plant having a capacity of 7.5 mgd could produce tertiary treated water at a cost of about $250 per million gallons.

If impoundment must be employed for wastes because of zero tolerance of discharge into receiving waters, as is the case in Tahoe, and may be the case in many places in the future, one actually has two alternatives: (1) treatment followed by impoundment; and (2) impoundment followed by treatment. The first alternative is that employed at STPUD; the second alternative is that employed at Saint Helena. Reflection will indicate that where land is inexpensive and climate permits, impoundment followed by treatment should intrinsically be more economical than the alternative, because one obtains buffer capacity, flow regulation, and fail-safe advantages from the impoundment, which at the same time is a most economical unit. Because of lower costs, one may then apply fundamentally different, more economical and more complete methods of treatment to impoundment effluents than can be applied to raw sewage.

The conclusions are thus suggested that where climate permits, complete waste treatment is attainable in ponds. It is also suggested that where ponds can be operated, which includes more than 60% of the world's habitable surface, a properly designed system involving deep impoundment followed by algal growth and removal is a more simple, reliable, and economical method of attaining a high degree of secondary and tertiary (i.e. complete) waste treatment than are the systems currently applied.

The potential economy, dependability, and simplicity in operation and maintenance of the pond system described are believed to be such that wastes could be economically transported substantial distances to reach 'pondable areas' where land costs would permit their use in waste treatment and reclamation. For example, in the San Francisco Bay area, piping, and even pumping of wastes to pondable inland areas, may be economically justified in view of a need for fresh, low-nutrient water for release into the upper reaches of the Bay. On the other hand, export of wastes from cities near pondable areas to more congested areas to achieve low unit costs may not be economically feasible, should the low unit costs of the Saint Helena-type system be applied in cost effectiveness calculations. Indeed, the potentially lower costs and high degree of waste treatment that may be obtained in systems of the Saint Helena type could, if proved to be generally applicable, greatly modify the planning of many of the world's major waste management schemes in the warm and arid regions of the world.

ACKNOWLEDGEMENTS

Most of the work reported here was supported in part by the various U.S. pollution control authorities over the years. Support for preparation of this review was provided in part by the Ecology Development Corporation, Washington, D.C. and by the University of California, Berkeley. I am indebted to Mrs. Joan Montoya for typing this manuscript and to Dr. C. G. Golueke of the University of California for subjecting it to a detailed review.

REFERENCES

1. South Tahoe Public Utility District *Advanced Wastewater Treatment as Practiced at Tahoe.* Project 17010 ELQ Water Quality Office, Environmental Protection Agency, Room 1108, Washington, D.C. 20242.
2. Oswald, W. J., 'Light Conversion Efficiency of Algae Grown in Sewage', *Transactions Amer. Soc. of Civil Engrs.*, Paper 3395–128 Part III, pp. 47–83 (1960).
3. Oswald, W. J., C. G. Golueke, R. C. Cooper, H. K. Gee and J. C. Bronson, 'Water Reclamation, Algal Production and Methane Fermentation in Waste Ponds', 1st Inst. Conf. Water Pollution Res., London, pp. 119–157 (1962).
4. Parhad, N. M., 'Studies on the Microbial Flora in Oxidation Ponds', Ph.D. Dissertation, Central Public Health Engineering Research Institute, Nagpur 3, India, 151 pages (1970).
5. Golueke, C. G. and W. J. Oswald, 'Harvesting and Processing Sewage-grown Plantktonic Algae', *Journ. Water Poll. Cont. Fed.* 37:4, pp. 471–498 (1965).
6. Sobsey, M. D., 'Interactions of Poliovirus with Biological Components of Algal-Bacterial Wastewater Treatment Systems'. Dissertation submitted in partial fulfillment of requirements for the Ph.D Degree, University of California, Berkeley (1971).
7. Oswald, W. J., 'Growth Characteristics of Microalgae Cultured in Domestic Sewage: Environmental Effects on Productivity'. *Proceedings IBPIPP Technical Meeting Productivity of Photosynthetic Systems, Part I. Models and Methods,* Trebon, Czechoslovakia (1969).
8. California Department of Water Resources, *Bioengineering Aspects of Agricultural Drainage: Removal of Nitrates by an Algal System.* Calif. Dept. of Water Resources, 1416 Ninth St., Sacramento, Calif. 133 pages (1971).

9. French Petroleum Institute, 'State of Development of the I.F.P. Algae Process at Dec. 1970', Report to FAO/WHO/UNICEF Protein Adv. Group. French Petroleum Institute, Res. 18730–1A (1970).

10. Oswald, W. J., 'Ecological Management of Thermal Discharges', Presented at Symposium on Beneficial Uses of Thermal Discharges, Amer. Soc. Agronomy, New York, 17 pages (1971).

11. Oswald, W. J. and H. B. Gotaas, 'Photosynthesis in Sewage Treatment', *Trans. Amer. Soc. Civil Engrs, 122*:73 (1957).

12. Oswald, W. J. and C. G. Golueke, 'Large-Scale Production of Algae', in *Single Cell Protein*, Ed. by Mateles and Tannenbaum, MIT Press, pp. 271–305 (1968).

13. Hintz, H. F., H. Heitman, W. C. Weir, D. T. Torell and J. H. Meyer, 'Nutritive Value of Algae Grown on Sewage', *Journ. of Animal Science, 25* No. 3, pp. 675–681 (Aug. 1966).

14. Jewell, William, J. and P. L. McCarty, *Aerobic Decomposition of Algae and Nutrient Regeneration*, Technical Report No. 91, Dept. of Civil Engineering, Stanford University, 282 pages (1968).

15. Foree, E. G. and P. L. McCarty, *The Decomposition of Algae in Anaerobic Waters*, Technical Report No. 95, Dept. of Civil Engineering, Stanford University, 202 pages (1968).

16. Ludwig, H. F., W. J. Oswald, H. B. Gotaas and V. Lynch, 'Growth Characteristics of *Euglena gracilis* Cultured in Sewage'. *Sewage and Ind. Wastes 23*:11 (Nov. 1951).

17. Zabat, Mario, 'Kinetics of Phosphorus Removal by Algae'. Dissertation submitted in partial satisfaction of the requirements for the Degree of Doctor of Philosophy in Engineering, University of California, Berkeley (1970).

18. Porcella, D. B., P. Grau, C. H. Huang, J. Radimsky, D. F. Toerien and E. A. Pearson, Provisional *Algal Assay Procedures,* SERL Report No. 70–8, Sanitary Engineering Research Laboratory, College of Engineering, University of California, Berkely, 179 pages (1970).

19. Maloney, T. E., *Algal Assay Procedure*, National Eutrophication Research Program Environmental Protection Agency, Corvallis, Oregon (1971).

20. Toerien, D. F., C. H. Huang, J. Radimsky, E. A. Pearson and J. Scherfig, Final Report, Provisional *Algal Assay Procedures,* SERL Report 71–6, Sanitary Engineering Research Laboratory, 211 pages (Oct. 1971).

21. Oswald, W. J., H. B. Gotaas, R. J. Hee, *Studies of Photosynthetic Oxygenation* I. Pilot Plant Experiments, IER Series 44, No. 9, 192 pages (1958).

22. Oswald, W. J., 'The Influence of Physical Environment Upon the Overall Efficiency of Light Energy Conversion by *Chlorella* in the Process of Photosynthetic Oxygenation'. Dissertation submitted in partial satisfaction of the requirements for the Degree of Doctor of Philosophy, University of California, Berkeley (1957).

23. Shelef, G., J. C. Goldman, W. J. Oswald, M. Sobsey, Joan Harrison, H. Gee and R. Halperin, *Kinetics of Algal Systems in Waste Treatment: Ammonia-Nitrogen as a Growth-Limiting Factor and other Pertinent Topics.* Final Report, Part II. Federal Water Quality Administration, U.S. Dept. of Int., Grant No. 17010 DZQ. Sanitary Engineering Research Laboratory, University of California, Berkeley, Calif. (Sept. 1970).

24. Shelef, G., 'Kinetics of Algal Biomass Production System with Respect to Light Intensity and Nitrogen Concentration', Dissertation submitted in partial satisfaction of the requirements for the Degree of Doctor of Philosophy, University of California, Berkeley (1968).

25. Shelef, G., W. J. Oswald and P. H. McGauhey, 'Algal Reactor for Life Support Systems', *Journ. San. Eng. Div., A.S.C.E.* SA1 *7105*, pp. 91–109, (Feb. 1970).

26. Goldman, J. C., D. B. Porcella, E. J. Middlebrooks and D. F. Toerien, 'The Effect of Carbon on Algal Growth — Its Relationship to Eutrophication', Occasional Paper 6, Utah Water Research Laboratory, College of Engineering, Utah State University, Logon, Utah 84321 (April 1971).

27. Oswald, W. J., A. Meron and M. Zabat, 'Designing Waste Ponds to Meet Water Quality Criteria', pages 184–194, in 2nd International Symposium for Waste Treatment Lagoons, 217 Nuclear Reactor Center, University of Kansas, Lawrence, Kansas 66044 (1970).

28. Meron, A., 'Stabilization Systems for Water Quality Control', Dissertation submitted in partial satisfaction of the requirements for the Degree of Doctor of Philosophy, University of California, Berkeley, 318 pages (1970).

29. Smith R., *Cost of Conventional and Advanced Treatment of Wastewaters*, U.S. Dept. of the Interior Federal Water Pollution Control Administration, Advanced Waste Treatment Branch, Cincinnati Water Res. Lab., Cincinnati, Ohio (1968).

Discussion *by* J. Benjamin Sless
Environmental Engineering Laboratories
Technion – Israel Institute of Technology, Haifa

The applicability of the AGP concept for predicting algal concentrations and nutrient uptake in ponds seems to be questionable. This concept is based on the following fundamental assumptions: all micro-biological processes in ponds are, or can be, aerobic and all nutrients present in inflowing wastes are made directly available for algal growth. Ideally therefore, all energy in the form of nutrients is, or can be, conserved within the ponds.

In the "high rate" ponds studied by Gotaas et al[1] and Oswald et al[2], the daily changes in dissolved oxygen concentrations were identical to those found in typical facultative ponds[3,4], except in the case of "high rate" ponds operated at 5 – 10 cms depth. Complete oxygen deficit for 6 – 10 hours at night was the rule, not the exception, in both types of pond. In the "high rate" ponds, loss of ammonia nitrogen from pond liquids was frequently recorded.[2] This was also evident in studies in Israel in facultative pilot ponds, in addition to precipitation of carbonates and phosphates.[3] In all cases high pH values resulting from vigorous algal growth was the cause.

The evidence therefore suggests that microbiological processes in stabilization ponds cannot be exclusively aerobic and that nutrients are dissipated. In consequence, algal growth cannot, in principle, attain the predicted AGP based on inflowing nutrient concentrations.

The degree of light penetration is central to any discussion on potential algal yields. Particular attention was paid to this feature in studies carried out by the Environmental Engineering Laboratories of the Technion, Haifa (EELT).[3] Among numerous experiments two series of ponds, each 80 cms deep, were operated in parallel. One series consisted of 2 ponds, each with liquid retention times of 20 days, with an organic loading of 4 – 5 gm $BOD/m^2/day$ in the primary pond ("low loading"). The other series consisted of 3 ponds, each with liquid retention times of 8 – 10 days, with an organic loading of 8 – 10 gm $BOD/m^2/day$ on the primary pond ("high loading").

Representative results of analyses of algal and total suspended solids concentrations, and the depth of light penetration are shown in Fig. 1. The depth of light penetration refers to the liquid depth at which a light intensity of 250 lux (about 25 foot candles) was measured at the time of peak light intensity, i.e. 10^{00} – 14^{00}. This corresponds to the light intensity found to represent compensation level for algal photosynthesis and respiration in "high rate" ponds.[5]

The figure shows that, in general, the percentage of algal to total solids, and the depth of light penetration, increased progressively with cumulative retention time in summer and winter, while algal solids increased with retention time in winter, and decreased in summer. Increase in light penetration was due partly to infestation by rotifers which reduced small algae and turbidity of non-algal origin. Also, the (assumed) compensation level of light intensity was generally recorded at 30 – 40 cms depth, i.e. about one half of the liquid depth, and algal yields per unit illuminated surface area were 5 and 10 $gm/m^2/day$ in winter and summer respectively, in the high loaded primary pond.

Conversion of areal to volumetric yield gives values of 4 and 8 $gm/m^3/day$ in winter and summer. Such yields compare very favourably with average annual yields of 27 $gm/m^2/day$ (40 tons/acre/year) in "high rate" ponds of 30 cms depth;[2] and 9 $gm/m^3/day$ on a volumetric basis. Here it may be noted that these ponds were operated under conditions considered optimal for algal growth, involving stirring and recirculation, unlike the EELT ponds.

The foregoing shows that interrelationships between algal and total suspended solids, light penetration and algal yields in serially operated ponds of 80 cm depth differ radically from those in single – cell "high rate" ponds. For instance, in "high rate" ponds, non-algal turbidity was essentially zero at retention times in excess of 3 or 4 days, with organic loadings of about 20 $gm/BOD/m^2/day^2$. In the EELT ponds, with half of such organic loadings, non-algal solids accounted for 10 – 30% of total suspended solids, and presumably turbidity, at retention time of up to 20 days, and infestation of the ponds by rotifers was a complicating factor in summer.

In other EELT studies on highly loaded field ponds (15 – 50 gm $BOD/m^2/day$), a net increase in algal concentrations always occurred with cumulative retention time.[3] The general inference is that in very high loaded primary ponds of series, severe light restriction occurs, so that more light becomes available for algal growth in succeeding ponds, whereas in low loaded systems, algal predators and/or detritus feeders invade secondary or tertiary ponds of series in summer, leading to increased light penetration, without net increase in algal solids.

A possible explanation for the surprisingly high yields, on a volumetric basis, of primary ponds in the EELT model pond system (Fig. 1) is that algae in lower pond levels were periodically exposed to light, through movement into the upper levels. The fact that light penetration invariably declined logarithmically with depth and the generally homogeneous vertical distribution of algae support this view. Unfortunately the dearth of dual data on light penetration and suspended solids in field ponds prevents extrapolation of results of EELT studies. Nevertheless, it seems likely that the effective depth of light penetration even in apparently stagnant ponds, is several times greater than indicated in formula no. 2a presented by Professor Oswald. Alternatively, this vital factor could be a function of

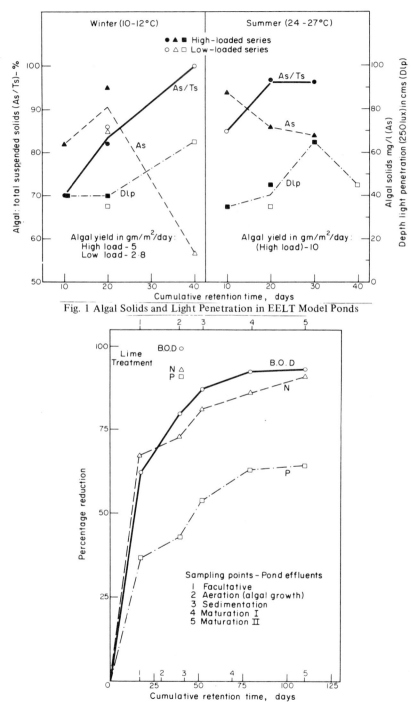

Fig. 1 Algal Solids and Light Penetration in EELT Model Ponds

Fig. 2 Percentage reduction of B.O.D. C and N in Saint Helena Ponds as function of cumulative retention time.

organic loading, which in turn is a function of sewage strength and retention time. Clarification of these questions surely merits careful study.

The concept of nutrient subtraction prior to waste discharge to ponds, as applied in the primary unit (facultative pond) of the Saint Helena system[6] could be particularly helpful in countries such as Israel, where sewage strength in urban areas is very high (BOD of 300 – 800 mg/l). In this connection, application of the formulae presented by Professor Oswald to waste treatment in an Israeli city such as Jerusalem (population 250,000) would involve the use of algal ponds with a liquid depth of 17 cms for carbon and nitrogen assimilation alone, and a land area of at least 300 hectares (750 acres) for winter operation (i.e. 800 people per hectare). This amply confirms the conclusions reached by Oswald and Gotaas[5] concerning the shallow depths and extensive land areas required for nutrient assimilation by algae, where the BOD of wastes exceeds 300 mg/l.

An interesting feature of the Saint Helena system was the overall degree of nutrient reduction. This was very impressive in regard to BOD, but somewhat less so in regard to nitrogen and phosphorous. In general the system seemed to be excessively large, unless die-off of pathogens was the principal objective.

This is borne out by Fig.2, which shows BOD, nitrogen and phosphorous reduction as a function of cumulative retention time. Point No.1 represents the effluent of the primary unit where the main reduction in BOD and nutrients occurred. Point No.2 represents the effluent of the algal growth pond, with a cumulative retention time of 18 days. Beyond this retention time, little or no further reduction in BOD was evident. Reduction in nitrogen proceeded slowly but progressively from a retention time of 18 – 110 days, while phosphorous reduction remained unchanged after 75 days liquid retention. The fact that appreciable reductions in nitrogen and phosphorous occurred within the algal growth pond show that nutrients were lost; otherwise nutrients would have been converted to algal cells, according to the AGP concept, so that no losses would have been evident.

Although the total reduction of nitrogen and phosphorous in the system as a whole was 60 – 90%, the concentrations of these nutrients in the final effluent were around 4 mg/l. Such effluent characteristics would debar their use for discharge to receiving waters, unless the dilution rate were extremely high. In brief, the degree of treatment was far from complete. In contrast, application of lime to the effluent of the algal growth pond gave dramatic results: almost 100% reduction in BOD and over 90% for nitrogen and phosphorous, corresponding to 1.4 mg/l total N and 0.6 mg/l total P.

Such effluent characteristics undoubtedly represent an advanced degree of treatment. But again, we must consider the land area required. Pond treatment plus lime treatment in the Saint Helena system required an area equivalent to 1400 people per hectare (4000 people for 2.8 hectares). This represents a decided improvement in regard to the area required for partial nutrient assimilation in purely algal ponds, but the area would still be excessive in regions of high land values.

An alternative to lime treatment of the effluent of an algal growth pond of the Saint Helena type is disinfection (e.g. by chlorination), so as to render the effluent suitable for unrestricted agricultural irrigation. This would be particularly useful in arid countries such as Israel, where stabilization pond effluents are currently used for restricted irrigation.[7]

In considering the obvious advantages of waste treatment in pond, we must not overlook the disadvantages. Apart from the large land areas required, chemical or physico-chemical treatment is essential, if advanced or tertiary purification is to be achieved. The cost of combined pond and chemical treatment may well exceed that of other types of advanced waste treatment, even in countries where the climate is optimal for use of ponds.

REFERENCES

1. GOTAAS, H.B., OSWALD, W.J., and GOLUECKE C.G. (1954) Algal bacterial symbiosis in sewage oxidation ponds. Univ. Calif. Inst. Eng. Res. RG 2601(C3) and RG 2601 (C4).
2. OSWALD, W.J., GOTAAS, H.B., GOLUEKE, C.G., and KELLEN, W.R. (1957) Algae in waste treatment. Sew. and Indust. Wastes 29.4 : 437–454.
3. SLESS, J.B., (1967) The role of algae in oxidation ponds in Israel. Thesis submitted for the degree of Doctor of Philosophy, Hebrew University, Jerusalem (in Hebrew).
4. MARRAIS, G.v.R. (1966) New factors in the design, operation and performance of waste stabilization ponds. Bull W.H.O. 34:737–763.
5. OSWALD, W.J., and GOTAAS, H.B., (1957) Photosynthesis in sewage treatment. Trans. Am. Soc. Civ. Eng. 122:73–97.
6. OSWALD, W.J., MERON, A. and ZABAT, M. (1970) Designing waste ponds to meet water quality criteria in 2nd International Symposium for Waste Treatment Lagoons, Univ. of Kansas, Lawrence, Kansas pp.184–194.
7. WACHS, A.M., AVNIMELECH, Y. and SANDBANK, E. (1970) Effect of the irrigation with stabilization pond effluents on the concentration of nitrates in underground water. Technion – Israel Inst. of Techn., San. Eng. Labs. Ann. Rep. Res. No. 013–386.

Reply

Professor Sless has brought out several factors that were perhaps not made entirely clear in the original presentation.

It was my intent in the paper to point out that there are two types of ponding systems which may be designed to achieve a high degree of primary, secondary and tertiary treatment, viz[1] shallow, mixed, aerobic ponds in which oxidation occurs and the products of oxidation are assimilated by algae. While it is true that conventional ponds as they are currently designed in Israel and elsewhere cannot be exclusively aerobic or will tend to dissipate ammonia into the air, our experience has been that virtually all inorganic carbon and nitrogen can be assimilated by algae in the very shallow ponds designed according to equations (1) and (2) and that as the algae are removed either by alum flocculation and floatation or perhaps by lime precipitation or perhaps by autoflocculation nitrogen and phosphate are also removed. The algae must be harvested and removed from the pond if complete tertiary treatment is desired. Because of the large area of land required, systems of this type cannot be justified unless the algae removed are of value. On the other hand, if waste-grown algae should prove to have a substantial value as a protein supplement in animal feed or for other purposes as is quite likely, one would wish to optimize to convert a maximum of nutrients to algae, and shallow ponds would be required. The large amount of land required should not be a problem in this case because the productivity of the land involved would be far more than could be obtained from an equal area of land devoted to conventional agriculture. The major deterent to use of shallow ponds other than the need to develop a market for algae is the problem of seasonal changes in the weather and predation of algae by several types of organisms such as Daphne which establish themselves in shallow ponds.

The second type of system (i.e. the St. Helena system) involving deep primary ponds followed by high-rate ponds and settling ponds or chemical treatment requires much less land than the first alternative and is not dependent on recovery and sale of algae to be economical. As Professor Sless has emphasized, removal of the algae with lime or alum is necessary to obtain a degree of treatment that may be termed "complete." His concern that, because of land use, such systems may be less economical than other types of tertiary treatment is understandable. Obviously in land treatment as land values increase a point may be reached where the cost of land is excessive. Our studies of the economics of pond systems indicates, however, that where the system can be applied land costing in excess of $20,000 per acre may be utilized competitively with available alternatives. There are few places in the world where land substantially less costly cannot be obtained within a few kilometers of the civic center.

This data requested by Professor Sless was not available when the paper was written, but it has since been obtained and is given in Table 1.

TABLE 1. Nitrogen Transforms, Saint Helena, California, January 5, 1972*

No.	Sample	pH	Temp. $^\circ$C	Orgn. mg/l	NH3N mg/l	NO5 mg/l	Total mg/l	ΔN	Removals* ΔCum.	Cum. %
1	Sewage	6.6	18	14.5	4.1	0.3	18.9	–	–	–
2	Pond 1–Eff.	6.9	11	3.63	5.0	0.27	8.90	10.0	10.0	53
3	Pond 2–Eff.	7.6	8	1.52	4.6	0.28	6.40	2.5	12.5	66
4	Pond 3–Eff.	7.2	8	2.58	5.1	0.22	7.90	1.5	11.0	58
5a	Pond 4–Eff.	6.7	9	0.96	0.5	0.84	2.30	5.6	16.6	88
5b	Pond 4–Filt.	6.8	–	0.31	0.4	0.83	1.54	–	17.4	92
6	Pond 5–Content	6.9	8	2.70	0.5	1.60	4.80	2.5	14.1	75

*Not corrected for decrease in volume.

The data show that had algae been removed as was done by filtration of pond 4 (sample 5b) final nitrogen would have been less than 2 mg per liter. The fact that pond 3 effluent actually contained an increase in nitrogen compared with pond 2 effluent simply reflects the fact that pond 3 after accumulating settled algae for several years is beginning to recycle nitrogen from decomposing, accumulated algae and should be cleaned. If it is not cleaned, the groundwater recharge capabilities of ponds 4 and 5 will be increasingly impaired due to clogging by regrowths of algae. The need to periodically clean pond 3 was recognized when the St. Helena system was designed. However, the remarkably high removals of nitrogen attained during the first four years of operation caused this requirement to be temporarily overlooked.

J.C. Barnard, South Africa

Did predators occur in the ponds? Were seasonal changes in the microbial population observed?

Reply

Predators were mainly confined to ponds 3, 4 and 5. Pond 1 had some Daphne the first few years but has had none the past two years. Pond 2 (the mixing pond) has never had Daphne in substantial numbers. Since there has been only limited support for the St. Helena studies we have not closely followed microbial populations. Green algae including Chlorella, Scenedesmus and Euglena have predominated in ponds 1 and 2 the year round. Blue greens have been increasing in numbers in ponds 3, 4 and 5 as the amount of nutrient being recycled in pond 3 increases. During the summer of 1972, blue-green algae blooms were sufficiently intense in pond 3 to stimulate a move by the local board to clean the pond. However, the actual cleaning will probably not be done until late in 1973.

W. Engelbart, Germany

Our proposal for the treatment of the organic wastes from a city of 1,000,000 persons and the cooling water from a power plant serving the community is to use bacterial ponds in cold climates where sunlight is insufficient for algal ponds. A specially mixed 50-stage pond system of about 500,000 in surface area divided by broad dams should be an excellent second stage treatment system and be sufficient to cool water from a 600 MW nuclear power plant. Because heat transfer predominantly takes place by evaporation only about 100,000 m^3d are needed as a minimum to avoid enrichment of salts. Because even in severe winters suitable conditions can be maintained, more than 1,000 tons of carp can be produced for every 10,000 m^2 of pond used in this stage as well as to aerate the fish pond and to treat the organic wastes from it.

The key to this proposal is the use of 2,500 m of a line aerator. This device generates the needed surface stream in the ponds and at the same time it can supply more than 60 tons of O_2 a day. Removal of well stabilized sludge is required after 2 − 5 years. The energy demand of this system will be less than 1 MW. Pretreatment of sewage should be mechanical with first stage biological treatment of rotating pipe filter rings. However, the effluent from a normal sewage plant or the water from a small creek could be used.

Reply

I appreciate the reminder that photosynthetic oxygenation must be replaced by mechanical aeration in cold climates and his proposal that wasted heat from power plants may be used to maintain biologically acceptable temperatures in ponds as well as to cool the power plant. The production of carp is also an attractive byproduct of his proposed system. I question the need for mechanical pretreatment of the sewage entering such a system. A deep, long-detention primary pond heated by the power plant should be much more economical and would eliminate the need to deal with continuous sludge handling.

N. Allen, South Africa

Could the additional information on the St. Helena plant given by Professor Oswald be included in the proceedings?

Reply

This request has been fulfilled in conjunction with my response to the discussion by Professor Sless.

THE DEVELOPMENT AND EFFECT OF CONSTRUCTION AND OPERATION COST IN BIOLOGICAL SEWAGE TREATMENT PLANTS

E. SICKERT
Baudirektor,
Baubehörde, Hauptabteilung Stadtenwässerung,
2000 Hamburg 36, Federal Republic of Germany

INTRODUCTION

In recent years hardly a country had remained unaffected by the wage and price spiral. The theme of this discussion is to make clear some important factors that influence the construction and operation cost of relatively large treatment plants and their dependence on differing cost factors, with a view to finding ways to bring down cost furthermore and to achieve even more economy in running the plant.

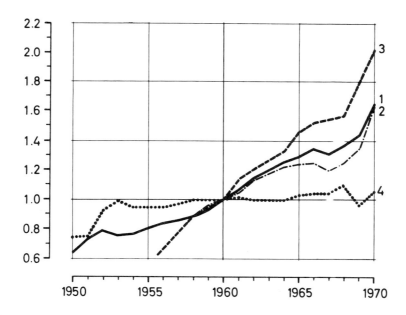

Fig. 1. Development of cost in Germany 1950 – 1970 valid for sewage treatment plants.
1. Construction cost for housing
2. Construction cost for Or. C. bridges
3. Wages of a skilled worker on a treatment plant
4. Cost of electricity for bulk consumers

Fig. 1 illustrates the changes that have taken place in the construction costs of treatment plants as well as the cost of wages and energy. Whereas the cost of energy, during the last twenty years, has almost remained stationary, wages costs have increased by 100% and construction costs by 65% in only ten years.

171

CONSTRUCTION COST

The factors that go to influence the construction cost of a sewage treatment plant are:
(a)   The size of the treatment plant, expressed as the number of persons served or the quantity of influent sewage.
(b)   The desired degree of purification.
(c)   Other factors such as local conditions, plant equipment, etc.

Size of treatment plant

The most influential factor that determines the local cost of a sewage treatment plant is the size of the treatment unit. Fig. 2 shows the interdependence between specific construction cost (cost based on one population equivalent PE) and the population equivalent connected to the treatment works.

For the sake of simplicity, the relationship of specific construction cost is based on a full biological treatment plant for 100,000 PE as 1.0. It is to be inferred, that for the price conditions obtaining in 1970, one such plant cost on an average 70 DM/PE.

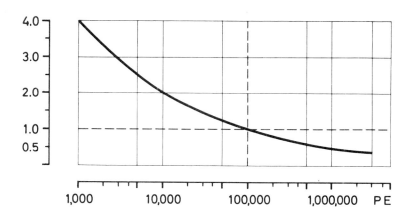

Fig. 2. Relations between specific construction cost for biological treatment plants and PE served.
Specific cost for a plant for 100,000 PE = 1.0.

However, the size of a treatment plant and with it the cost of construction are dependent not only on the population it serves, but also on the quantity of the incoming sewage. Because a similarity of relationship is to be expected, as illustrated above, as in the case of population equivalent served, no diagram is shown. However, in view of increasing water consumption and industrial effluents such as occurs in big cities, this criterion should be given its due weight and it would be more correct to take into account both the population equivalent and the quantity of incoming sewage simultaneously. This has already been done in the case of the determination of part of the operation cost. However, at this stage it is sufficient to mention that by increasing the quantity of sewage the construction cost increase is only gradual. This means that the specific construction costs based on quantity of incoming sewage tend to go down more rapidly than is the case on a population equivalent basis. For biological treatment plant SCHMIDT derived the following cost relationship.

$$\text{Total construction cost} \quad y = A \cdot x^{0.58} \text{ (x in PE)}$$
$$\text{Total construction cost} \quad y = B \cdot x^{0.49} \text{ (x in } m^3/h)$$

The small cost increase where the quantity of incoming sewage is considered is explained as follows. An increasing quantity of influent results in the increased size of primary and final sedimentation tanks, whereas increasing pollution means comparatively costly additional aeration tanks and sludge handling equipment have to be provided for.

In any case it should be realized that it is possible to build large sewage treatment plants with lower specific construction expenditure.

In Fig. 2 the cost curves for different purification processes are not given because, excepting for oxidation ditches, the specific costs do not show much deviation from one another. It may be seen that on an average the activated sludge plants are about 10% under and the trickling filter units 10% over the curve. According to SCHMIDT, from the point of view of construction cost, oxidation ditches for a population equivalent of up to 5000 could be estimated to be about 30% less than the cost curve actually indicates.

Degree of purification

The second most influential factor affecting cost is the degree of purification desired. An overall idea is obtained from illustration 3, from a publication of PROF. DR. MÜLLER-NEUHAUS. When the cost basis for a mechanical treatment plant is taken as 1, the cost for biological treatment rises gradually till the degree of purification of 80% is reached and thereafter the rise is steep. According to MÜLLER-NEUHAUS, with mechanical treatment, doubling of cost occurs at 80% purification, and a trebling between 93% and 94% purification.

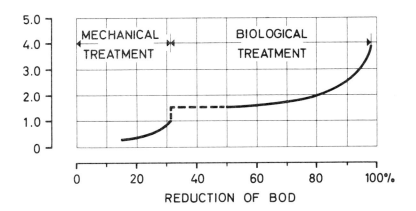

Fig. 3. Construction cost for biological treatment plants in relation to the degree of purification, after PROF. DR. MÜLLER-NEUHAUS.

Recent investigation by PROF. DR. V. D. EMDE show that the rise of costs may be smaller. Thus he found that for a plant serving a community of 50,000 PE giving an effluent BOD of 80 mg/l, the quality could be upgraded to 25 mg/l BOD with only an additional investment of 20%. For a plant of 2,500,000 PE upgrading the effluent from 40 to 25 mg/l BOD would require an increase in construction cost of about 18%.

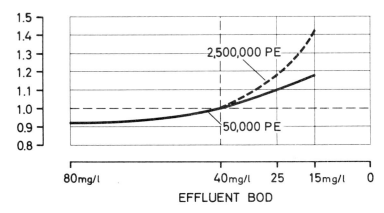

Fig. 4. Construction cost in relation to the degree of purification, after PROF. DR. V. D. EMDE, for activated sludge plants of different size.

It is to be inferred without going deeply into cost relations, that increase in the degree of purification has less effect on the increase of construction cost for small plants than for large treatment plants. The reason for this is that part of the construction cost, which is independent of sewage and sludge handling such as administration building, roads, connections etc., and which more or less stay constant when the plant is being upgraded to higher BOD reduction is, in the case of a small treatment unit, larger than in the case of larger treatment plants and vice versa. Also, increasing the mass (more concrete etc., for the bigger volume of tanks and basins) in the case of larger treatment plants is not as conducive for cost reduction as in the case of smaller treatment plants.

Other cost factors

The combined effect of the 'other cost factors' can be equal to, or even greater than, the influence of size and degree of purification on the total construction cost of the treatment plant. Local factors (land acquisition, preparation of site, cost of amenities, subsoil and ground water conditions, connections to the public traffic systems, the degree of free capacities of the building industry in the period when tenders are called for, the proximity to larger cities or other densely populated areas etc.) especially in relation to the standards set by the planning, contracting, and supervising engineers play a great role in determining the cost of a treatment plant. In addition, the lay out, the general outfit, the construction technique employed play a significant role either in reducing or increasing the cost of construction. BUCKSTEEG, in an analysis of 250 treatment works, has found out that in case, of treatment units for less than 50,000 PE there could be a cost deviation, due to the 'other factors' by as much as 50% in excess of average construction cost (10% of the investigated treatment units had even larger deviations). That means that the specific cost based on the price situation in 1970 for treatment plant, for example for 50,000 PE, could vary between

$$1.3 \cdot 70 \pm 50\% = 91 \pm 45 \text{ DM/PE}$$

simply on the influence of the 'other factors'. In the case of large units these 'other factors' lose their almost dominating aspect and fall to approximately ± 30%. In addition to the 'other factors' mentioned the process used for sludge disposal plays a not inconsiderable part in the determination of the total construction cost where large sewage treatment plants are concerned.

RUNNING COST OF TREATMENT PLANTS

The total annual cost of a treatment plant consists of:

(a) Capital cost   (depreciation or annulment of construction cost and interest)

(b) Operation cost

The operation cost could be divided into 3 categories.

1. Personal cost
2. Cost of repairs and materials including cost of upkeep, but without cost of energy
3. Cost of energy.

Operation cost

Smaller treatment units do not only carry a higher specific operation cost but a higher specific operation cost as well. This can be seen from Figures 5 and 6 which show an analysis of the annual and the operation cost of biological treatment units. The curves have been worked out from dates of different authors and Hamburg plants.

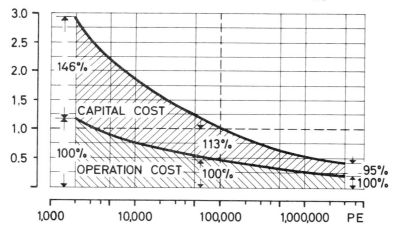

Fig. 5. Relations between specific annual cost for biological treatment plants and PE served. Specific cost for a plant for 100,000 PE = 1.0.

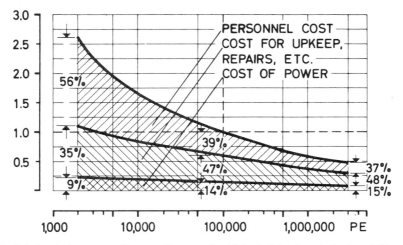

Fig. 6. Relation between specific operation cost for biological treatment plants and PE served. Specific cost for a plant for 100,000 PE = 1.0.

For the sake of easy comparison the illustrations are again based on 1.0 for a plant of 100,000 PE. For the plant under consideration and for the conditions obtaining in 1970 the average specific annual cost worked out to be 10.60 DM per PE and per annum, in which the capital cost have been calculated with 8% p.a. of construction cost. The average specific operation cost alone were 5.0 DM per PE and per annum.

These curves show average values and therefore the same reservation should be applied as in the case of construction costs. Deviations are possible both above and below the average which can be larger than in the case of construction cost. The reasons for these violent deviations may be found in the following factors.

The method of disposal and amount of remaining material to get rid of.

Method of purification, possibly in relation to topographical conditions (e.g. gravity flow trickling filter plant).

Purification efficiency.

Equipment installed.

Utilized capacity and age of the plant.

Operation management.

Quality of the staff.

Special local conditions.

Without discussing these factors in detail it can be stated that not only the specific constructions cost but also the specific operation cost diminish considerably with increasing number of PE served; but it can be pointed out that with the increase in the size of the plant the specific capital cost tends to diminish more steeply than the specific operational cost does or, expressed differently, with the increasing size of a treatment plant the operation cost as against the capital cost tend to play a key role.

Illustration 6 shows by what magnitude the individual factors pertaining to operational cost could change in treatment plants of differing sizes. It is to be noted that the cost of personnel in larger treatment plants fall considerably, but they still account for not an insignificant percentage of the total operational cost.

Personnel cost factor

Fig. 7 shows the increasing role personnel cost may play in the determination of the operational cost in future. The changes in cost during the past 20 years for a Hamburg treatment plant serving a PE of 100,000 is illustrated here. The calculation is based on an applied stationary loading, the same number of staff, and constant amortization rates; the last mentioned, however, is calculated on the value of a new plant, i.e., taking into account the increase in the construction cost index. It shows that in this case the personnel costs have become the determinating factor of the operation cost whereas the ·share of cost of energy has gone down tremendously. It is logical to expect that in the future the same development will persist as in the past two decades and therefore the percentage of personnel costs will go on increasing, whereas the importance of cost of energy and amortization in relation to personnel cost will furthermore go down.

Whereas in the past much attention has been given in research of those factors which concern the requirements of energy, i.e. BOD removal per unit or oxygen input per unit and which from the point of total economy, as shown in Fig. 7, are less significant, little research has been done on the operational field such as work cycle, work organization and requirement of personnel in treatment plants.

Recently LONDONG has made a comparative study of personnel cost in 60 treatment

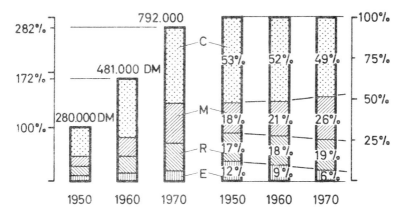

Fig. 7. Development of annual cost of a biological treatment plant serving 100,000 PE and its changes in composition during the last 20 years.

C Capital cost, 7.25% of construction cost for a contemporary plant
M Personnel cost
R Cost for repairs, materials, upkeep, etc.
E Cost for electrical power

plants belonging to the Emschergenossenschaft, along with those belonging to the Lippeverband and the findings by MICHEL of 1600 American treatment units, and has arrived at a set of orientation data for the personnel requirement in treatment units for the first time in the Federal Republic of Germany. The results obtained, being valid for

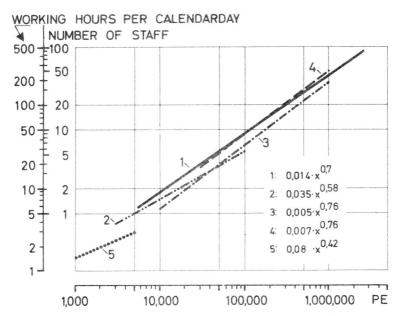

Fig. 8. Demand of personnel for municipal treatment plants in Germany.
1. Activated sludge plants with separate sludge treatment, but without power generation
2. Trickling filter plants with separate sludge treatment, but without power generation.
Personnel being used on treatment plants of the organizations 'Emschergenossenschaft' and 'Lippeverband'
3. Activated sludge plants with separate sludge treatment, but without power generation
4. Activated sludge plants with separate sludge treatment and with power generation
5. Activated sludge plants with aerobic sludge stabilization.

municipal works, are shown in illustration 8, and concern activated sludge plants without power generation and plants with trickling filters. The figures found in the Emschergenossenschaft and Lippeverband for their activated sludge plants with and without power generation, and also biological units with aerobic sludge stabilization, have been added for comparison.

LONDONG has summarized in the form of an equation based on average population equivalent (PE_m) the important factors that most influence the demand for personnel such as pollution load, quantity of influent, and degree of utilizations.

$$PE_m = \frac{1}{3} \cdot \frac{\text{design BOD load/d}}{54 \text{ g BOD/PE} \cdot \text{d}}$$

$$+ \frac{1}{3} \cdot \frac{\text{present BOD load/d}}{54 \text{ g BOD/PE} \cdot \text{d}}$$

$$+ \frac{1}{3} \cdot \frac{\text{present inflow/d}}{200 \text{ l/PE} \cdot \text{d}}$$

In the graphical representation of Fig. 8 it is shown that as a result of having 42 hours working time a week the average productivity per person per calendar day is about 5 hours only instead of 6 due to vacation, sickness and holidays.

The 'demand of personnel' curve depending on the magnitude of the different influencing factors would rise or fall as already indicated by the low values obtained in the curves of Emschergenossenschaft and Lippeverband. The large number of treatment plants these associations have to administer has the advantage of a central administration, which makes possible a more rational utilization of manpower than in the case of communities with individual treatment units to care for. Beyond this, it is the practice in such associations to carry out maintenance work of periodic occurrence such as horticultural maintenance, painting etc., through external agencies on the grounds of economy; major repairs are carried out either through external agencies as well or through a central workshop.

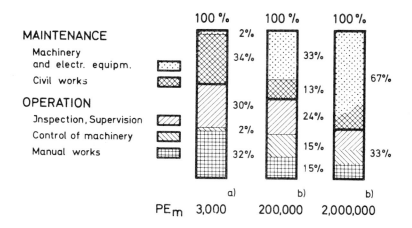

Fig. 9. Analysis of working hours on activated sludge plants of different size.
(a)   Activated sludge plants with sludge stabilization.
(b)   Activated sludge plants with separate sludge treatment and power generation.

Illustration 9 shows an analysis of expenditure on personnel concerning treatment plants of differing sizes. Due to the labour to be done the hours worked have been divided into two categories: 'operation', i.e. control of process machinery, operational manual work, supervision of processes, and 'maintenance', i.e. inspection, servicing and repair. Here the considerably high rate for the proper upkeep of the plant (maintenance) can be recognized; this percentage would be even higher where the work due to the external agencies would have been done internally. In small activated sludge units manual process work, process supervision and civil work maintenance, which here means mostly cleaning of tank walls and channels, claim approximately the same magnitude of the total share. The share of hours spent on large biological treatment plants having their own power generation shows the high degree of mechanization and automation there. It should not go unmentioned here, thanks to the specially qualified staff of the bigger plants, that even complicated maintenance jobs are carried out to a large extent by own personnel.

CONSEQUENCES AND CHANCES FOR HIGHER ECONOMY

Just as water economy demands a certain purification standard to be achieved in sewage treatment works, the general economy requires sound economic principles to be observed. From this it follows that besides assuring an optimum operational safety and protecting the high standards already obtained in treatment, the annual cost of running treatment plants should reach a minimum.

After having shown the important cost factors and their inherent nature to change it is but logical to examine the possibilities that exist to bring about a higher degree of overall economy. It should be realized that the enormous increase in cost of construction as well as operation has been brought about in recent years in the first place by an increase in the cost of wages, to a lesser degree by increase in cost of material, and least of all by higher energy cost, and this tendency may be expected to persist in the future also.

The essential ingredients towards further cost reduction therefore is to apply of all possibilities of operational and constructive concentration and simplification at the stage of planning and design itself and, secondly, to lay emphasis on a rational operation procedure with extensive automation combined with improved organization. As a consequence it should be taken into account that, in certain cases, a fair increase in energy consumption may result.

Concentration

One big treatment plant is more economical than many smaller ones. This knowledge has been verified by the construction of bigger and centralized treatment works. Where the collection of sewage to be treated in one big unit necessarily involves too expensive a transport sewerage system, 'concentration' with the following facilities would be advantageous.

(a) Common sludge treatment in one plant. The transport of sludge could take the form of a pumping main and pumping station or in certain cases through tankers. Examples are treatment plants Farmsen and Volksdorf in the city of Hamburg.

(b) Operation and supervision of widely scattered treatment units from one central switchboard station. However, this method assumes a certain degree of automation in the treatment units concerned.

In Hamburg at the moment investigations are underway to supervise and operate the treatment plant at Stellinger Moor, 500,000 PE, during night time, weekends, and holidays through television and remote control from the main treatment plant at Köhlbrandhöft, 2,750,000 PE, which is at a distance of 10 kms and has to be supervised continuously by operating personnel.

(c)   Coordinated operation of a group of treatment plants as practised in cities with a number of treatment units and by drainage authorites. One form of coordinated operation can be seen in a central workshop where specialized craftsmen are available.

(d)   Coordination of all operating devices including those used for reporting and measurement, into one switching panel instead of at decentralized locations in order to reduce the amount of construction and to minimize the supervising staff.

(e)   The arrangement of electrical operating devices, which otherwise are installed in the immediate neighbourhood of the machines that are to be operated, into centrally situated and well air-conditioned control rooms. This does result in a considerable prolongation of the useful life of the electrical equipment and, reduces the frequency of repairs and maintenance work.

(f)   Common location of all the different units in the treatment plant. When possible one should avoid locating pumps, screens, and grit chambers separately from other process units – as in Hamburg is the case with pumpwork Hafenstrasse and treatment plant Köhlbrandhöft by the river Elbe separating the plants because of high personnel and other costs.

(g)   Blockwise arrangement of tanks. A combination, for example, of aeration and final sedimentation tanks, and in some cases inclusive of primary sedimentation tanks in one block or a combination of grease trap and aerated grit chamber in one block reduces cost, spaces and work involved in operation.

Simplification

Increasing mechanization and automation, especially in the bigger treatment plants, actually means great simplification in operation, but unfortunately not in plant equipment due to the greater number of more complicated operations and control devices. Nevertheless it is possible to achieve, apart from a simplification in the operation, simplification in the sphere of civil and mechanical construction so as to keep down investment and operation cost.
Examples:

(a)   Simpler construction such as level floors instead of floors sloping towards hoppers, avoiding haunches, using plain walls and when possible many similar constructions, among others usage of prefabricated constructions, will still bring a construction cost reduction.

(b)   The construction of fewer, but all the same bigger units (tanks and basins) tend to reduce both construction and operation cost. A good example is the design for the new main treatment plant in Vienna. This partial biological plant, with the capacity of 12 m^3/s, possesses only 4 rectangular primary sedimentation tanks each of 5300 m^3 volume, 4 aeration tanks each 8900 m^3 and 18 final sedimentation tanks each 3600 m^3. Consisting of a larger proportion of cheap bottom surface and a smaller proportion of costly wall surfaces, the cubic-meter price is brought down considerably. The plant has only 2 return sludge pumps and the excess sludge need be drawn off only at two points.

(c)   The arrangement of fewer and at the same time larger mechanical units results in operational advantages. The treatment plant Köhlbrandhöft brought into operation in 1961 with a capacity of 3.3 m³/s, was equipped with aeration tanks consisting in all 190 Kessener brushes, which rest on 235 bearings and driven by 45 motors. On grounds of considerable operation upkeep and maintenance cost the new aeration tanks of capacity 4.6 m³/s are equipped with only 16 surface aerators. Each cone has 3.6 m diameter and a capacity of 250 kg O₂/h. Altogether, including one reserve, they cost about 600,000 DM. It is hardly probable that other aeration systems with such provisions could be more economical in both installation and operation cost.

(d)   Limitation of the number and size of machines, if possible all uniform in manufacture, leads to easier maintenance and reduces the cost of spares.

(e)   The omission of primary sedimentation – in the present meaningful for treatment plants serving up to 200,000 or 300,000 PE – reduces not only the investment cost but, due to simplification, the personnel cost as well.

(f)   A look into the future may show that due to the ever dwindling cost of power, internal power generation in treatment plants may no longer prove economical for plants serving 100,000 PE but only for those for 1 million or more. However, the limit to the overall economy of a power station largely depends on the cost of power available from electrical undertakings. In Fig. 10 the operational data pertaining to the internal power generation are shown, and also the cost of power drawn from the Hamburg electricity works in the case of the treatment plant at Köhlbrandhöft. Full economy will be restored as soon as the enlargement under construction is in operation.

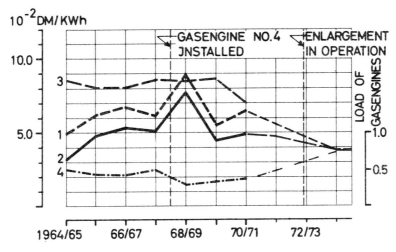

Fig. 10. Cost of power generation at Köhlbrandhöft, 2,750,000 PE.
Capital cost have been calculated on cost of contemporary plant.
1.   Absolute cost of internal power generation
2.   Cost of power generation, sold electricity being subtracted
3.   Cost of power if taken from the electricity works
4.   Degree of usage of the four gas engines of 950 kw each.

(g)   Simplification of plants as well means the installation of machines which are robust and at the same time require little maintenance, even if it has a slight disadvantageous effect on the process itself.

In order to return sludge to the aeration tanks as quickly as possible rectangular final sedimentation tanks with endless-chain scrapers are preferred. It has been observed, after experience with the Hamburg treatment units, that the endless-chain scrapers as against the mobile scraper should be rated much lower. Therefore it is advisable to install endless-chain scrapers only in places where there is no alternative possibility, for example in two-storey sedimentation tanks because of the high cost of repairs and attendent personnel cost.

Screw pumps, due to their robustness, reliability, and low maintenance cost should be installed wherever possible. It is even possible to sacrifice a standby unit at all places where no great threat to safety exists and temporary alternative arrangements could be provided. Therefore the first stage at Köhlbrandhöft is equipped only with one screw pump of 2 m^3/s discharge for the return sludge lift. This screw pump in ten years of operation was off service for only 2 weeks for repairs.

Because of good experience, a new pumpwork of 10 m^3/s maximum capacity and 20 m pumping head under plan for the Köhlbrandhöft plant would be of 3 pump sets only (2 screw pumps one after another) and of a capacity of 3.3 m^3/s each.

Inspite of their low efficiency the single shovel pumps and the turbulence flow pumps are to be preferred for pumping sewage and sludge so as to avoid clogging and consequently personnel cost. Almost the same advantages should be valid for simple air lift pumps.

(h)   At the design stage it is of fundamental importance to design units that are easy to repair, that means that especially those parts which ask for frequent maintenance, such as machines, that are easily accessible. Also designs should permit easy reassembly of equipment. Because of this, according to experience gained at the Hamburg treatment plants, external stirring of sludge by pumps, which consumes more electrical energy, has been found more economical than stirring by screw blade mixers inside the digesters.

It is conducive to a reduction of work on repairs if the building block principle is also kept in mind in case of mechanical and electrical equipment. This means that units such as surface aerators, including drives and motors or pumping units, wherever possible should be replaceable as a whole by a spare unit so as to have no interference with the operation as such during repair. It must be added here that the variety of machines such as electrical motors, drives, pumps, valves, blowers, a.s.o. should be limited so as to keep the inventory of spare units or parts as small as possible.

Improved organization

Rationalization has brought about through the years noteworthy success. The larger sewage authorities have proved themselves to be the forerunners in this field as is evident when one compares their operation cost with those of plants belonging to municipalities. In general, further reduction of personnel cost through rationalization should be possible and would be the goal to be reached in the future. This could be realized in addition to 'concentration' and 'simplification' by organizational measures.

(a)   Although it is desirable to have two categories of personnel such as operators and maintenance teams due to the difference in content of work on a treatment plant there are great advantages, with a few exceptions of specialists, of universal employability and exchangeability of personnel, as has been shown in the German treatment units. Because of this flexibility in the overall service planning it is possible to keep down the demand of personnel, e.g. where substitution of personnel is required.

(b) Universal employability and in addition specialization of part of the co-workers, brings economy not only in the larger treatment plants but also in the smaller ones when these are grouped to have bigger maintenance units.

In the neighbourhood of big cities — as in Hamburg — smaller treatment units are subordinated to a bigger one. In the case of sewage authorities which have many small and medium sized units under their care, the treatment units in some cases are divided into several districts with one central unit in each district. In this manner, as in the example of 'Lippeverband', 2 to 4 treatment units, each serving 1000 to 10,000 PE, are looked after at specified hours only by one trained treatment worker equipped with tools and transport. Depending on the location, 6 to 12 treatment plants form a sewage plant district. The central treatment unit is the headquarters of a district with a foreman as head, equipped with a small workshop, spare stores, and a field laboratory.

(c) It may seen from Fig. 9 that due to extensive mechanization and automation in the larger plants room for further economy in personnel cost, at least in the operation, is indeed small. A reduction in running cost, however, could still be conceived in the field of maintenance. For example, it is possible to work intensively after the principle of preventative maintenance as already practised with success every 6–8000 hours of operation in the case of gas engines. By using this system breakdowns can be avoided to a large extent and the safety of plant operation considerably be raised. Attendance to breakdowns should be immediate and as in certain cases this makes uneconomical use of manpower, preventative maintenance might provide the best possible use of personnel rather than the employment of external agencies. This presupposes the existence of a sufficient number of suitably qualified personnel on the staff.

(d) Through the increasing usage of fully automatic analytical equipment there exists the possibility of utilizing personnel in a more rational way in laboratory work.

(e) Due to design gaps and operational gaps treatment plants do not always give the results expected of them. In order to minimize design gaps a type of organization suggests itself in that the planning and designing team is associated with those constructing and operating the plant. Through cross checks, i.e. checking designs by operation personnel and vice versa, the number of deficiencies and insufficiencies can be kept to a minium. There is only one way to overcome the operation gap, namely to educate and qualify one's own personnel. For this job the leading staff of treatment plants could be approached as well as their colleagues of the design team About ten years ago the city of Hamburg provided a precise description of work, educational demands included, for operation workers and foremen, which is recognized by the ministry concerned, for well graded payment as well, so that there is no dearth of qualified personnel in this field. Great advantage can be seen in training staff on those plants they are going to work on or have been working.

In recent years the ATV has developed a multistage programme for qualifying and training treatment personnel in the Federal Republic of Germany whose further development will probably satisfy the increasing demand for qualified personnel in this field. It is unavoidable to equate these qualified technicians and foremen with corresponding personnel in skilled work (handicraftsmen) and industries. This is necessary to increase the attractivity of this profession. Thanks to the efforts of ATV the day is not far off when this training and regulation will be recognized for the FRG as a whole.

Automation

Today every large and modern treatment unit is, more or less, mechanized and provided with automation. One finds here, side by side:

> Manual control, e.g. of unimportant valves
> Local control, e.g. of screens
> Remote control, e.g. of power stations
> Automated control, e.g. of desludging
> Automated plant, e.g. O_2-control in aeration tanks.

The deciding factors for the choice of this or that mode of control for the various operational areas in a treatment plant among others are economy (cost/benefit), reliability, and effectiveness.

It should be stated here that the factors mentioned last have been responsible for preventing extensive automation up till now. Not that the control devices were too expensive or unreliable, but the one and only reason is that the measuring instruments using chemical and biochemical parameters for control were simply unsuitable for rough plant operation. One thinks for instance of automatic measuring devices for O_2, PV, COD, TOC, NH_3 suspended solids etc. They are available, but their susceptibility and complicated nature make them, for the present, less suitable for full and extensive use in such a way as to be able to build a fully automatic control system entirely relying on them. One notable exception is perhaps the automation of oxygen supply in activated sludge plants which have been installed or planned in recent times in many treatment plants and have led to considerable savings in electrical energy. It should be remembered, at these treatment plants, apart from the difficulties encountered with measuring devices as such, there arise other practical problems from them awaiting solution. Considerable reliability could be placed on measuring devices, which measure physical parameters such as water level, flow of water etc. Tested and reliable instruments are available for this purpose and the high degree of automation in our treatment plants achieved so far is based on such instruments. Founded on this knowledge we could see advantages in replacing the control of our surface aerators at Köhlbrandhöft by oxygen measuring devices actuating a local control system. By means of parabolic weirs that would raise the surface in the aeration tank and hence the emergence depth of the aerators according to the inflow, combined with change of pollution load, the oxygen input is controlled automatically without any complicated instrumentation. In order to be on the safe side as far as the oxygen level is concerned this simple system may result in a higher oxygen consumption when there happens to be dilution by rain or at weekends, but this should not be important from the point of view of cost especially when, as it is generally the case in large units, internal power plant is available and the surplus power could not be sold to the electricity works.

There is a good scope for further automation in the area of sludge treatment. There are tested instruments available which could measure exactly the quantity as well as the CO_2-content of the digester gas. The feeding of digesters depending on the gas quality at the moment — as a measure of the loadability of each digester itself — leads to an optimal utilization of sludge treatment plant thereby doing away with expensive reserve units.

SUMMARY

Regarding the continuously increasing load caused by rising cost an attempt has been made to point to the different factors influencing the construction and operation cost of treatment plants, to fix the limits of the influence and to show how higher economy can be achieved. Particular weight has been put upon analysis of operation and maintenance because operating sewage treatment plants is a constant burden on the public purse and therefore should be kept as low as possible. The many aspects which not only have to be taken into account when operating a plant but even when planning and designing have been discussed from an exclusively practical point of view. Because of this the interdependence of these points of view have purposely been shown not in an abstract, mathematical form but in a simple way, supplemented by suggestions and examples for further cost reductions. Of course this list is not complete, but from many years of practical experience it is my view that many of the problems of today can be solved and there are fundamental possibilities for improvement.

REFERENCES

1. Schmidt, U., Über die Kosten der biologischen Abwasserreinigung, Veröffentlichungen des Instituts für Siedlungswasserwirtschaft der T. H. Hannover, Heft 13 (1964).
2. Bucksteeg, K., Korreferat zu DÖNGES: Bau- und Betriebskosten von Kläranlagen, Gewässerschutz − Wasser − Abwasser, Veröffentlichung der T. H. Aachen, Heft 4, 75−89 (1971).
3. V. D. Emde, W., Wiener Mitteilungen, Veröffentlichung der T. H. Wien, Band 4, Q-1-36 (1969.
4. Imhoff, K., Taschenbuch der Stadtentwässerung, München (1969.)
5. Schoenenberg, H., Korreferat zu DÖNGES: Bau- und Betriebskosten von Kläranlagen, Gewässerschutz − Wasser − Abwasser, Veröffentlichung der T. H. Aachen, Heft 4, 91−99 (1971).
6. Londong, D., Betrieb und Instandhaltung von Kläranlagen, Gewässerschutz − Wasser − Abwasser, Veröffentlichung der T. H. Aachen, Heft 4, 109−126 (1971).
7. Londong, D., Über den Personalaufwand beim Kläranlagenbetrieb. Korrespondenz Abwasser, 8, 141−144 (1971).

Discussion *by* Wilbur Torpey,
Rutgers University New Jersey

The writer's experience indicates that as plant capacity is increased to about 100 M.G.D. both capital and operating costs decrease significantly. Above that capacity costs decrease quite slowly. Moreover, the overall project cost can be affected adversely if the plant size is increased inordinately due to the requirement of having to construct a larger, deeper, more expensive intercepting sewer system to the plant.

In contrast to the authors' recommendation for eliminating primary settling tanks, the writer has found that they should be retained in an activated sludge plant in that they have a beneficial effect on plant performance. The primary settling tanks are normally employed to remove the raw settleable solids, before their exposure to the subsequent biological environment, which practice conserves their compactability and thus serves to reduce the ultimate sludge volume. The separation of the settleable solids from the wastewater reduces the energy requirements for dissolving oxygen. Moreover these tanks perform the function of capturing greases and oils prior to their entry into the secondary system.

One method of reducing plant capital and operating costs is by eliminating the conventional grit and screenings chambers and transferring these functions to appropriately designed primary settling tanks. Thus grit and screenings are passed through the main pumps and collected by the primary settling tanks. The settling tank underflow, containing the concentrate of grit and screenings, is elevated above the ground surface by the use of a torque flow or other non-clogging pump. The concentrate is then screened at a level sufficiently high so that the continuous overflow has hydraulic head to satisfy the needs of the subsequent cyclone degritter. The degritted flow which is, in effect, a dilute primary sludge is conducted to a separate thickening tank system, there to undergo volume reduction jointly with the waste activated sludge at specific areal loading rates, as to both solids and liquid. Such prethickened mixed primary and activated sludge can be subjected to high rate digestion in tank volumes of only about one quarter the size required by the conventional two stage digestion system.

Integrating functions in the primary settling tank has the beneficial side effect of being able to elevate these "expensive to handle" grit and screenings solids to the surface using a fluid, rather than some mechanical vehicle, thereby largely reducing the need for operating personnel. This flow sequence advances the concept of working on the concentrate which represents only a small fraction of the flow, and so avoiding having to supply expensive facilities for handling of the main flow of wastewater. This basic design, employing the primary tanks for multiple purpose, will reduce substantially both capital and operating charges.

Although there are many processes which use different modes of contact between the wastewater and the activated sludge, the use of the step aeration process, which provides for the multiple addition of wastewater to the flow of return sludge through the aerator, is recommended for reducing capital and operating costs. Step aeration is designed to use only one half of the volume of the aerator that would be required conventionally. Its efficiency derives from the ability to hold an adequate amount of activated sludge solids under air to cope with the organic loading rate while maintaining a low concentration of activated solids passing to the final tank for separation and recycle. The use of the step aeration process not only effects a marked reduction in capital costs but affords an operational tool capable of dealing with transient changes in the settling capacity of the activated sludge floc by providing facilities for shifting wastewater flow addition to maintain continuity in treatment efficiency. That is to say, when the floc settling velocity decreases, the flow addition is shifted towards the effluent end of the aerator thus diluting the aerator effluent. After the activated floc recovers sufficient density the flow addition is shifted slowly backwards.

Experience indicates that the effect of rising labor costs on operating costs can be dealt with effectively through compact plant design layouts which minimize operating stations. Many repetitive operating tasks can be eliminated by such means as non-clog pumps, time and level controls, using visual and adible alarms where necessary.

Reduction of the number of plant personnel, doing repetitive operation functions, is a sound objective. In large plants, even those supplied by a combined sewer system, the number of men on each operating shift might be as low as two, with provision for paid holdover or call-in off duty personnel. One of these two men would be paid as a working supervisor. It has been observed that the installation of sophisticated automated equipment frequently results in providing leisure time for the operating personnel rather than fully eliminating need for their presence at particular locations and only adds to the already inflated charges for treatment. Where possible, a policy can be pursued during the course of the plant operation wherein operating personnel can be gradually shifted to the daytime maintenance function by the installation of positive means for eliminating repetitive tasks at particular locations in the plant.

The three categories of maintenance that must be performed in the plant namely (1) equipment breakdown repairs, (2) intermediate servicing and repairs of equipment, (3) major electrical, mechanical and structural repairs and replacements. It is mandatory that the first be done by on-site plant maintenance personnel. Breakdown maintenance demands that an adequate supply of spare parts be kept on hand. This is an unavoidable liability because, by its very nature, breakdown maintenance requires purchase and storage of unnecessary materials. When doing intermediate servicing and repairs of equipment, as in the case of multiple plants, consideration should be given to a program using plant maintenance personnel supplemented by work crews rotating between plants, supplied largely by materials purchased according to needs. Doing major equipment and structural repairs requires a multiplicity of skills. Moreover such replacements cannot be stored. For these reasons the contractual method of accomplishing this objective is more effective and economical than using plant personnel even though the costs for preparing the contracts and for administrative overhead and profit are included.

Reply

In my paper I said that for the reason of simplifying plants one could omit primary sedimentation for plants serving up to 200,000 or 300,000 P.E. A very good example for this thesis is the Blumenthal-plant near Vienna which proved to be very economical. Omitting primary sedimentation tanks is only possible when using low loaded activated sludge plants.

With highly loaded activated sludge plants you have to have primary tanks having at least a detention time of 40 min. Operating primary tanks with 12 to 20 min detention time only for the last 11 years at our large treatment plant at Köhlbrandhöft proved that this detention time is far too small. In the subsequent biological part we had much trouble.

It is not clear why Americans use compressed air aeration.

Mr. Cohen referred to the Denver plant, in which approx. 500,000 m^3/d are treated and that for the aeration equipment 9 mill US$ were spent. For reason of comparison I like to emphasize that for 16 huge aeration cones (ϕ 3.60 m), which are enough for treating 320,000 m^3/d and which will shortly be installed in the enlargement of Kohlbrandhoft we only spend 600,000 DM i.e. approx. 200,000 US$

ASPECTS OF SLUDGE DISPOSAL IN GREATER LONDON

S. H. DAINTY

Director of Public Health Engineering, Department of Public Health Engineering, Greater London Council, 10 Great George Street, London SW1, United Kingdom

INTRODUCTION

The Greater London conurbation lies on an estuary and all its sewage treatment plants are virtually surrounded by urban development. Pressures increase for greater utilization of land and critical eyes are continuously being cast at areas reserved for further sewage treatment or for emergency purposes. The condition of the River Thames flowing past the seat of government is always under the spotlight — the clamour for its continued improvement goes on. Greater London covers an area of some 1300 square km (600 square miles) and outside those boundaries planning authorities resist further encroachment.

The technological explosion in industrial fields continues to impose greater loads on the waste disposal services while world-wide attention is being drawn to the risk of pollution in the open sea. With such pressures on all sides it is imperative that the waste disposal services shall match the problems imposed by the most efficient service possible.

It became obvious immediately that a paper on the water pollution problems of an urban authority dealing with flows from some 8 million people would demand far more space than is admissible. A careful selection of possible subjects had to be made. In doing this it became apparent that most problems in sewage purification involved either directly or indirectly the treatment and disposal of sewage sludge. I have therefore chosen the sludge problem as the common theme on which to base this paper. The subject is not a new one and is generally acknowledged to be the real headache of the service.

Treatment and disposal of domestic sludge in an agricultural environment presents relatively few problems. The availability of adjoining land areas can provide the medium for the final breakdown of the sludge to a beneficial use. Cost of this final disposal is then of little significance. But as the rural community expands into larger units the simplicity of sludge disposal deteriorates. The factors which dictate this situation are many. Population expands and with it the demand for larger disposal areas. Industrial activity introduces into sewage flows materials less acceptable to the land. The co-operation of the farming community becomes less certain and adverse publicity can quickly affect disposal facilities. The demand for higher living standards generally involves the removal of all odour nuisance and the production of better effluents. Further increase in sludge output follows. Finally, urban development may well absorb the treatment plant itself and pressures for the release of boundary lands become acute.

London has experienced all these phases of a growing community and the impact on sewage treatment and sludge disposal is a subject of continuous study. The operating costs of large sewage works are lower than those of smaller works and in general the abandonment of smaller works in favour of large compact plants can be shown to be economically viable and to reduce such problems to a minimum. Land values become an important consideration and whatever the size of plant the method of treatment and disposal of sludge may dominate practical and financial feasibility study. Sludge treatment may be carried out cheaply on large areas or be a costly operation on small areas but in either case the demands of amenity or land restrictions may call for the complete removal of sludge from the works at an early stage.

Removal of sludge to a distant location for disposal introduces transportation problems and the possible choice of medium into which the sludge can be assimilated. If all the options are to be kept open then the quality of the sludge is an important factor for consideration. Trade waste control is an essential element in ensuring this and the future may demonstrate that more rigorous controls are necessary if land or sea is to be the final recipient of the sludge.

The risk of serious trouble from industrial discharges increases with the escalating sophistication and complexity of science and technology. Every day more and more drugs, pesticides, chemicals and processes are evolved which constitute a potential threat to the stability and efficiency of the main drainage system. Incredibly small concentrations of certain preparations could totally inhibit sewage purification. The need, therefore, is for increasing vigilance and care combined with education of all concerned and an extension of present control and disposal facilities. At present in the U.K., industrial or toxic wastes which cannot be accepted into the sewers must be disposed of by the producer himself or by private firms which offer a service of industrial waste disposal. [That is apart from radio-active materials, the disposal of which is carefully and properly controlled by a Government inspectorate under separate legislation.]

The majority of such wastes, which in the GLC area (Greater London Council) totals nearly 10,000 tonnes/week, are disposed of to land tip. A relatively small proportion is incinerated and some of the most toxic substances are dumped at sea. A GLC Working Party and a Government Committee have recently and independently concluded that much greater control needs to be exercised over the disposal of these wastes. No doubt such control will be a considerable help in minimizing still further the very real hazard of major damage or pollution from this source.

We regard the breakdown of all sludge by anaerobic digestion as a basic need to provide an innocuous material suitable for disposal, but transportation of this material requires thought. Dewatering of the sludge will reduce its volume considerably but it is open to question whether such a process is justified where sludge can be pumped cheaply by pipeline to a distant site or tanked or piped conveniently to sea.

Before commenting further, one general point needs to be emphasized. On the scale of the Greater London operation breakdown of the service from whatever cause must be met by planned procedures. The size of the community at risk and the scale of nuisance created demand attention to these situations and it may well be that in the ultimate, decisions have to rest not so much on performance but on the repercussions of breakdown. In line with this consideration is the perpetual risk of industrial unrest and the consequent reliability of all operations. Works capacities at all sites whether urban or rural should be designed to meet such eventualities.

THE FACTORS WHICH INFLUENCE VOLUME AND QUALITY OF SLUDGE

Primary sedimentation

There appears to be a small section of opinion in the sewage treatment field — possibly originating from a theoretical, academic approach — which suggests that primary sedimentation, in advance of an activated sludge plant, is unimportant and could in fact be dispensed with. This sort of radical approach must not be discarded out of hand or accusations of being reactionary or old-fashioned could well be true! But except for very small works, where the activated sludge process would probably not be the best answer anyway, the elimination of primary sedimentation is considered to be unsound for the following reasons:

(i) The purpose of a sewage treatment plant is to reduce the impurity load which would otherwise pass direct to a watercourse and create unacceptable conditions in it. The reduction of impurity load needs to be carried out without nuisance and as cheaply as possible. The cost of removal of a unit of impurity (BOD for example) by sedimentation is, at a large works, only about two-thirds of the cost of removal of a unit of impurity by biological or chemical treatment, e.g. by an activated sludge plant. The required capacity of a biological (or chemical) treatment plant may not be directly proportional to the strength of the sewage imput, but it is a principal factor in its determination. The elimination of good sedimentation would at least double the impurity load on the activated sludge plant and it is estimated that the extra cost of the activated sludge plant would approximate to the 'saving' expected by elimination of the primary sedimentation stage. Downing[1] states that the minimum period of aeration will be roughly proportional to the 5-day BOD of the sewage applied.

(ii) The biggest single change which would result from an elimination of sedimentation would be the enormous increase in surplus activated sludge. The 'saving' of primary sludge would be relatively insignificant. Consider a works serving one million people: The quantity of primary sludge at 4% solids would be about 25 l/s (0.5 mgd). This includes re-settled surplus activated sludge which would originate as about 50 l/s (1 mgd) containing 0.6% solids; without primary sedimentation there would be 100 l/s (2 mgd) of surplus activated sludge at 0.6% – a nett increase of 75 l/s (1½ mgd). As activated sludge is much more difficult to dewater, digest and handle when thickened than raw sludge, it is inevitable that a considerable increase in sludge disposal costs would be incurred. It is also relevant that activated sludge possesses the unfortunate characteristic of high oxygen demand and consequent rapid deterioration to septicity when not aerated. The risk of smell nuisance would therefore be increased by the elimination of primary sedimentation.

(iii) The performance of an activated sludge plant is improved and can be more efficiently controlled if the impurity load imposed is kept constant. Moreover it appears that satisfactory automation of sewage treatment processes (almost inevitable in the future) will be assisted by, if not dependent upon, a high degree of flow and load balancing, to maintain as near as practicable, even and stable conditions.

There can be no doubt that an efficient primary sedimentation stage performs not only load balancing and load reduction but also acts as an important buffer against intermittant or shock loads.

(iv) Primary sedimentation tanks are usually designed to remove grease and scum as well as suspended solids. If such material is allowed to pass forward freely to the activated sludge plant, additional problems, cleaning and costs will be incurred.

If it be claimed, not that sedimentation should be eliminated, but merely that its efficiency is relatively unimportant, then it is suggested that in practice the reverse is the case.

If waste liquors and surplus activated sludge are intended to be settled out in the sedimentation tanks, any inadequacy of sedimentation or sludge disposal rapidly reacts on the biological treatment plant and a spiral of increasing inefficiency and difficulty can easily, and usually does, result! In one instance, inadequate sludge digestion and disposal facilities resulted in a build-up of sludge in the sedimentation tanks. This rapidly went septic, resulting in a poor tank effluent and greater volumes of surplus activated sludge; this overloaded the sedimentation tanks still further and so the 'spiral' continued, accompanied by undeniable odour nuisance over wide areas of residential character. Drastic measures were called for until improved sludge handling and disposal facilites reduced the pressure on the sedimentation plant and allowed it to function normally.

More recently, at another major works, similar conditions arose due to (a) out-of-condition activated sludge which, in a state of partial nitrification, characteristically had a very low density (SDI) and abnormally poor settling properties, and (b) large sedimentation tanks which were not equipped with individual scrapers and had poor sludge withdrawal facilities. In these circumstances, sludge again built up in the sedimentation tanks to such an extent that the tank effluent was adversely affected, the load on the aeration plant increased and the surplus sludge increased in quantity whilst it reduced still further in quality. Again extensive smell nuisance resulted and, in this case, the situation was only retrieved (with the co-operation of the River Authority) by the 'spilling' of all surplus activated sludge to the river for several days. Good sedimentation minimizes the production of surplus activated sludge and helps to maintain an activated sludge of good characteristics — so minimizing subsequent treatment and disposal costs and creating a self-improving situation.

Activated sludge separation

Good design is of first importance to the successful operation of an activated sludge plant. In London we have experienced time after time the need for the designer to be aware of the need for the design to promote production of activated sludge of good settling characteristics. Whether the plant is required to produce a nitrified effluent or not, tank capacities, sludge return facilities and power or air supplies must allow for the creation of a stable, active and good settling sludge. If the operator is given these facilities with plenty of reserve in hand, problems of activated sludge separation will be minimized and probably will never be acute.

On a number of occasions, difficulties at GLC works have arisen due to the poor condition of the surplus activated sludge at times when partial nitrification only existed in the aeration plants. Usually, this has been the result of 'pressing' a plant to produce a nitrified effluent when in fact it was not designed or equipped to do so. Maybe nitrification could be attained during favourable summer conditions, when enhanced temperatures accelerate biological activity and the incoming sewage load is reduced by a proportion of the population being away on holiday; but during the autumn, as the loading and the colder weather return, nitrification falls away and the activated sludge condition deteriorates.[2] Because of the extremely critical situation which arises so quickly in these circumstances and is so difficult to correct, we have found it necessary at one large works to run the two halves of the plant quite differently — seeking nitrification in one half whilst actively discouraging it in the other. Under such an arrangement it is obviously essential that the returned activated sludges should also be kept separate because of their different biological content. Wherever practicable, we would urge, for convenience and flexibility of operations, the separation of the activated sludge returns by units — so that any temporary malfunctioning plant can be nursed back to health one unit at a time, by reducing imput load, increasing air or power, or increasing returned sludge rate, without being 'reinfected' by sludge returned by a 'common' supply. The above problems are, of course, especially difficult and significant at large works, where there may be millions of gallons (many millions of litres!) of surplus activated sludge requiring disposal each day.

At large works there is a temptation to build large tanks; it is more economical than building a greater number of smaller tanks — but troubles are more likely. Problems of excavation and ground water increase with the floor slope so again seeking economy, there is a temptation to reduce the floor slope. The large radius and the gentler slope both tend to increase the average time of collection of the sludge at the centre, and if the scraper blades are inefficient the risk of dentrifying or deteriorating sludge is further increased.

Whilst flat-floor final tanks certainly can be efficient units when fitted with suction scrapers — which ensure that the tank floor is kept clean with a slow rate of revolution of the scraper — the most reliably efficient final settling tanks in GLC works are considered to be those which are only about 18 metres diameter, 2—5 metres side wall depth and 30° floor slope. At the Mogden works such tanks were originally fitted with ring-beam scrapers but for various reasons a few tanks have been operated quite successfully without any scraping whatever. Side wall scrapers are now normally fitted having single or double chains dragged over the sloping floor to discourage any temporary hang-up of sludge.

Automatic sludge separation

The volume and water-content of sludge passed to digestion and disposal should be minimized for maximum efficiency. However monotonous and time-consuming may be the duties of 'tank attendants', who are responsible for controlling the withdrawal of sludge from primary sedimentation tanks, the importance of this job being done well cannot be in doubt; and if this time-consuming job is to be eliminated and automation substituted, it must be with little or no loss of efficiency.

In considering the case for abandoning manual operation one must bear in mind not only the fallibility but also the value of human control — in being able to make judgements and take action appropriate to abnormal circumstances, which automative devices may not be able to do.

Sedimentation is a continuous process, although the volume of water and the amount of solid matter to be separated varies diurnally and with weather conditions etc. If automation is to be successful, the withdrawal of sludge should approximate to the rate at which solid matter is deposited. Otherwise there will be a build-up of sludge in the tanks or a withdrawal of uneconomic volumes of watery sludge.

It may be easier to achieve a matching rate of withdrawal if the flow through the sedimentation tanks is fairly constant. Separate sewerage systems and large catchment areas will help in this respect but, if sludge withdrawal is to be based on some form of time control, it is suggested that flow balancing prior to sedimentation will be necessary. Even then, unless some method of sludge density control is also exercised, there will be an increase in the average water content, and hence volume, of the sludge withdrawn, requiring increased sludge digestion capacity.

There are basically two ways in which overriding control on the volume withdrawn, depending on the thickness of the sludge, can be exercised. First, by centralized visual remote control using T.V. cameras at each withdrawal chamber. Secondly by some form of resistivity,[3] optical[4] and audio automatic device. In the last category, GLC experiments with ultrasonic equipment have met with encouraging success and if the field trials now proceeding confirm the results obtained in the earlier experiments, a considerable step forward will have been made. It appears likely that water-contents can be assessed reliably to within 1% so that it would be possible to override a time-control system to prevent a sludge wetter than between, say, 97 and 98% being withdrawn.

It is unlikely that any automated system will result in greater efficiency — expressed as smaller, drier volumes of sludge withdrawn for treatment and disposal. Indeed it is likely that some increase in volume of sludge, requiring increased sludge treatment and disposal capacities, must be weighed against any savings in manpower and wages achieved by automation or remote control. There will undoubtedly be further progress soon in this field, accompanied by pressures to eliminate the boring and unpleasant job of manual de-sludging. But, at large works, the risks of total automation, the relatively small proportion of costs attributable to manual control of sludge withdrawal together with the probable necessity to balance flows and extend or amend other related parts of the works, suggest that progress towards full automation should be made with caution. For reliability's sake, remote visual control is preferred at the present time. The automated control of returned activated sludge is much easier because there is no necessity to limit the volume so carefully. Water content of the returned sludge is neither so variable nor so critical as with raw sludge withdrawal. Also, where an automated activated sludge plant, based on D.O. control is concerned, automatic withdrawal and proportioning of returned activated sludge to mixed liquor is most desirable.

D. A. D. Reeve[5] suggested that primary sludges might be withdrawn automatically, without particular reference to the water content — to be subsequently thickened for sludge treatment and disposal. The dangers here are that (i) the process of thickening and its supervision may cost as much as manual de-sludging and prove to be an amenity problem and (ii) the recirculation of increased volumes of liquors reduce overall plant performance.

SLUDGE PUMPING

There are many advantages, now well known and generally accepted, of large works and regional schemes. One of the difficulties which attend large scale sewage disposal operations, however, is the distance between the various operations and this applies particularly to sludge 'handling'.

One of the earliest regional schemes in the world (certainly in the U.K.) was the West Middlesex main drainage system constructed between 1931 and 1936: 110 km (70 miles) of new trunk sewers serving over 1 million people in 435 square km (170 square miles) discharge to the Mogden works which replaced 28 former small works in the area. Mogden was situated close to the upper reaches of the tidal Thames but in a rapidly developing residential area: suitable therefore for the discharge of a high quality effluent, but not for the storage and disposal of huge quantities of sludge. So it was essential to develop a separate sludge disposal works 11 km (7 miles) away at Perry Oaks, adjacent to the then small, quiet Heathrow Airport.

Raw sludge, at Mogden as elsewhere, does not usually pose many pumping problems providing (a) rags and grit have been removed adequately, (b) suction lines are kept short, (c) scum lines are short and scoured frequently.

In these circumstances, centrifugal pumps have proved entirely satisfactory and rising mains, usually to digesters, can be quite long — say 1 km (half a mile). Pumping heads seldom exceed 15 or 20 m (50 or 60 feet).

The pumping of digested sludge is a different matter, however, especially when considerable distances (as with Mogden to Perry Oaks) are involved. The GLC and its predecessor have now had over 35 years' experience of this activity and on the basis of this experience now have similar pipelines carrying sludge from

(i) Deephams to Rammey Marsh — 1963 — 6 km (3½ miles)

(ii) Deephams to Beckton — 1971 — 19 km (12 miles)

(iii) Riverside to Beckton — 1968 — 5 km (3 miles).

In each case, and for the original Mogden to Perry Oaks installation, there are twin pipes, 250 or 300 mm (10 or 12 ins) diameter, spun/cast iron with fully valved crossovers about every 2 km.

In one instance, a considerable length of main is in flange-jointed pipe, suspended in the soffit of a 3 m diameter main sewer. A few years ago, some of the pipe hangers corroded and gave way, thereby casting doubt on the success of the system. Repair of the damaged main proved to be a difficult and slow operation as it was only possible to gain access into the partially-separated sewer for two or three hours each morning. The remainder of the pipelines are buried about 1.5 m below ground, with lead-filled spigot and socket joints. Reciprocating pumps driven by variable speed D.C. motors were installed initially, capable of lifting 40 litres per second against a head of 60 metres. There was never any evidence of scaling or coating in the pipes even though the average velocity of the sludge was only 0.5 m/s (max 1 m/s and min 0.25 m/s).

There were, however, numerous occasions when blockages occurred. These were generally due to digester cleaning operations, when there was a great risk of pockets of scum or grit being drawn through the pumps and into the pipelines, increasing friction, slowing down pumping rates and ultimately blocking the line. On these occasions, leakages from 'blown' pipe joints were frequent side-effects. The flexibility in pumping conditions provided by frequent crossover connections has usually provided the facilities for clearing blockages. An added facility is to be able to pump from either end of the mains and if necessary to pump clear water. Washouts are provided at all low points and air-valves at all high points. The latter were originally manually operated but are now of the automated ball-valve type. Gas generation in the pumps and main can be a problem and was one of the reasons for the widely held view that positive displacement pumps were more suitable for digested sludge pumping — apart from their value in clearing partial blockages! This concept has been confounded to a large extent during the last few years, when two-stage centrifugals installed initially as an experiment at Mogden, have performed consistently, reliably and much more economically than the ram pumps did. The digested sludge is normally about 97.5% water, the combined static and friction head about 50 m (75 lb/sq in) and the quantity about 42 l/s (800,000 gallons per day), delivered in 15 or 16 hours. Variations from these average values, obtained in experimental test runs, are given in the table below:

Solids Content %	Velocity m/s	Head Loss m of sludge/m of pipe
5	0.58 1.07	0.013 0.015
2.5	0.43 1.10	0.00078 0.0046

Between 1966 and 1968 the GLC commissioned Consulting Engineers to undertake a feasibility study[6] for pumping digested sludge from Mogden to Crossness (45 km across London) and from Crossness to the North Sea (a distance over land of nearly 110 km, and submarine of 14.7 km.

For the main section, designed to convey 330 l/s (6.2 mgd) in 16 hours, twin 700 mm S.I./steel mains were proposed with only one intermediate pumping station at Darenth between Crossness and the North Foreland (near Margate). The following table gives the principle pumping data. The scheme has not been proceeded with, although as intimated in the following section, this does not imply any dissatisfaction with or apprehension about the viability of sea disposal in general.

	Rate/ Quantity l/s	Velocity m/s	Head m
Crossness to South Darenth Twin 27" diameter pipes			
16 hour (normal) duty	536.6	.726	88.8
12 hour duty	715.6	.964	99.4
Pumping Effluent	564.5	.763	90.3
Cleaning	564.5	.763 & 1.525	98.8
South Darenth to Coast Twin 27" diameter pipes			
16 hour (normal) duty	536.6	.726	54.2
12 hour duty	715.6	.964	111.6
Pumping Effluent	564.5	.763	62.2
Cleaning duty	564.5	.763 & 1.525	74.6

METHODS OF SLUDGE TREATMENT AND DISPOSAL

Sludge digestion

The most widely adopted method of sludge treatment is the digestion process. The population served by digestion in the U.K. is 10 times that of any other process for sludge treatment. The principle reasons for digesting sludge are:

(i)   To reduce the amount (weight) of solid matter requiring final disposal.

(ii) To render the sludge relatively inoffensive.

(iii) To reduce risk from pathogens.

(iv) To reduce the fat/grease content of the sludge.

(v)   To take advantage of the by-product methane — usually as a fuel for power generation.

(vi) To assist subsequent dewatering.

Reasons (ii), (iii) and (iv) are particularly relevant where ultimate disposal is to land as a soil conditioner, digestion being virtually essential in these circumstances.

The GLC are committed to sludge digestion in spite of the apprehension expressed in many quarters about the future of the process. In fact, no serious difficulties have so far been experienced at GLC plants and the survey by Swanwick[7] indicates that the number of instances of inhibition due to detergents may have been exaggerated. Certainly where difficulties have arisen, the use of stearine amines has been very effective in correcting the situation.

The water content of digested sludge is usually quite high and most methods of final disposal require some dewatering if the disposal method is to be reasonably economic.

Dewatering

The thickening of raw sludge prior to digestion increases the efficiency of the process but is not widely favoured in the GLC works because of odour nuisance arising therefrom. However, the thickening of surplus activated sludge has been tried on pilot plant scale and appears promising, in reducing considerably the volume of sludge requiring treatment and disposal without accompanying smell nuisance. A final decision on the installation of a major plant at Crossness on the (i) flotation by air using polyelectrolytes or (ii) electrolytic flotation method, has not yet been made.

Lagoons, the most commonly employed method of separation, require about 5 weeks of quiescent settlement to achieve a reduction of water content from 97.5 to 96%, and about a year to reach 92%, when dealing with a well-digested mixed sludge.

A process has been developed recently by the GLC and reported by B. R. Brown and L. B. Wood[8] which involves aeration of the sludge prior to settlement. This process enables the sludge to be dewatered by 40% of its volume after only 3 or 4 days settlement, and is now operating on a works scale at the GLC's Crossness works.

Mechanical treatment

The alternative to the above methods of dewatering is a mechanical system, such as vacuum filtration, pressing or centrifuging. About 5 years ago a GLC Working Party considered all such methods and found them all too expensive. While existing methods of disposal remain available, we regard mechanical means of dewatering as contributing an unnecessary expense.

Heat treatment or wet oxidation are also considered expensive because another stage of treatment becomes necessary, i.e. only a relatively small part of the total treatment and disposal is achieved by them.

Drying beds are not so expensive as most mechanical methods and produce a drier product but in spite of efforts to mechanize, their operation in application and removal of sludge are not regarded as avenues for extension in the future. The tendency where appropriate is to lagoon sludge over a period of 3–4 years, thereafter to remove the sludge directly to land at a water content of some 85%.

Incineration

Incineration of sludge is only possible after dewatering to an autothermic level of about 70%. It is therefore again not considered attractive on economic grounds. The joint incineration of sludge and refuse, although practised in some plants in Europe and America, has in our opinion not reached the required standard of development. Problems of mixing, smell and most of all, suitable proportions of sludge to refuse, still persist and until such problems can be solved economically, application on a large scale is unacceptable.

The most attractive alternative to wet digested sludge disposal to land (as a soil conditioner) or to sea appears to be the use of waste heat from refuse incineration to dry the sludge prior to its utilization on land. It is surprising and disappointing that the few existing plants on these lines in Sweden have not yet been followed by greater attention, refinement and improvement to derive benefit from this attractive concept.

Utilization

The recommendation in 1954 by the Ministry of Housing and Local Government that sewage sludge should be digested and disposed of to agricultural land, preferably in the liquid form, was generally re-affirmed by the Jeger Working Party in 1970.

About one third of London's sludge is so utilized and because we think this basic concept of utilization is right and proper, consideration is being given to how this proportion may be increased without significant increased cost to the GLC. An assured outlet to land is, however, essential and this can involve vast areas outside London's boundaries.

Disposal to sea

Since 1885 – when sedimentation tanks were originally built at the outfalls of Sir Joseph Bazalgette's intercepting sewers, constructed 20 years earlier – ships have plied up and down the River Thames taking London's sludge to the dumping grounds in the Black Deep and the Barrow Deep – over 80 km from the works and about 16 km from the nearest land. This represents 70% by weight of the total solid matter derived from all GLC's disposal plants.

During the whole of the intervening 80 years there has been no evidence of significant troubles or nuisance from this operation, which is nowadays carried out with the co-operation of, and in consultation with, the Port of London Authority (within whose harbour boundaries the sludge is dumped) and the Ministry of Agriculture, Fisheries and Food who operate a voluntary system of monitoring and approving dumping wastes at sea.*

There are two methods of dumping sludge at sea: (a) by ship and (b) by pipeline. The comparative economics of these two methods must be tempered by consideration of reliability, flexibility, permanence and geographical factors. But in general, a pipeline will be higher in capital cost and lower in operational costs than shipping. Costs will evidently be lower for larger quantities dealt with and for shorter distances – but from London, where over 200 l/s (4 mgd) of sludge (initially at 2.5% dry matter) is shipped to sea as 14,000 tonnes/day at about 3.5% dry matter – the total cost of shipping equates to about £5 per dry ton, or £900,000 per annum. The capital cost of a new pipeline system has been estimated at between £19 and £20 million in 1968.

Where large quantities are concerned, the use of several ships, running continuously, is justified. This reduces the effect of an accident or unplanned maintenance to a vessel compared with the case of only one or two ships being required normally. A pipeline does not require dewatering of sludge to be carried out – this may be considered an operational advantage. On the other hand, any significant breakthrough in dewatering techniques may well have a marked effect on the economics of shipping – or for that matter on sludge disposal techniques as a whole.

Although the quantity of sludge requiring disposal is continuing to increase, and at 4½

---

* Negiotiations are in progress to control sea dumping by legislation.

million tons per annum is now double the quantity in 1963, it is relevant to note that the proportion of the sludge which is digested prior to disposal is now much greater (70%) than formerly (25% prior to 1959) and by 1976 will be 100%.

Digested sludge (screened if necessary) derived from predominantly domestic sewage, holds no threat of marine or coastal pollution when dumped in a reasonable depth of water, and well away from land and shellfish waters. There appears to be no evidence otherwise and reports from Glasgow[9] — where it is suggested that anglers' catches have improved — and Los Angeles — appear to support the evidence available of the outer Thames estuary. Some of the most recent evidence stems from the feasibility study[10] referred to in a previous section. The third stage of that study concerned the submarine pipeline outfall and included a detailed assessment by the Hydraulics Research Station, the Water Pollution Research Laboratory and the GLC's Scientific Adviser of the diffusion, ultimate destination and effects of sludge discharged through an outfall several miles from the coast. Surface and sea-bed drifters, sludge labelled with radio-active tracers and chemical and biological examinations all played their part in the study, which concluded that 'long term' changes in the concentration of dissolved oxygen, due to the accumulation of digested sludge solids on the sea bed, can be largely discounted because:

(i) Surveys of the sea bed before and towards the end of the pilot experiment failed to provide evidence of changes other than those which could be attributed to seasonal factors.

(ii) The investigation carried out by the Hydraulics Research Station showed that the possibility of extensive accretion occurring in the vicinity of the proposed outfall was remote.

(iii) The oxygen consumption rate of settled digested sludge is low.

Also, 'the concentration of dissolved oxygen in the sea is unlikely to be seriously depleted'.

On the beaches of the Kent coast, 'no significant increase in pollution of the beaches either chemical or bacteriological was detected'.

It is suggested therefore that with the immense dilution which the sea affords[11] the dumping of biodegradable sludge from a population of some 8 million — even into the relatively small and shallow North Sea — does not represent a real hazard to life or amenity.

It is, of course, essential that if sludges are destined for disposal either at sea or on land, efficient trade effluent control already referred to must be exercised to ensure that heavy metals and other toxic substances are limited to concentrations in the sludge (as well as the effluent) which may be regarded as valuable trace elements instead. The temptation must be resisted for traders to regard the presence of a sludge main to sea as a suitable back-door for the disposal of unwanted wastes.

A uniform approach to trade effluent control by all members of the E.E.C. (European Economic Community) is also most desirable. Otherwise those nations not exercising control and not charging the producer for the cost of dealing with his waste will be at a financial advantage. Moreover, there will be an incentive, especially for the multi-nation firm to cheapen his production costs and thus gain advantage, by concentrating work where trade effluent control and charging are not enforced. The problem of disposal of industrial and toxic sludges (excluded from the sewers) is a very different matter from 'good' sewage sludge.

Here is the danger area to which attention must be directed most carefully: where persistent pesticides, mercury, cadmium and other heavy metals, arsenic and other poisons and sophisticated new products constitute a very considerable threat of pollution on land or at sea. These wastes are at present dealt with by the trader himself or by a waste disposal contractor. It is suggested by the Technical Committee on the Disposal of Toxic Solid Wastes in their Report of 1970 that the new authorities (due to emerge in 1974) should be responsible for the authorization of disposal arrangements for all such wastes produced in their areas. Most of these wastes are at present tipped (on land), but control as recommended, together with the legislation now planned to flow from the N.E. Atlantic Convention (Calo 1971) whereby approval will be required for *any* dumping at sea, should safeguard our seas to a very great extent. It is hoped that the principle will be applied on a world-wide scale.

REFERENCES

1. A. L. Downing, A. G. Boon and R. W. Bayley, 'Aeration and Biological Oxidation in the Activated Sludge Process.' Journal of I.W.P.C. 1962–1–66.
2. A. L. Downing, H. A. Painter and G. Knowles, 'Nitrification in the Activated Sludge Process.' Journal of I.W.P.C. 1964–2–130.
3. C. B. Townend, 'Recent Middlesex Developments in Mechanization and Automation of Sewage Plant Operation.' Journal of I.W.P.C. 1961–4–273.
4. G. Williams and R. A. Ownsworth, 'Single Photo-electric Sludge-Level Detector.' Journal of I.W.P.C. 1967–3–283.
5. D. A. D. Reeve, 'Automation of Sewage Treatment Processes', paper presented at I.W.P.C. Conference, Brighton 1971.
6. R. F. Pearson, E. V. Finn and D. R. Miller, 'Study for disposal of digested sewage sludge from the Greater London Sewerage area into the North Sea by pipeline.' Journal of I.C.E. March 1971.
7. J. D. Swanwick, D. G. Shurben and S. Jackson, 'Survey of the Performance of Sewage Sludge Digesters in Great Britain.' Journal of I.W.P.C. 1969–6–639.
8. B. R. Brown and L. B. Wood, 'Some experiments on the dewatering of digested and activated sludge' presented at I.W.P.C. (Metropolitan & Southern Branch) March 1971.
9. D. W. Mackay and G. Topping, 'Preliminary Report on the effects of sludge disposal at sea.' Effluent and Water Treatment Journal, 19 Nov. 1971.
10. S. H. Dainty, R. F. Pearson, L. H. Thompson and E. V. Fin, 'A Regional Scheme for the Disposal of London's Sludge to Sea.' I.W.P.C. Symposium, Bournemouth, May 1970.
11. A. Key, 'Water Pollution in Coastal Areas – Where do we go from here?' I.W.P.C. Symposium, Bournemouth, May 1970.

Discussion *by* G.C. White, U.S.A.
"The Related Problems of Odor, Septicity and Corrosion in Large Collection Systems"

One of the most trying problems of these large systems is the disposal of sludge. Closely akin to this are the collective problems of odor, septicity and corrosion, all of which find their origin in the collection system. The occurrence of odor problems is usually related to sulfide build-up and this contributes to the corrosion of concrete. Septicity in raw sewage is also related to sulfide build-up; however, septicity can occur for other reasons but with very little sulfide generation. It is well known that a septic sewage has very poor sedimentation characteristics which results in poor overall performance in the primary treatment process and further upsets the digestion process of raw sludge.

Sulfide Generation

The basic problem in prevention of sulfide build-up and septic conditions is one of oxygen depletion in the flowing sewage.[1,2]

Methods of Control

Free-flowing Sewers

The build-up of hydrogen sulfide in free-flowing sewers can be controlled by a rigid program of intermittent shock dosing, using sodium hydroxide.[3] This attacks the slime layer on the submerged surface of the pipe, which reduces the oxygen up-take of the slime layer.

Force Mains

Force mains can be treated by the threshold air injection method. Theoretically the oxygen added by air injection is just sufficient to offset the oxygen up-take in the force main.

There are always situations where additional treatment or alternate solutions are desirable. This review addresses itself to the chlorination of the flowing sewage and the scrubbing of foul air in ventilation systems.

Treating the Sewage Flow

The Case for Chlorine

Chlorine is useful because it can destroy hydrogen sulfide instantaneously.

Destruction of hydrogen sulfide by chlorine eliminates two problems at once: odor and corrosion. If hydrogen sulfide is eliminated by chlorination, odors from other compounds will also be controlled. Elimination of septicity by adding chlorine also removes the odors resulting from the septicity because the settling characteristics of the sewage are restored by prechlorination.

Chlorine applied to sewage reacts first and instantaneously with inorganics, such as sulfides and ammonia compounds. After these initial reactions the remaining chlorine will enter into a series of complex reactions with organic compounds. In sulfide destruction and septicity control it is not necessary to add additional chlorine to satisfy the organic reactions. The way to determine the correct dosage of chlorine is by laboratory chlorine demand studies. This is not practical when the plant is in the design stage so we must rely on experience. It is not necessary to chlorinate to any specific residual for odor or septicity control, however dosage is related to the severity of the problem. For a moderately septic sewage and dissolved sulfides up to 2 ppm an equipment capacity of 15-20 ppm of peak flow should be adequate. For a grossly septic sewage such as one containing domestic sewage two to three hours old with a thirty percent mixture of cannery wastes, 30 to 40 ppm of chlorine might be required for proper control. For the case of hydrogen sulfide destruction chlorine reacts stoichiometrically, 2.2 ppm chlorine being required per ppm of sulfide as hydrogen sulfide.

The stoichiometric relationship requires 8.32 ppm of chlorine to reduce sulfides to sulfates. It is interesting to note that the Los Angeles County Sanitation District found by actual field tests that it required 4 mg/l of chlorine to overcome 1 ppm of sulfide build-up in a sewer line.[3] The chlorine requirement cannot be accurately predicted except for existing plants where chlorine demand studies can be made.

Treatment of Ventilation Systems

There are many instances where it is desirable to remove odors from the ventilation system of an outlying pumping plant or a treatment plant separately or in conjunction with treatment of the sewage flow. Various methods have been used. Some of the most ambitious have been those where sewers were purged with forced draft ventilation, and the exiting air forced up through a spreading field with natural dirt as the filtering medium.

Outlying pumping plants or siphons in a large system can become a severe odor nuisance due to the presence of odiferous compounds in the exiting air of the ventilation system. As the flow of sewage varies in the sewer lines they must "breathe". This breathing can be a problem at a pumping station.

Scrubbing with Chlorine Solution.

Assume, for example, that the ventilation system of a modest sized pumping plant is approximately 5000 CFM. Let us further assume that it may be necessary to scrub out of this system up to 10 ppm of H_2S. This can be done effectively with chlorine solution applied to the scrubbing water. The removal of hydrogen sulfide is accomplished by dissolving it in the aqueous solution of a scrubbing tower where it is destroyed by the free available chlorine in the scrubbing water. The water is recirculated continuously with one complete turnover every five to seven minutes. Chlorine solution is added from a gas chlorinator as needed to maintain a sufficient excess to convert the dissolved sulfides to sulfates. The alkalinity destroyed in the recirculating water by the addition of chlorine is restored by adding caustic. This assures maintenance of pH near the neutral zone thereby eliminating the possibility of molecular chlorine being aerated out of solution.

Packed Towers Using Secondary Effluent

Quite recently Hicks[4] of New Zealand found that foul air could be de-odorized by being scrubbed in a packed tower using secondary sewage effluent. Pilot plant tests of this system have shown phenomenal results.[5] The treated effluent is put through a vertical tower packed with media having a surface area of approximately 30 sq.ft. per cu.ft. The ratio of tower height to diameter is on the order of 5 to 1. The media builds up a coating of slime about 1/32 inches thick. This zoological slime coating creates a high ORP situation which is conducive to high transfer rates of malodorous gases in the foul air. It appears that this system can remove not only hydrogen sulfide, but also all the other malodorous compounds found in sewage odors. Operation of test units have shown hydrogen sulfide removals as high as 25 ppm in the foul air. Secondary effluent will work without recirculation while primary effluent requires a 9 to 1 recirculation rate. The BOD_5 loading must be very light, i.e. 0.5 lbs per cu.ft. The organic carbon source that enhances the proliferation of the nitrifiers is in this BOD_5 loading. The critical contact time is only 10 seconds.

Presently prototypes of the test towers are being designed for both San Diego and Los Angeles, California.[5] The one for Los Angeles will handle up to 30,000 CFM of foul air, heavily laden with H_2S. These installations are adjacent to their respective treatment plants and are treating air that comes from the collection system.

This innovation in odor control may help solve odor problems in both large and small systems as there does not seem to be any physical limitation.

REFERENCES

1. POMEROY, R.D. and BOWLUS, F.D. "Progress Report on Sulfide Control Research," Sew. Wks. Jour. 18, 597 (1946)
2. WHITE, G.C. "Handbook of Chlorination," Van Nostrand Reinhold, New York 1972.
3. Private communication, Chas. Carry, Los Angeles County Sanitation District, 1972.
4. HICKS, R., "Odor Control in Wastewater Treatment Systems", paper presented at the International Air Pollution Meeting, Wash. D.C. (1970).
5. Private communication, T.V. Lutge, Brown and Caldwell Consulting Engineers, Alhambra, Calif. (1972).

Reply

Although Mr. White did not deal specifically with the subject of the paper, the aspects he presented had a severe import on the efficiency of treatment plant and the sludges which resulted from it.

The performance of sewerage systems had been given the greatest attention in London for many years. Nevertheless the earliest systems particularly in inner London suffered from the piecemeal design both historically in the last century. He attributed the absence of serious problems of septicity to the general climatic conditions which applied throughout the year. Evenly spread rainfall characteristics giving frequent scouring of sewers were helpful; temperatures never reached consistently high levels and the two factors together resulted in the discharge of relatively fresh sewage for treatment. The malaise of odour nuisance was rare in London. It coincided normally with exceptional temperatures and dry conditions. It was apparent that these two local factors must be given high priority in the development of any large scale sewerage scheme.

In the other areas where large regional schemes of sewerage and treatment had been produced in recent years conditions were even more satisfactory. Great attention had been paid to minimum velocities and to adequate ventilation of sewers. Provision had been made to augment minimum flows by river water should it be necessary; chlorination stations had been provided in the system to counter tendencies towards septicity. In neither case had use of these been required and the chlorination stations were now adopted for other purposes.

Corrosion problems had been negligible – recently a sludge main suspended in the soffit of a trunk sewer had been inspected in detail and no corrosion effects were evident in the main but careful attention needs to be given to the protection straps and anchor bolts and more particularly to those where movement of the main is allowed for at bends etc.

In regard to force mains, he assumed that Mr. White had referred to long rising mains involving, under certain conditions, extended times of retention. He could recall no mains of this type in the London system where gravity mains were generally supplemented by pumping stations having short vertical force mains discharging into a high level gravity system.

C.F. Seyfried (Germany)

Referring to the discussion, E.G. White stated that odours could be avoided by ensuring that the sewage received adequate aeration, by cleaning the sewers regularly with high pressure water and by preventing the admission to the sewers of trade effluents that had an offensive odour.

Odour nuisance could occur in sewers that were under pressure. Experiments had been carried out in Hannover for removing odours in pressure mains. The addition of sodium nitrate, chlorine, sodium hypochlorite proved to be effective only in large quantities that would have escalated costs. The chlorine dosage recommended by Dr. White for the raw sewage was of limited value except to the chlorine manufacturer.

To avoid septic conditions in long pressure mains only aeration under pressure gave satisfactory results.

Activated sludge was also found to be effective as a deodourising agent under these conditions. Foul air can either be admitted to an aeration tank or scrubbed in a tower with the air entering at the bottom flowing past the sludge dropped in from the top. This method has also proved to be useful in removing odours from the disposal of condemned meat and animal waste.

R.W. Goodwin, U.S.A.

What are the advantages and disadvantages of the suction secondary clarifier of the rectangular compared with the circular type?

What are the relative merits of waste activated sludge return to the primary clarifier compared with the combining of both the underflows? Apparently the former is a British practise as mentioned by L.B. Escritt. Does this technique include a primary clarifier modified with raking arms to function as a semi-thickener? The US has had excellent success with the latter technique of combining the primary and secondary underflows to a thickener prior to digestion. What is the author's view on sea disposal of undigested sludge?

Reply

The author has no experience of the performance of rectangular suction secondary clarifiers. They were not common in the UK.

The relative merits of returning surplus activated sludge to the primary settling tanks compared with combining it with primary sludge and dewatering were dependent, principally, on the degree of thickening achieved by the respective methods. Until recently technically sound and economic methods suitable for the large-scale dewatering of G.L.C. surplus activated sludge were not available. The sludge was returned to the primary settling tanks which were not usually fitted with thickening paddles or arms.

The employment of one of the activated sludge thickening devices described in the paper would enable the thickened sludge to be sent directly to the digestion tanks. It was expected that the solids content of the raw primary sludge would then increase and that the amount of surplus sludge would decrease resulting in an overall reduction in the volume of sludge produced. In addition, but of equal importance, the unstable conditions described in paragraph 2b of the paper should no longer occur. He was opposed to open-air primary sludge thickeners on the basis of odour nuisance.

As regards the disposal of undigested sludge to sea, he could only repeat that the G.L.C. were committed to sludge digestion for reasons already given. The disposal of undigested sludge to sea was aesthetically unacceptable and could prejudice public opinion against all sludge disposal to sea.

M. Bernard, France

What are the dimensions of the Crossness Works tanks?

What are the power requirements and separation periods of the aerators?

What are the volatile solids content of the digested sludge and the dry solids content of the sludge before and after aeration-settling process?

What is the main equipment in the tanks?

Reply

The dimensions of the tank at Crossness Works used for the experiments on dewatering digested sludge by aeration and subsequent settling were 60 ft by 100 ft with a sludge depth of about 9 ft. It was an existing flat bottomed rectangular tank and at no time has it been established that it was of optimum shape.

The tank contained two floating surface aerators each powered by 20 h.p. motors. Liquor was removed by a floating decanting arm and sludge through twelve 12 inch diameter outlets in the tank bottom.

An aeration period of 12 hours and quiescent settlement for 3 days followed by decanting of the supernatant liquor increased the dry solids from, typically, 25% to 4%. The volatile solids content was unchanged at about 1.5%. The results are fully reported in reference 7.

W.F. Garber, USA

The author points out that the treatment and disposal of wastewater solids is perhaps the greatest remaining problem now present for most works and that such solids will not simply disappear but must go to land, sea or air and really only to land or sea since incineration inevitably returns most of the components of environmental significance to the sea or the air. In Los Angeles over 90% of the Hg and Pb reaching the ocean now comes from the air. However, in the USA wastewater treatment authorities are almost precluded from any avenue but land disposal. Under the lash of heavy publicity by so called environmental groups and apparently regardless of excellent scientific data, the Environmental Protection Agency has by fist allowed only one avenue of disposal to be considered, that is land. This in the face of the fact that through high energy use the total environmental impact is greater. It is suggested that the wastewater treatment authorities in other countries be alerted to this and enter the public opinion field to prevent such political override of scientific data.

Reply

The author agreed with Mr. W.F. Garber on the need to cultivate public opinion. The G.L.C. ensured, by means of Press releases, Works Open Days, visits to Works, lectures, films and exhibitions, that its role in protecting the environment from pollution was well publicised.

In the UK it was generally considered that the disposal of sludge to land was both environmentally and economically sound providing that close control over the methods and rate of disposal was maintained and vigorous control was exercised on the disposal of industrial wastes.

ACTIVATED SLUDGE TREATMENT
OF SMALL WASTE VOLUMES

BRIAN L. GOODMAN
Director, Technical Services, Ecodyne Corporation
Smith & Loveless Division, Lenexa, Kansas

INTRODUCTION

Small activated sludge plants have been used in the United States for the treatment of domestic, commercial and industrial wastewaters for over forty years. For the past twenty-five years, the designs of these small capacity (<1.0 MGD) systems have featured low process loading (<0.1 lb. BOD_5/lb. MLSS) and clarifier surface overflow (<600 gal/ft^2/day) rates. Of the several variants of the basic activated sludge process, only extended aeration and two stage aeration flow sheets have been commonly employed in

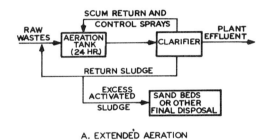

A. EXTENDED AERATION

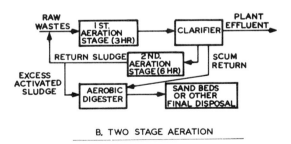

B. TWO STAGE AERATION

Fig. 1. Small plant flow sheets

the United States for the treatment of small daily volumes of wastewater, see Figure 1. Presently, some 10,000 such treatment systems are in operation in the United States. Several thousand others are located in other parts of the world. Small activated sludge plants are treating wastewaters North of the Arctic Circle, South of the Antarctic Circle, on many Pacific and Atlantic Ocean islands and in very nearly every country of the world. In view of the demonstrated usefulness of small activated sludge plants, it is indeed unfortunate that highly conservative design constraints have been widely imposed, first by custom and later by regulation, which seriously limit the capabilities of these plants. In 1947, it was noted that if sludge was not wasted from the activated sludge plant at

205

TABLE A.

Common Design Criteria for Small Extended Aeration Plants (U.S.)

1. **AERATION TANKS**

    A.   Volumetric BOD_5 Loading $\leqslant 15$ lbs/1000 ft.^{3}

    B.   Detention Time   24 hours

2. **CLARIFIERS**

    A.   Detention Time   4 hours

    B.   Surface Overflow Rate $\leqslant 300$ gal/ft^2/day

    C.   Weir Overflow Rate $\leqslant 5000$ gal/ft^2/day

3. **AIR REQUIREMENT**   1500 ft^3/day/lb BOD_5 removed

4. **SLUDGE RETURN RATE** $\geqslant$ Av. Daily Influent Flow Rate

NOTE: Detention Times and Rates all based on Design Average Influent Flow Rate

TABLE B.

MLSS Concentration (mg/l) along a Cylindrical Aeration
Tank (Total Length, 29 feet)

		Distance from Inlet (feet)			
DAY	5.8	11.6	17.4	23.2	AV.
0	2350	2220	1980	2450	2250
1	2520	2170	2250	2190	2283
2	2270	2090	2310	2220	2223
4	2130	1850	2170	2210	2090
7	2260	2080	2080	1820	2060
9	2490	2420	2380	2410	2425
14	2540	2320	2570	2830	2565
15	2620	2750	3050	3090	2878
16	2670	2770	3390	3030	2965
18	3390	3560	3470	3380	3450
22	3460	3230	3460	3810	3490
23	3610	3790	3830	3720	3738
24	3650	3650	4030	3750	3748
Av.	2766	2685	2844	2839	2782

East Palestine, Ohio, the mixed liquor suspended solids concentration increased at an ever decreasing rate until a point was reached at which the solids accumulation rate was just offset by the loss of solids in the plant effluent.[1] At equilibrium, the solids accumulation and oxygen uptake rates were sufficiently low that an adequate effluent quality for many discharge conditions was achieved as judged by the regulations then in effect. Since sludge wasting was not practiced, this activated sludge process variant was termed 'total oxidation'. This concept had obvious appeal in the treatment of small daily wastewater volumes since the disposal of excess solids presented a major problem, particularly if a large number of very small treatment facilities were contemplated. Increasing utilization and study of small 'total oxidation' systems coupled with an increasing attention to effluent quality, soon made it clear that mixed liquor suspended solids level control through sludge wasting must be practiced if 90% BOD5 and suspended solids removal was to be reliably achieved. As mixed liquor suspended solids operating levels were decreased in order to achieve reliable process performance, the aeration period was extended in order to maintain low (Lf) values and, consequently, low solids accumulation rates. By the late 1950s, the customary design criteria for these extended aeration plants had been largely formalized as regulations which have since remained essentially unchanged. The same situation exists with regard to design criteria for two stage aeration plants. Table A presents these formalized criteria as they presently exist in the United States; however, it should be appreciated that there exists a fair degree of variability and some conflict in these regulations from jurisdiction to jurisdiction. Similar regulations exist in other parts of the world.

During the past fifteen years, extensive researches have been reported which have established a rational basis for the design of small activated sludge wastewater treatment systems.[2,3,4] The process models are supported by the data presented here. It is, therefore, contended that existing design criteria should be revised to reflect established process fundamentals. The extended aeration design criteria will be utilized here in the development of this thesis.

EXTENDED AERATION

Complete Mixing

Small (2000 — 100,000 GPD) extended aeration wastewater treatment plants commonly consist of a horizontal cylindrical, rectangular, square or circular aeration tank together with any of a variety of final clarifiers. Primary clarification is not often employed. Solids return from the final clarifier is either by pump or by so-called 'gravity' sludge return, see Figure 2. Process models assume complete mixing of the aeration tank contents. Of the various aeration tank configurations commonly employed, it might be supposed that the horizontal cylindrical with sludge return at its effluent end would be least likely to display complete mixing as evidenced by uniform mixed liquor suspended solids concentration, residual dissolved oxygen concentration and oxygen uptake rate throughout the tank contents. Such an aeration tank was selected for test. It was ten feet in diameter, twenty-eight in length and had a center water depth of 9.75 feet. The plant clarifier was of the gravity sludge return type. Aeration and mixing of the tank contents resulted from the release of air at a depth of 6.25 feet from air diffusers spaced forty-eight inches on center along one side of the tank. Influent domestic waste was continuously introduced at one end of the tank and exited at the opposite end. The data of Table B resulted from the analysis of mixed liquor samples collected along the length

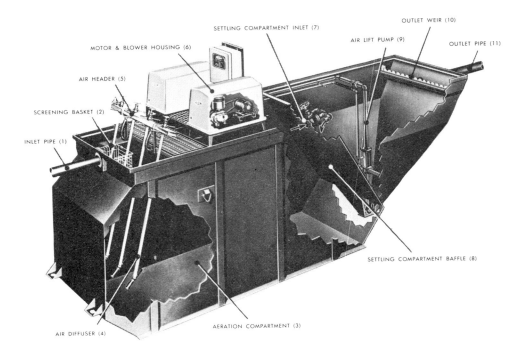

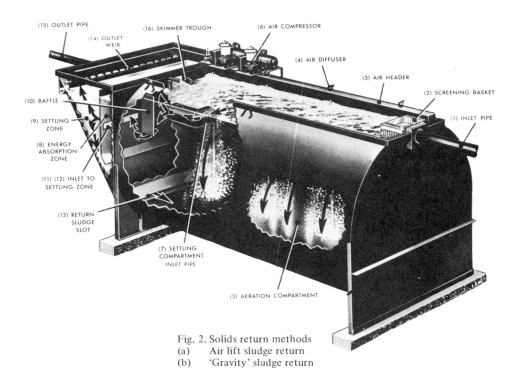

Fig. 2. Solids return methods
(a)   Air lift sludge return
(b)   'Gravity' sludge return

of the tank. These data indicate an essentially uniform suspended solids concentration in the tank with only a slight apparent tendency toward higher solids concentration toward the effluent end. This nominal variance simply indicates the effectiveness of backmixing. The contention that this tank was, for all practical purposes, completely mixed is further supported by the dissolved oxygen and oxygen uptake rate profiles of this same aeration tank, see Table C. These findings are not surprising in view of the fact that for small, cylindrical, rectangular, square and round aeration tanks, complete mixing conditions are routinely demonstrated.

TABLE C.

Dissolved Oxygen and Oxygen Uptake Rate Profiles
Cylindrical Aeration Tank

Distance from Inlet Feet	Residual Dissolved Oxygen at Peak Load mg/l	Oxygen Uptake Rate mg/l hour
0.67	0.6 ⎤	13.0 ⎤
4.67	0.7 ⎬ 0.60 Av.	12.0 ⎬ 12.3 Av.
8.67	0.5 ⎦	12.0 ⎦
12.67	0.6	—
16.67	0.8 ⎤	11.5 ⎤
20.67	0.6 ⎬ 0.63 Av.	12.5 ⎬ 12.0 Av.
24.67	0.5 ⎦	12.0 ⎦

BOD Removal

The removal of degradable organic matter in completely mixed activated sludge systems has been extensively studied and reported on by McKinney.[2,3] The rate of removal (K_m) can be determined from the relationship:

$$\frac{F_i - F}{t} = K_m F$$

Where:

F_i = Influent BOD_5, mg/l

F = Effluent Soluble BOD_5, mg/l

t = Aeration Period, days

Figure 3 presents the data of McKinney[3] together with that being reported here. These data suggest that the degradable organic matter, measured as BOD_5, is metabolized at the rate of approximately 203 mg/l oxygen demand per mg/l oxygen demand remaining per

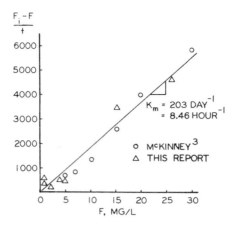

Fig. 3. Value of K_m at 20°C

day at 20°C. It should be recognized that the precision of the standard BOD test[5] is not great at low values of effluent soluble BOD_5. Since the reaction rates discussed here have been reported as approximately being doubled by a 10°C temperature increase,[3,6] they can be corrected to observed temperatures (T) through the relationship:

$$K_T = K_{20} (1.075 \, T - 20)$$

Solids Accumulation

The synthesis and subsequent oxidation of cell mass, measured as volatile suspended solids, has been determined[3,6] from a data plot of degradable organic matter removed (measured in terms of BOD_5) per unit of volatile suspended solids versus the aeration period employed. The synthesis of cell mass, measured as volatile suspended solids, has been determined to be approximately 0.7 mg/l per mg/l BOD_5 removed.[3,7] Thus, the active mass (M_a) synthesis rate would be 0.7 K_m. For low sludge age (t_s) values, the rate of cell mass oxidation (K_e) due to endogenous respiration has been evaluated as 0.48 day^{-1}.[3] Available data suggests that this rate is decreased to 75% of its former value with each doubling of sludge age.[8] Utilizing these values, K_e for any combination of observed t_s and T values can be determined from the relationship:

$$K_e = [0.48 - (0.12 \, \mathrm{Log}_e \, t_s)] \, 1.075 \, T\text{-}20$$

Measured as volatile suspended solids, 24% of the active mass oxidized remains as inert residue (M_e). Other inert materials also accumulate in activated sludge systems. About 0.1 ($M_a + M_e$) is inert inorganic matter. Also, influent inert organic and inorganic materials accumulate in the system. For domestic sewages in the United States, these inert fractions (f_i) appear to average, in total, 55% of the influent suspended solids.

The total solids accumulated (M_T) in a completely mixed, extended aeration plant is equal to the sum of the active, endogenous and inert masses.

$$\Delta M_a \quad = \quad \frac{M_a}{t_s}$$

$$\frac{M_a}{t_s} \quad = \quad K_sF - K_eM_a$$

$$M_a \quad = \quad t_s(K_sF - K_eM_a)$$

$$M_a \quad = \quad \frac{K_sF}{\dfrac{1}{t_s} + K_e}$$

$$M_e \quad = \quad 0.24\, K_eM_a\, t_s$$

$$M_i \quad = \quad f_i\,(SS\ inf)\frac{t_s}{t} + 0.1\,(M_a + M_e)$$

$$M_T \quad = \quad M_a + M_e + M_i$$

Clarification

The loss of suspended solids in the plant effluent can be approximated if the clarifier efficiency factor (f_e) is known, see Figure 4.

$$M_{Te} \quad = \quad f_e\, M_T$$

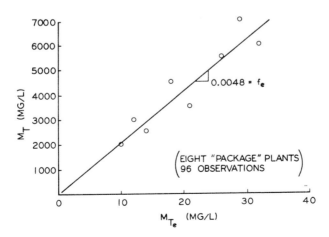

Fig. 4. Effluent suspended solids as a fraction of total mass

Full Scale Test Data

Test No.	Qa (gpd)	t (Days)	ts (Days)	T (°C)	f_e	K_m (Day^{-1})	K_e (Day^{-1})	Influent BOD (Mg/l)	Influent SS (Mg/l)
							General	Influent	
1	8,500	1.18	28	26	0.0048	313	0.12	190	240
2	8,500	1.18	28	26	0.0048	313	0.12	240	260
3	8,500	1.18	28	22	0.0048	235	0.09	200	260
4	8,200	1.22	20	14	0.0048	132	0.08	310	480
5	10,000	1.00	25	20	0.0042	203	0.094	145	165
6	15,000	1.00	40	25	0.0030	291	0.054	214	173
7	18,000	1.00	15	17	0.0055	163	0.125	303	274
8	36,000	0.50	20	18	0.0072	176	0.104	110	108

Test No.	Mixed Liquor MLSS Observed	MLSS Calculated	O$_2$ Uptake Observed	O$_2$ Uptake Calculated	Effluent SS Observed	SS Calculated	BOD$_5$ Observed	BOD$_5$ Calculated
1	4700	4559	-	-	14	22	10	3
2	5000	5201	-	-	18	25	3	3
3	5700	5047	-	-	18	24	3	4
4	6000	6389	-	-	33	31	7	6
5	3396	3566	-	-	12	15	6	3
6	7130	6948	-	-	20	21	8	4
7	3660	4014	11.8	12.4	16	22	6	6
8	3898	4015	7.9	8.9	32	29	14	6
Av.	4936	4967	9.9	10.7	20	24	7	4

TABLE E.

Calculated Values

General				Mixed Liquor				Effluent		Loading
t	t_s	K_m	K_e	MLSS	O_2 Uptake		ΔM	SS	BOD_5	L_f
(Day)	(Days)	(Day^{-1})	(Day^{-1})	(mg/l)	(mg/l/hr)	(mg/l/day)	(lbs/1000 gal. treated)	(mg/l)	(mg/l)	(lbs. BOD_5/lb. MLSS)
1.00	15	203	0.155	3053	8.6	204	0.170	15	3	0.07
0.67	10	203	0.204	3085	12.5	309	0.173	15	5	0.10
0.33	5	203	0.287	3268	23.8	654	0.180	16	6	0.18
0.17	3	203	0.348	3937	42.7	1312	0.186	19	9	0.30

NOTE: Influent— SS = 240 mg/l, BOD_5 = 200 mg/l, T = 20°C, f_e = 0.0048

Surface overflow rate controls the gross design of small activated sludge system clarifiers. Design surface overflow rate must be based on the values of L_f, peak flow (Q_p, gpd) and maximum M_T at which the system will be operated. Figure 5 relates the settling rate (V_s, ft/hour) found associated with various values of M_T[9]. Thus, the important clarifier design considerations can be correlated by the relationship:

$$A_s = \frac{Q_p}{V_s(180)}$$

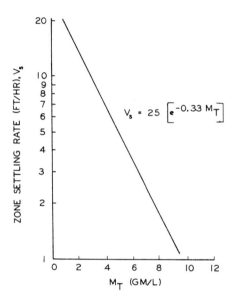

Fig. 5. Relationship of V_s to M_T

Effluent BOD and System Oxygen Requirement

The effluent BOD_5 concentration is determined as the sum of the unmetabolized influent substrate BOD_5 (F) and the five day oxygen equivalent of the degradable fraction of the effluent suspended solids.[3,10]

$$BOD_{5e} = F + 0.84\, M_{Te}\, \frac{M_a}{M_T}\, 0.76$$

The system oxygen requirement (lb/day) is the sum of the synthesis (O_{us}) and oxidation (O_{uo}) oxygen demands.

$$O_{us} = 0.5\,(F_i-F)\,8.34\,Q_a$$
$$O_{uo} = 1.415\,M_a\,V\,(8.34)\,(0.76)\,K_e$$

VERIFICATION OF MODEL

Table D reports the results of eight full-scale plant studies. The wastes treated were of domestic origin. The analytical methods employed were those specified by STANDARD METHODS.[5] Calculated values resulted from computations which employed the relationships presented in the preceding section of this paper. Each treatment plant was operated .at the specified sludge age and aeration period for a minimum of thirty days. Steady state influent flow was maintained. Composite sampling of the influent and effluent was utilized. Grab samples of the mixed liquor were taken. The mean of the daily values is reported in each category.

Generally, good agreement between observed and calculated values will be noted. Differences between observed and calculated effluent BOD_5 values are thought to be the result of the low precision associated with the BOD test[5] and the effect of nitrification occurring during the test in some instances. Considering the order of agreement, one could realistically expect it is submitted that the process model presented is verified by the data reported.

PRACTICAL APPLICATION OF THE MODEL

Presently, aeration tank detention times of about twenty-four hours are specified in many jurisdictions for small, completely mixed, activated sludge wastewater treatment plants. In terms of volumetric loading rate, this is about $12.5 - 15.0$ pounds BOD_5 per 1000 cubic feet of aeration tank volume if primary clarification is not practiced (based on common U.S. domestic raw wastewater values).

The model presented was employed to generate the data of Table E. From these data, it is clear that volumetric loadings of $2 - 4$ times or more than those presently utilized could readily be employed without adverse effect on plant performance.

In order to further test the model, a four month test was conducted utilizing an eight hour aeration period and a sludge age of ten days. The treatment system consisted of a 12,000 gallon horizontal cylindrical aeration tank and a gravity sludge return type clarifier (see Figure 2) with a surface area of 65 square feet. The wastewater treated was of domestic origin. The mean influent BOD_5 and suspended solids concentrations were, respectively, 184 and 160 mg/l. The mean observed values and calculated values for this test system were:

ITEM	OBSERVED	CALCULATED
MLSS, mg/l	4300	4696
M_{T_e}, mg/l	15	17
BOD_{5_e}, mg/l	9	6
O_2 Uptake, mg/l/hr	19	23

SUMMARY

A verified process model has been presented which is capable of predicting the performance of small, completely mixed, activated sludge wastewater treatment plants. This model indicates that considerably higher process loading rates than those commonly employed could be used without adversely affecting plant performance. Predicted performance, at a loading rate some three times that commonly employed in the United States in connection with small activated sludge systems, was closely approximated by the observed values. It is suggested that a considerable savings in treatment plant cost could be realized if higher process loading rates were adopted.

From an operating cost standpoint, it should be noted that the solids accumulation rate is increased only about 9%, pounds of solids accumulated per thousand gallons of waste treated basis, when the aeration period is decreased from twenty-four to four hours. On the same basis, the system oxygen requirement is reduced to about 60% of its twenty-four hour aeration period value.

Thus, in an era when required land area, capital costs and operating costs are of such importance, the opportunity to significantly reduce these factors should not be overlooked. This is especially true since it appears that such reductions can be made without sacrificing process efficiency.

REFERENCES

1. Anon. 1961. A study of aerobic digestion plants in Ohio 1959–1960. *Ohio Department of Health*, Columbus.
2. McKinney Ross E. 1963. Mathematics of complete-mixing activated sludge. *Transactions ASCE* 128:497.
3. McKinney Ross E. and Ooten Robert J. 1969. Concepts of complete-mixing activated sludge. *Bulletin of Engineering and Architecture No.60* University of Kansas, Lawrence.
4. Goodman Brian L. 1971. *Notes on Activated Sludge*. Smith & Loveless, Lenexa, Kansas.
5. 1971. *Standard Methods for the Examination of Water and Wastewater*. 13th Edition, Amer. Pub. Health Assn., New York.
6. Reynolds Tom D. and Yang J. T. 1966. Model of the completely-mixed activated sludge process. *Proceedings* 21st Industrial Waste Conf. Purdue Univ., Lafayette, Indiana.
7. Wuhrmann K. 1954. High-rate activated sludge treatment and its relation to stream sanitation. *Jour. Water Poll. Control Fed.* 26, 1.
8. Goodman Brian L. Unpublished data.
9. Goodman Brian L. 1966. *Package Sewage Treatment Plant Criteria Development, Part I: Extended Aeration*. National Sanitation Foundation, Ann Arbor.
10. McCarty Perry L. and Brodersen C. F. 1962. Theory of extended aeration activated sludge. *Jour. Water Poll. Control Fed.* 34, 1095.

Discussion *by* S.E. Jørgensen

To look after small plants requires at least 10 hours per week, not taking the sludge treatment into consideration. Biological treatment cannot be automated to the same extent as the physical-chemical treatment and therefore the cost of labour will always be relatively high for small biological treatment plants. I should like to hear the author's opinion about it.

Only physical-chemical treatment plants offer a satisfactory solution to the problems of small communities. This can be achieved by precipitation by calcium hydroxide followed by treatment on activated carbon, or precipitation by calcium hydroxide or aluminium sulphate followed by an ion exchange, using a combination of a cellulose ion exchanger, which removes organic matter and an anion exchanger removing phosphate, nitrate and some organic matter — and if desired zeolite for removal of ammonia.

I have been concerned with a plant in Sweden which is able to treat 70,000 gallons per day by precipitation and ion exchange. It gives a phosphate reduction of 99%, and suspended matter is reduced almost 100% and the BOD_5 is reduced by 95%. The running cost included labour is approximately 5 cents/m^3 or 18 cents/1000 gallons without regeneration of the chemical and 3,5 cents/m^3 or 13 cents/1000 gallons, when the chemicals are regenerated. The investment cost is reasonable. The plant costs 30,000 $. It is fully automated and is not sensitive to toxic material or sudden changes in loading. It requires a space of 3 x 3 x 2.5 m^3.

Admittedly the plant described by Dr. Goodman produces no sludge. In the plant I have mentioned this problem was solved by drying, which is a very expensive solution (the figures given above are exclusive of drying). On the other hand communities with 1200 inhabitants or less should more easily, than larger communities, be able to find area for sludge disposal.

Reply

Attempts to automate biological treatment plants have encountered significant difficulties due to the difficulty associated with the proper maintenance of the prime sensors. Also, as the complexity of the treatment system increases the opportunity for system failure increases. If sufficient complexity is introduced the probability of failure approaches certainty. The same considerations also pertain to physical-chemical treatment systems.

As suggested, biological treatment systems have failed in some instances. On the other hand, there have been many notable successes. For instance, a complete-mixing, two-stage, activated sludge plant at Addison, Illinois, U.S.A. achieves an effluent containing 7 mg/l suspended solids and 4 mg/l BOD_S. Dual media filtration of this effluent reduces these concentrations to 3 mg/l suspended solids and 2 mg/l BOD_S. I would suppose that chemical-physical treatment plants could not be considered failure free either and that process failure at either type of plant would, in most cases, be mainly attributable to operator failure either in an operations or maintenance sense.

In any case, I would not presume to contend that either system was suitable for all applications. I would prefer to view all of the processes and equipment available to us as potential candidates for incorporation into specific designs. Our problem is to make judicious selections from among the candidates in arriving at the optimum design solution to each individual treatment problem. In short, there is no universal solution.

Dr. Jørgensen implies that the achievement of zero solids accumulation in the biological system described herein is possible. Unfortunately, such is not the case. Some excess solids are always accumulated in biological treatment systems even with very high sludge ages. Indeed, I must point out that my paper contains an error in Table E where the $\triangle M$ values should have read 1.7, 1.73, 1.8 and 1.86 lb/1000 gal. treated. Even if no biologically inert solids were entering the treatment system an inert residue of cellular origin would accumulate in the system as stated in the paper where this residue (e) was shown to equal 0.24 K_eMat_S.

Arne Rosendahl, Norway

In the author's system sludge is returned by gravity and this is not controlled adequately. Sludge may accumulate in the sedimentation unit. Sludge also often rises to the surface. With the usual method of returning sludge the top of the sludge blanket may be very high, resulting in sludge loss when the hydraulic load is increased. What are the author's views on this? Concerning small physical, chemical plants I cannot agree with Mr. Jørgensen that they are easier to operate and less expensive than extended aeration plants. Experience in Finland and Norway shows that BOD and P can be removed by more than 90% by simultaneous precipitation by a once daily manual addition of alum or iron into the aeration tank. This can be used with extended aeration plants for up to 1,000 to 1,500 persons.

Reply

The author cannot agree that the return of sludge from the final clarifier described "is not controlled adequately." Sludge return from the clarifier described is by positive hydraulic means.

Sludge is driven from the clarifier by directed liquid flow. It is not "pulled" from the clarifier. Further, the sludge does not fall from the sludge return slot into the aeration tank. In short, sludge return is not passive, it is positive. The author acknowledges that so-called "gravity sludge return" clarifiers of somewhat similar appearance have failed to perform in the manner suggested by Dr. Rosendahl's question. It was for this reason that the "positive hydraulic sludge return" clarifier described here was developed. Sludge return from this unit is quite rapid and sludge losses to the effluent are quite low as indicated by the data presented (see: "Practical Application of the Model").

Barry Storch, UK

What is the significance of figures of 24 hours detention in aeration and 4 hours in settlement in Table A? Aeration tank loadings can be increased without detriment to performance but only if the sludge return and air supply are abundant. The complete mixing of the test plant is questioned since it is probably that complete mixing was only achieved in the small plant, with a very small throughput of low biological activity. It is suggested that if the size is increased and detention decreased the short-comings of this vital aspect will become apparent.

The mathematical connection between effluent quality from the clarifier and total solids in aeration is questioned since no connection appears to exist between solids in aeration and overflow rates upon which the author's calculations depend. At the high loading of 36,000 U.S.g.p.d. the clarifier overflow rate was constant at 550 gals/ft/day. One would expect the results to be good under these conditions. The similarity between observed and calculated results is therefore probably fortuitous. Can any reason be given for the low air requirement of 900 cu.ft. per 1lb BOD/day at the higher loading?

Reply

Table A was presented *only* to indicate the common design criteria imposed by the majority of U.S. State Regulatory Agencies. The author agrees fully with Mr. Storch's contention that aeration tank loading can be increased without process efficiency degradation. Indeed, that contention forms the central thesis of the author's paper.

Complete mixing activated sludge plants utilizing aeration periods of 3−5 hours and having capacities of as little as 2,000 US GPD have been extensively studied by the author who observed consistent BOD removals of 90 percent or greater.

The reader's attention is directed to Figure 4 of this paper in connection with the question raised concerning the clarifier efficiency factor. Also, the relationship presented in Figure 5 should be noted in this same connection. Since the sludge solids zone settling rate is shown to be a function of solids concentration (other factors being equal), it is not surprising that clarifier efficiency should be shown to depend on solids concentration (again, other factors being equal).

As to the matter of the lower oxygen demand at the higher loading rates, it should be noted that sludge age (t_s) decreases as the aeration detention time (t) decreases, see Table E. Thus, sludge wasting is at a more rapid rate and a smaller proportion of the endogenous oxygen uptake demand is exerted in the treatment system.

Ballay, France

Has the author been able to verify that the value of K_m used in the model, describing the removal of BOD is independent of the concentration of activated sludge? The author indicated that the sludge excess at a works in which the sludge was activated 4 hr per day was only 9% greater than at a works where it was aerated 24 hr per day. But to maintain a stable concentration of activated sludge, removal will have to be six or seven times more frequent than with continuous aeration. Does not the author think that the advantages of operating under conditions which provide fresh conditions compensate for the plant capacity required to attain these? What kind of sludge treatment does the author envisage for small plants with prolonged aeration and with activated sludge aerated occasionally?

Reply

The substrate removal rate, K_m. can be defined as:

$K_m = kX_v$ where : k = substrate removal rate per unit of mixed liquor volatile suspended solids, mg/l per mg/l per unit time. X_v = Mixed liquor volatile suspended solids, mg/l.

Thus, K_m is a constant while both k and X_v vary, k being highest when X_v is lowest. These relationships have been shown to hold for a very wide variety of wastes and waste combinations.

All of the studies reported herein employed continuous aeration, separation and sludge return means. The author does not advocate intermittent aeration. Such aeration is considered to be and has been frequently demonstrated to be highly detrimental to process efficiency.

J.A. Greives, USA

Has the author any experience of combining activated sludge and trickling filters as one process?

Reply

The author has studied biological wastewater treatment units which might best be described as "Fixed Activated Sludge" processes. Such a treatment unit has been in service on board the river tow boat M.V. Missouri for nearly a year. Effluent from this unit is recycled for use as toilet flush water. Continuous recycle periods of up to 90 days have been achieved. Effluent quality is equal to that achieved by advanced wastewater treatment units (BOD_s and suspended solids less than 5 mg/l). These "FAS" units represent a combination of activated sludge and trickling filter fundamentals.

Treatment system employing a trickling filter followed by complete-mixing activated sludge are also well known and frequently utilized in the treatment of high strength industrial-commercial wastes.

PARTIAL TREATMENT OF DOMESTIC SEWAGE BY HIGH-RATE BIOLOGICAL FILTRATION

A. M. BRUCE and J. C. MERKENS

Water Pollution Research Laboratory, Stevenage, Herts, United Kingdom

INTRODUCTION

It is well established that the biological filtration process may be employed either for complete purification of sewage or for any specified degree of partial treatment, given a suitable type of medium and the appropriate rate of application of the sewage in relation to other operational factors such as temperature. In theory, at least, use of the process for partial treatment at high loadings will produce maximum removal of BOD per unit volume of medium per day[1] without increasing operational costs. If, therefore, the mass of BOD load removed, rather than a high degree of purification, is the important requirement, high-rate filtration is often an attractive proposition. In practice, it is finding widespread use, particularly for relieving BOD overload on existing sewage works but also for other purposes; it is claimed[2] to offer substantial cost savings compared with alternative methods.

The trend towards high-rate filtration has undoubtedly been influenced strongly by the availability of proprietary plastics media of which a fairly wide variety of types is now marketed. For optimum design, the engineer requires reliable data on the relative performance capabilities of the different types of media, including the relatively cheap conventional mineral aggregates, and he also generally requires information on the way the other main design and operational variables affect performance.

To this end, long-term pilot studies on high-rate filtration of settled domestic sewage were started at the Laboratory in 1968, the main investigation involving a comparison of performances of different types of medium over a range of loadings. Preliminary results from these studies were published earlier[3]. The main part of this paper is devoted to summarizing the results obtained over a 3½-year period of investigation.

Reported briefly also are results of a shorter-term subsidiary study on high-rate filtration of comminuted unsettled domestic sewage on an open-structured plastics medium. This system of treatment may have certain cost advantages.

All the studies of high-rate filtration have included observations on the rates of production of 'humus' sludges and their dewatering characteristics, since these factors may often influence markedly the economics of a given treatment system The results are dealt with separately in the final part of the paper.

DETAILS OF INVESTIGATION

For the main investigation of partial treatment of settled domestic sewage, six pilot filters were operated at the Laboratory under various conditions for a 3½-year period. Each filter contained a different type of medium two being mineral aggregates of coarse

TABLE 1

Type and specific surface area of medium in each pilot filter, and nominal rates of application of settled sewage during successive phases of operation

Filter reference No.		F25	F26	F27	F28	F29	F30
Type of medium		Slag (75–125 mm)	'Flocor'(1)	'Crinkle'(2) close' ('Surfpac')	'Cloisonyle'(3)	'Surfpac'(2) Standard	Basalt (75–125 mm)
Specific surface area (m²/m³)		40*	85†	187†	220†	82†	40*
Phase	Period	Volumetric loading of sewage (m³/m³ d)					
I	≠ September 1968– June 1969	6	6	6	6	6	6
II	August 1969– July 1970	6	12	12	12	12	6
III	September 1970– August 1971	3	16	16	16	6R1	6
IV	October 1971– February 1972	2.25R1	4.45R1	9R1	9R1	6R2	6

* Determined by paint-dipping technique
† From manufacturer's literature
≠ Actual start-up in July 1968
R1 Recirculation of effluent (1:1)
R2 Recirculation of effluent (2:1)

(1) ICI Ltd   (2) Hydronyl Ltd   (3) Satec Ltd

grade (75—125 mm) and four being commercial plastics packings. Three of the latter were orderly formations of plastics sheets, while the other comprised vertical septate tubes. A detailed description of the filters and their media has been given elsewhere[4]. Fig. 1 outlines the main features of an individual pilot unit; some essential information about the media and operating conditions is given in Table 1.

The study comprised four successive phases of operation; the first three lasted about 12 months each in order that seasonal effects on performance could be observed. The fourth phase covered about half a year, including the late summer and winter. Throughout each phase settled sewage was applied to the filters at selected nominal rates (± 5 per cent) with no diurnal variation in flow.

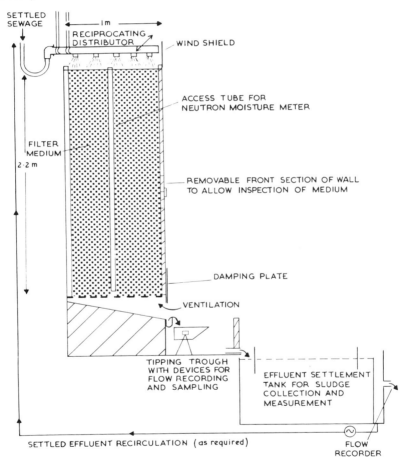

Fig. 1. Schematic diagram of single pilot high-rate filter and settlement tank.

Briefly, the scheme of investigation was:

Phase I All six filters were operated at the same sewage loading (6 m^3/m^3 d) to obtain a direct comparison of the relative efficiencies of the various media.

Phase II The sewage loading to the four plastics media was increased to 12 m^3/m^3 d, while that to the filters containing the mineral media remained at 6 m^3/m^3 d to provide a control. Subsequently, Filter 30 (basalt) continued to serve as control throughout the other phases of operation.

Phase III The sewage loading on three of the plastics media was raised to 16 m³/m³ d in order to compare performances at, and assess tolerances to, very high BOD loadings. The sewage loadings to F25 (slag) and F29 ('Surfpac' Standard) were reduced to 3 and 6 m³/m³ d respectively with, in the latter case, employment of 1 : 1 recirculation of settled effluent. Here the main objectives were (a) to test the prediction that, despite the very different types of medium involved, the degree of purification by the two filters would be comparable because sewage loadings were proportional to specific surface areas, and (b) to observe the effect of recirculation.

Phase IV Four of the filters were operated at sewage loadings which were lower than those in Phase III and, in the case of three of the media (slag, 'Flocor', and 'Crinkle-close'), the loadings were proportional to their specific surface areas; recirculation of effluent at 1 : 1 ratio was also employed. The main objective was to confirm, or otherwise, the expected similarity in the quality of effluent from each medium when operated at similar loads per unit specific surface area. It was also desired to establish whether a 'standard' of 150 mg/l for BOD of the settled effluents could be achieved in each case as predicted from previous results. The effect of an increased rate of recirculation on F29 was also observed.

Monitoring of performance

In all phases, composite samples of the settled sewage feed and of the filter effluents, respectively, were obtained over 20-h periods three times each week. Samples of effluent were settled quiescently for 30 min in 10-l vessels with a take-off point 100 mm below the surface. Appropriate sanitary analyses, normally 5-day BOD, COD (dichromate value), suspended solids (SS), anionic detergent, and total organic carbon, were carried out on the sewage and settled effluents. Determinations were also made of SS and organic carbon in samples of unsettled effluents.

Samples for bacterial counts (specifically for *E. coli* Type I and faecal streptococci) were collected over 1-h periods each week and processed immediately, using the membrane filtration technique[5].

Routine measurements were made of the degree of accumulation of biological film in the filters using a neutron moisture probe[6].

SUMMARY OF RESULTS FROM MAIN INVESTIGATION

A comprehensive summary of the conditions of operation and of the mean performances of the filters in each of the four phases is presented in Table 2. Most of the variability in the performance parameters, as indicated by the standard deviation, for a particular filter in a given phase is actually attributable to seasonal variations in purification efficiency. This is demonstrated in Fig. 2, which shows the monthly mean percentage BOD removals achieved by each filter in each phase together with the mean BOD and temperature of the sewage.

For ease of identification in Fig. 2, the results for the control (F30) have been plotted separately from the rest, and those for the other five filters have been segregated into two groups so as to avoid confusion of points. There is no basic significance in the groupings but it is convenient to be able to make a direct comparison of the removal efficiencies of F25 and F29 in Phases II and III when loadings were proportional to specific surface

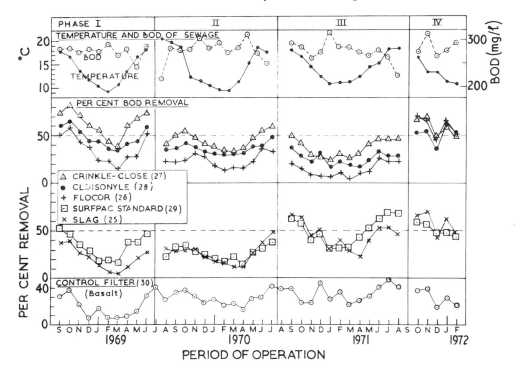

Fig. 2. Monthly mean temperature and BOD of settled sewage and percentage BOD removals achieved by each pilot filter during each phase of operation. (For sewage loadings to filters see Table 2).

areas, and also to compare directly the performances of the three other plastics media in Phases I to III when all received the same loading.

There was, generally (Table 2), a strong correlation at all levels between BOD and values for COD, organic carbon, suspended matter, and anionic detergent. Fractional removal efficiencies, in respect of COD and detergent, were consistently lower than those for BOD.

The results of bacterial counts, not detailed here, showed that the degree of removal of the pathogenic indicator organisms (referred to earlier) was normally similar to or less than that for BOD removal. Compared with ordinary-rate percolating filters, the high-rate plants were not effective in removal of fine (bacterial size) suspended matter from the sewage.

Over the 3½-year period of observations none of the media gave any indication of clogging. The mean level of accumulation of biological film was well below 20 per cent saturation of voids, and film was evenly distributed with depth. Throughout, average thicknesses of film were greatest in the mineral aggregate media, with values of up to 2 mm: average thicknesses of film in the plastics media ranged from 0.1 to about 1 mm. In the two filters containing the media with the higher specific surface areas (F27 and F28) there appeared to be some small, seasonal variation in film accumulation; this may have been related to the 'grazing' activity of macrofauna, as occurs in lower-rate filters, since *Psychoda* larvae and nematode worms were present in the warmer weather. There was no significant seasonal sloughing of film.

TABLE 2

Mean (μ), and standard deviation (σ), of various parameters relating to operation and performance of six high-rate filters, containing different media, in each of the four phases of the investigation

(a) Composition of settled sewage applied to filters

Phase	BOD		SS		COD		Organic carbon		Anionic detergent (as Manoxol OT)		Temperature (°C)	
	μ	σ	μ	σ	μ	σ	μ	σ	μ	σ	μ	σ
	(mg/l)		(mg/l)		(mg/l)		(mg/l)		(mg/l)			
I	277	43	170	34	553	65	175	26	24.2	3.6	14.8	3.2
II	281	46	188	35	578	76	187	27	28.2	5.3	14.5	4.2
III	277	56	188	41	538	79	179	33	27.6	3.8	14.5	3.8
IV	288	54	172	26	564	82	177	28	31.0	6.7	14.7	2.4

(b) Operation and performance of filters

Parameter	Phase	25 Slag		26 'Flocor'		27 'Crinkle-close'		28 'Cloisonyle'		29 'Surfpac'		30 Basalt	
		μ	σ	μ	σ	μ	σ	μ	σ	μ	σ	μ	σ
Settled sewage hydraulic loading (m³/m³ d) R = recirculation of effluent at 1 : 1 ratio	I	6.1	0.4	6.0	0.5	6.0	0.4	6.1	0.5	5.9	0.3	6.1	0.4
	II	5.9	0.3	11.8	0.5	11.8	0.5	11.8	0.5	11.7	0.8	6.0	0.4
	III	3.0	0.2	16.0	0.9	16.0	0.9	16.1	0.8	5.5R	0.5	5.9	0.4
	IV	2.2R	0.4	4.5R	0.4	9.0R	0.7	9.0R	0.5	5.9R	0.7	6.0	0.1
BOD loading (kg/m³ d)	I	1.71	0.3	1.71	0.3	1.71	0.3	1.71	0.3	1.65	0.3	1.71	0.3
	II	1.71	0.3	3.30	0.6	3.30	0.6	3.30	0.6	3.30	0.6	1.70	0.3
	III	0.80	0.2	4.48	1.0	4.49	0.9	4.50	1.0	1.54	0.4	1.61	0.3
	IV	0.62	0.2	1.30	0.2	2.60	0.5	2.50	0.5	1.70	0.3	1.71	0.3
Per cent BOD removal	I	20.9	16.5	35.4	18.5	61.6	15.4	48.6	14.8	33.8	16.7	17.9	16.3
	II	24.6	17.7	23.2	16.5	42.2	14.2	35.9	11.8	24.8	14.0	26.5	14.2
	III	48.5	18.0	14.3	16.9	34.3	15.3	26.1	15.8	50.4	16.8	34.3	16.1
	IV	62.1	18.4	61.7	15.8	58.7	17.8	52.3	14.5	51.9	16.4	31.5	22.1
BOD of settled effluent (mg/l)	I	218	50	177	51	106	43	140	39	183	47	225	47
	II	213	55	216	52	164	44	179	38	211	46	206	46
	III	141	53	235	58	180	48	201	44	137	52	196	40
	IV	113	44	112	35	121	42	137	29	137	33	196	40
SS of unsettled effluent (mg/l)	I	190	42	196	68	160	57	179	58	215	70	219	82
	II	193	29	211	51	200	71	189	34	204	46	189	28
	III	181	45	214	74	200	57	225	78	192	92	187	67
	IV	148	26	152	29	175	33	215	62	158	45	186	25
SS of settled effluent (mg/l)	I	147	33	133	36	94	36	115	30	142	36	159	35
	II	147	31	156	30	125	31	129	27	150	27	151	26
	III	112	30	160	40	131	32	139	34	120	48	142	32
	IV	96	23	102	29	107	29	115	19	119	32	148	33
COD of settled effluent (mg/l)	I	449	78	390	72	280	74	331	64	407	73	477	72
	II	444	71	459	73	383	80	403	68	455	68	456	72
	III	322	73	460	82	385	79	413	72	324	87	409	71
	IV	265	48	268	49	282	53	303	36	314	61	419	51
Anionic detergent in settled effluent (mg/l as Manoxol OT)	I	20.4	2.7	19.0	2.6	14.2	3.1	15.8	2.8	19.2	2.5	21.9	3.5
	II	22.9	4.1	24.2	4.1	21.8	3.8	21.1	3.8	24.3	4.3	24.3	4.3
	III	18.7	3.6	24.9	3.5	23.1	3.3	22.4	3.3	19.8	3.5	23.9	3.8
	IV	17.3	5.8	16.8	5.4	17.6	5.5	16.6	4.3	19.5	5.3	27.8	6.2
Organic carbon in unsettled effluent (mg/l)	I	154	28	143	31	110	29	126	27	150	31	169	32
	II	157	25	174	30	148	30	149	22	166	26	160	26
	III	126	29	180	47	150	30	171	45	134	50	151	39
	IV	86	23	83	19	87	18	98	17	100	18	131	21

COMPARATIVE TREATMENT EFFECTIVENESS

As shown by Fig. 2, monthly average BOD removals ranged, overall, from just above 80 per cent to less than 10 per cent according to the particular sewage loading, the type of medium, and the season. For a given medium increased loading gave, as expected, decreased fractional BOD removal but increased BOD weight removal per unit volume of medium.

In general, BOD removals were closely related to specific surface area. The main anomaly was the rather lower efficiency in relation to surface area of the tubular medium (F28) although it should be stressed that this medium was second in order of effectiveness throughout the investigation.

The seasonal variations in performances of the filters usually corresponded fairly closely to the changes in sewage temperature, though there were some exceptions to the general rule. In particular, the performance of F26 in Phase II and of F30 in Phase III showed, in some months, an unexplained reversal of the normal temperature response.

The mean BOD removal efficiencies of F25 and F29 during Phases II and III, when each received a sewage load in proportion to the specific surface area of its medium were not significantly different (on the basis of a T-test). The effect of recirculation on performance of F29 in both Phases II and III was not significant, when account was taken of the performance of the control (F30) during the same periods. It is important to note that although operated under the same sewage loading throughout, the mean BOD

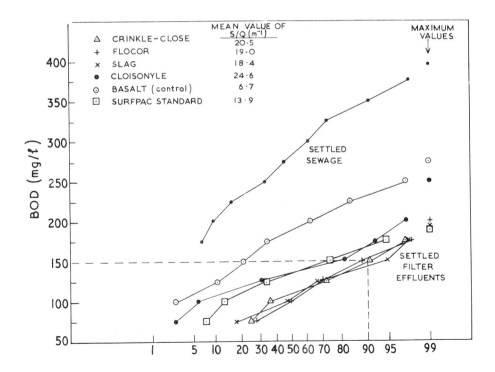

Fig. 3. Distributions of BOD values of settled sewage and of settled effluents from pilot high-rate filters during phase IV (Arithmetical probability plots).

removal efficiency of the control filter improved significantly from phase to phase. This may have been the result of 'maturation', or possibly differing weather conditions but the observation is still largely unexplained.

The results of Phase IV provide strong confirmation of the prediction (from the earlier results) that similar purification efficiencies would be achieved from the three different types of medium (slag, 'Flocor', and 'Crinkle-close') when loaded at equivalent rates per unit specific surface area of medium. Fig. 3 shows the probability plots of the BOD values of the sewage and of the settled effluents from all the filters during Phase IV. The close grouping of the results for the three filters (F25, F26, and F27) is evident and statistical tests confirmed that performances were not significantly different. It is also noteworthy that, as shown by Fig. 3, over 90 per cent of the BOD values of the samples of effluent from the three filters were of the anticipated 150 mg/l or less. In fact, performances of all the filters in Phase IV were rather better than expected on the basis of experience from the previous phases; this may have been due to the comparatively mild winter conditions.

MATHEMATICAL MODEL OF PERFORMANCE

In an attempt to explain, by a single mathematical expression, the overall spectrum of performances shown by the filters over the whole investigation, various theoretical and empirical approaches have been tested. So far, the most satisfactory model is based on first-order kinetics for rate of decrease of BOD with, in effect, two independent variables – volumetric loading of sewage and specific surface area – as the 'residence time' determinants[7-12]. In the most familiar integrated form the general equation is

$$\frac{L_E}{L_S} = \exp\left(-K_T S^a Q^{-b}\right) \tag{1}$$

where L_E and L_S are the BOD values of the sewage and of the settled effluent from the filter, respectively, S is the specific surface area of the medium (m^2/m^3), Q is the sewage volumetric loading (m^3/m^3 d), K_T (m d) is the exponential rate constant for the particular sewage at temperature $T^{\circ}C$, and a and b are constants.

The depth of filter is not included in the model, since it was not a variable in the studies; in any case, the present authors[4] and others[13,14] have shown that performance at a given sewage volumetric loading is not influenced significantly by depth over a range 2–5 m.

Examination of the data indicated that within the experimental range, it is appropriate to assign values of unity to both indices a and b, although it is recognized that values of less than one, and as low as 0.5, have been used in many other models of this type.

To account for seasonal variations in performance, it was necessary to incorporate a temperature function in Equation 1 in order to vary K_T with sewage temperature; for simplicity it was decided to employ a function of the type originally proposed by Streeter and Phelps[15] and later used by Howland[16] for biological filters, this being

$$K_T = K_{15} C^{(T-15)} \tag{2}$$

where K_{15} is the value of the rate constant at $15^{\circ}C$, and C is a temperature coefficient.

The temperature of 15°C, rather than the more usual 20°C, was selected as the reference temperature since it is closer to the annual mean temperature of the sewage used in the investigations, and also of that in the UK generally.

Substituting for K_T in Equation 1 gives the more definitive model

$$\frac{L_E}{L_S} = \exp\left(\frac{-K_{15}C^{(T-15)}S}{Q}\right). \tag{3}$$

With the aid of a computer, 'best-fit' values of K_{15} and C were calculated for each filter in each phase and also for groups of phases, using regression analyses incorporating all observed values of L_E, L_S, T, and Q, S being a constant for a given filter. Most regression analyses, except those for Phase IV, involved well over 100 sets of data.

The value of C, the temperature coefficient, was fairly consistent from filter to filter in a given phase, but varied to a greater extent between phases, being, on average, highest, at 1.14, during Phase I[3], and lowest, at 1.05, during Phase II. In a few cases, the estimated value of C for a particular filter and phase was less than 1.02 (e.g. F26 in Phase II, and F28 in Phase IV); this apparently very low temperature coefficient is probably mainly attributable to the previously mentioned occurrence of certain periods within the phase when performance actually improved while sewage temperature fell, thus tending to 'flatten' the normal response in the regression analysis. Such reversals may have been caused by many factors but no complete explanation is available. Published data on the effects of temperature on biological processes usually show a good deal of scatter and further studies are indicated.

In the present long-term study, the mean value of C for all phases and all filters was 1.08 which represents a 2.2-fold change in K_T from T = 10 to T = 20°C. The value is higher than that of 1.047 found by Streeter and Phelps[15] to describe biological oxidation processes generally, and that of 1.035 derived by Howland[16] for biological filters operated at a lower rate.

Fig. 4 indicates the values of K_{15} derived from the regression analyses for each filter

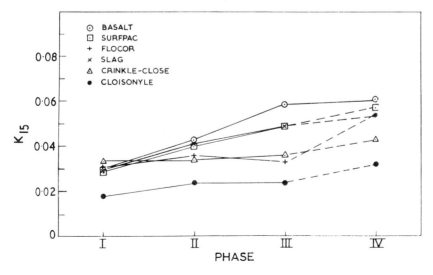

Fig. 4. Computed values of K_{15} for the various high-rate filters during each phase of operation.

and phase. As expected, from the earlier discussions, the value of K_{15} for the 'Cloisonyle' medium (F28) was consistently lower (by 30–40 per cent) than values for the other media; this reflects the relatively lower effectiveness of 'S' in the vertical tubular configuration rather than a real decrease in K_{15}.

The value of K_{15} for F30 (control) increased steadily from phase to phase. This observation is without obvious explanation except in terms of 'maturation' and, possibly, milder weather conditions. However, it should be stressed that at the relatively low level of BOD removal efficiency at which F30 operated, relatively small changes in L_E/L_S would effect large changes in K_{15}. Values of K_{15} for the remaining filters were reasonably consistent, although all showed an increase in Phase IV.

The effect of recirculation of effluent is uncertain, since no significant change in K_{15} was observed with those filters employing recirculation as compared with that for the control which did not have recirculation. A more definitive experiment is, perhaps, indicated to assess the influence of recirculation. The literature is directly contradictory on the matter; although the present theory suggests that recirculation could be detrimental to or, at least, have no influence on performance[14,17] some other authors[18, 19] indicate fairly substantial improvements – though these may possibly be attributable solely to improved uniformity of distribution of liquid over the surfaces of the filters[20].

The mean value of K_{15} for all filters (excluding F28) in all phases is 0.037, so that, together with the mean value of C given earlier, we may write Equation 3 as

$$\frac{L_E}{L_S} = \exp\left(\frac{-0.037 \times 1.08^{(T-15)} \times S}{Q}\right) ; \qquad (4)$$

this represents the average relation for the present investigation. Equation 4 may be regarded as fairly 'conservative' for design purposes and may need to be refined in the light of further investigation. It is of interest to note that results obtained from a 12-month period of operation of another pair of filters treating settled sewage at rates of 3 and 6 m^3/m^3 d give values of K_{15} of 0.053, and of C of 1.08. On the other hand, data from high-rate filters operated at Minworth sewage works[21] gave values of K_{15} from 0.024 to 0.039.

COMPARISON WITH OTHER MODELS

Fig. 5 shows, for the stated set of basic operating conditions of a hypothetical filter, the range of performance curves predicted by various published filter 'formulae' including that by Equation 3 for two values of K_{15}. It is seen that even with K_{15} equal to 0.053 the performance predicted by Equation 3 is not as good as that by most other models. There are many possible reasons for the very wide variation in predictions. Not all the published equations include all the variables indicated, and some have also been derived from investigations with types of sewage other than domestic, and the degree of pre-treatment may also have been variable. Other reasons have been suggested elsewhere.[12]

The very wide spectrum of predictions, however, highlights the difficulty often facing the design engineer in deciding on design requirements. While it is optimistic to expect that a single expression could ever describe all situations, it is important to achieve some reduction in the present rather wide confidence limits.

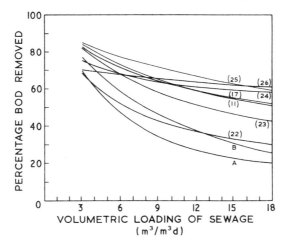

Fig. 5. Relation between volumetric loading of settled sewage and BOD removal efficiency as given by various mathematical models. Conditions (where applicable): depth of filter, 2m; specific surface area of medium, 100 m²/m³; temperature of sewage, 15°C. A, present paper (K_{15}=0·037); B, present paper (K_{15}=0·055); numbers on other curves are literature references.

FILTRATION OF COMMINUTED CRUDE SEWAGE

One unique advantage claimed in respect of some open-structured plastics media is their suitability for treatment of unsettled comminuted sewage, thereby eliminating the normal requirement for a primary settlement stage. Settlement is required only after filtration, and capital costs are therefore reduced. As a consequence, however, all of the sludge produced is of a secondary nature.

Studies of this type of filtration were carried out using a pilot filter (0.6 x 0.6 m in cross section and 2.4 m in depth) containing 'Flocor' medium (specific surface area, 85 m²/m³).

Comminuted sewage, of the same domestic origin as the settled sewage used for the other studies, was applied at a rate of 12 m³/m³ d for a 12-month period. Subsequently, the rate of application was reduced to 9 m³/m³ d for a 2-month period. Before passing on to the medium, the unsettled sewage passed through a plate perforated with holes of 10-mm diameter. The plate assisted uniformity of distribution and also served to remove coarse solids and rag which otherwise would have fouled the surface of the filter and caused clogging. Film accumulation was observed to be rather higher than that in filters treating settled sewage, with up to 20 per cent occupation of voids, but no clogging occurred.

The BOD removal efficiency of the filter was in line with expectation from the performance of the 'Flocor' pilot filter treating settled sewage (Fig. 6). However, the BOD of the comminuted sewage after settlement was some 10–15 per cent higher than that of ordinary settled sewage from the same source so that a slight disadvantage was incurred by comminution.

The general experience with pilot- and full-scale filters treating unsettled sewage has been that it is a feasible process but in view of the increased problems of blockages of the distribution system and occasionally of the medium, some form of fine free-screening or even short-term settlement seems to be worthwhile in order to minimize operational problems.

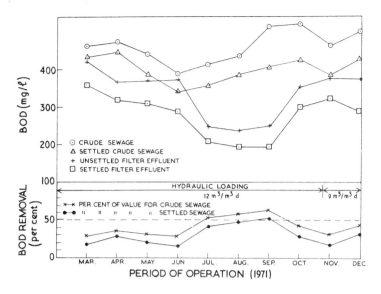

Fig. 6. Monthly average BOD values of sewage and effluents, and BOD removals achieved by a pilot filter treating comminuted crude sewage.

SLUDGE PRODUCTION AND FILTRABILITY

During the first three phases of the main investigation, weekly determinations were made of the quantities of humus sludge deposited, over a 24-h period, in the settlement tanks serving the pilot filters. Estimates were also obtained of the specific resistance to filtration of these sludges using the capillary suction time technique[27]. The coagulant demands and resistances to shear of the sludges were also determined regularly by the method of Gale and Baskerville[28].

The rate of humus sludge production by the various filters was fairly consistent in relation to BOD removal, and was generally in the range of 0.6-1 kg dry solids/kg BOD removed (Table 3). Taking all the filters and all the phases together, the average rate of sludge production was 0.77 kg/kg; much the same value has been found for other high-rate filters[20] but it is much greater than the rate of sludge production observed from a low-rate percolating filter (0.22 kg/kg), though rather less than that observed from an activated-sludge plant (0.8-1 kg/kg), in both cases for full treatment of the same sewage. The normal phenomenon observed with low-rate filters of a marked seasonal variation in sludge production was not noticed with the high-rate filters — as evidenced also by the relatively constant level of film accumulation in the respective media.

All high-rate sludges showed consistently high values of specific resistance to filtration (Table 3); even after the addition of coagulant, filtrability was not reduced to a level suitable for mechanical dewatering (i.e. 10^{12} m/kg), as shown by Fig. 7. The conditioned sludges were also susceptible to breakdown under storage and moderate shear forces. By comparison, the humus from a low-rate filter treating the same sewage had much better dewatering characteristics (Fig. 7) and was readily conditioned by coagulant.

Trial pressings of the conditioned humus sludges with a pilot filter press confirmed the difficulty in mechanical dewatering predicted by the filtrability measurements and a suitable cake could not be obtained even with a high dosage of chemical. However, much

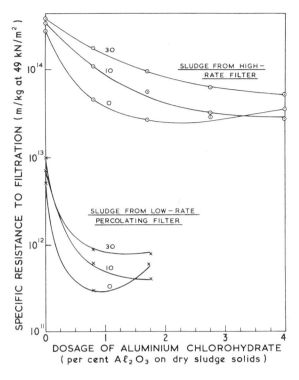

Fig. 7. Filtrability of humus sludges from high-rate and low-rate filters after addition of various amounts of conditioner and subjection to various degrees of shear. Average results for 12-month period. Values against curves indicate period of high-speed stirring (s) after coagulant addition.

TABLE 3

Mean rates of sludge production and filtrability of sludges from each filter during Phases I, II, and III

Results are means of weekly determinations

	Phase	F25 Slag	F26 'Flocor'	F27 'Crinkle-close'	F28 'Cloisonyle'	F29 'Surfpac' Standard	F30 Basalt
Sludge produced (kg dry matter/ m³ d)	I	0.28	0.45	0.61	0.47	0.43	0.24
	II	0.37	0.63	0.82	0.79	0.66	0.33
	III	0.29	0.72	1.04	1.24	0.59	0.37
Sludge produced (kg dry matter/ kg BOD removed)	I	1.05	0.81	0.63	0.64	0.75	0.93
	II	0.87	0.83	0.59	0.67	0.80	0.73
	III	0.74	0.97	0.65	1.00	0.75	0.61
Specific resistance to filtration of sludge (10^{12} m/kg) at 49 kN/m²	I	150	180	190	200	200	140
	II	150	180	250	180	230	170
	III	190	270	370	300	380	210

less difficulty in filter pressing was observed with the chemically conditioned humus sludge from the high-rate filtration of crude comminuted sewage, and a firm satisfactory cake was obtained. Even so, the chemical requirements and the pressing time were higher than required for a primary sludge produced by the same raw sewage.

It is important to note that others[29] have reported less difficulty with high-rate filter sludges; possibly, the situation may vary with the character of the sewage. Further investigations are required on this matter. Although in many cases the characteristics of the sludge will be of no particular concern, in many other cases it will be important to determine the likely effects on the economics of sludge disposal before deciding to employ high-rate filtration. The same would apply, of course, to other forms of high-rate biological treatment.

SUMMARY AND CONCLUSIONS

The points of most general importance which have emerged from the foregoing studies are, briefly, as follows:

1. The specific surface area of the filter medium was a major factor determining the performance of the high-rate filters.
2. Plastics-sheet media were of comparable efficiency to coarse mineral media in relation to specific surface area. The vertical tube configuration has been found to be somewhat less efficient than the other configurations in relation to specific surface area.
3. For high-rate filters operated in the open, BOD removal efficiency generally varied markedly with season, though the magnitude of the variation itself varied from one year to another, and was not solely dependent on sewage temperature. Design considerations should always allow for seasonal effects on performance.
4. Recirculation of effluent is of uncertain value, although it may usefully improve the uniformity of distribution of liquid over the medium. More definitive experiments are required to resolve the matter.
5. A simple empirical mathematical model, based on first-order reaction kinetics, has been derived from the data to relate BOD removal efficiency to sewage loading, specific surface area, and sewage temperature.
6. Treatment of comminuted unsettled sewage by a filter containing an open-structured plastics medium is feasible, but presents certain operational difficulties. It would appear generally worthwhile to effect some preliminary solids removal, by either short-term settlement or fine screening, before filtration.
7. Sludges produced by high-rate filters treating settled domestic sewage have been found to be extremely difficult to dewater satisfactorily by mechanical means even after full chemical conditioning. Sludges from filters treating comminuted crude sewage were found to dewater rather more easily, but not as easily as primary sludges from the same sewage.

ACKNOWLEDGEMENT

The authors' thanks are due to the many other members of the staff of the Laboratory who assisted in various ways with the investigation.

REFERENCES

1. Bruce, A. M., Merkens, J. C., and MacMillan, S. C. Research developments in high-rate biological filtration. *Instn. publ. Hlth Engrs J.,* 1970, **49,** 178
2. Summers, T. H. Plastics filter medium would make water recovery profitable. *Process Engineering,* March 1971, 58
3. Bruce, A. M. Some factors affecting the efficiency of high-rate biological filters. *Proc. 5th Int. Conf. Wat. Pollut. Res.,* 1970, II–14
4. Bruce, A. M., and Merkens, J. C. Recent studies of high-rate biological filtration. *Wat. Pollut. Control,* 1970, **69,** 113
5. Windle Taylor, E., and Burman, N. P. The application of membrane filtration techniques to the bacteriological examination of water. *J. appl. Bact.,* 1964, **27,** 294
6. Harvey, B. R., Eden, G. E., and Mitchell, N. T. Neutron scattering: a technique for the direct determination of the amount of biological film in a percolating filter. *J. Proc. Inst. Sew. Purif.,* 1963, 581
7. Howland, W. E. Flow over porous media as in a trickling filter. *Proc. 12th Industr. Waste Conf.,* Purdue Univ., 1958, 435
8. Sinkoff, M. D., Porges, R., and McDermott, J. H. Mean residence time of a liquid in a trickling filter. *J. Sanit. Engng Div., Am. Soc. civ. Engrs,* 1959, **85,** SA6, 51
9. Atkinson, B., Busch, A. W., and Dawkins, G. S. Recirculation, reaction kinetics, and effluent quality in a trickling filter flow model. *J. Wat. Pollut. Control Fed.,* 1963, **35,** 1307
10. Roesler, J. F., and Smith, R. Trickling filter model: Design and cost factors. *Ind. Wat. Engng,* 1969, **6,** 46
11. Eckenfelder, W. W., and Ford, D. L. Water Pollution Control: Experimental Procedures for Process Design. Jenkins Book Publishing Co., Austin and New York, 1970
12. Bruce, A. M., and Boon, A. G. Aspects of high-rate biological treatment of domestic and industrial waste waters. *Wat. Pollut. Control,* 1971, **70,** 487
13. Audoin, L., Barabe, J. P., Brebion, G., and Huriet, B. The use of plastic material as a medium for trickling filters treating domestic sewage. *Proc. 5th Int. Conf. Wat. Pollut. Res.,* 1970, Paper II–16
14. Askew, M. W. High-rate biofiltration: past and future. *Wat. Pollut. Control,* 1970, **69,** 445
15. Streeter, H. W., and Phelps, E. B. A study of the pollution and natural purification of the Ohio River. III. Factors concerned in the phenomenon of oxidation and reaeration. *Publ. Hlth Bull., Wash.,* 1925, No. 146, p.7
16. Howland, W. E. Effect of temperature on sewage treatment processes. *Sew. Industr. Wastes,* 1953, **25,** 161
17. Germain, J. E. Economical treatment of domestic waste by plastic medium trickling filters. *J. Wat. Pollut. Control Fed.,* 1966, **38,** 192
18. Stack, V. T. Theoretical performance of the trickling filter process. *Sew. Industr. Wastes,* 1957, **29,** 987
19. Hanumanulu, V. Performance of deep trickling filters by five methods. *J. Wat. Pollut. Control Fed.,* 1970, **42,** 1446
20. Bruce, A. M., and Merkens, J. C. Further studies of partial treatment of sewage by high-rate biological filtration. Paper presented to Institute of Water Pollution Control, East Midlands Branch, May 1972
21. Tomlinson, T. G., and Hall, H. Some factors in the treatment of sewage in percolating filters. Public works and Municipal Services Congress 1950, Final Report, 600
22. Schulze, K. L. Load and efficiency of trickling filters. *J. Wat. Pollut. Control Fed.,* 1960, **32,** 245
23. Lamb, R., and Owen, S. G. H. A suggested formula for the process of biological filtration. *Wat. Pollut. Control,* 1970, **69,** 209
24. Eckenfelder, W. W. Trickling filtration design and performance. *J. sanit. Engng Div., Am. Soc. civ. Engrs,* 1967, **87,** SA4, 33
25. Tucek, F., Chudoba, J., and Madera, V. Unified basis for design of biological aerobic treatment processes. *Wat. Res.,* 1971, **5,** 647
26. Galler, W. S., and Gotaas, H. B. Analyses of biological filter variables. *J. Sanit. Engng Div., Am. Soc. civ. Engrs,* 1964, **90,** SA6, 59
27. Baskerville, R. C., and Gale, R. S. A simple automatic instrument for determining the filtrability of sewage sludges. *Wat. Pollut. Control,* 1968, **67,** 233
28. Gale, R. S., and Baskerville, R. C. Polyelectrolytes in the filtration of sewage sludges. *Filtrn. & Separn.,* 1970, **7,** 37
29. Joslin, J. R., Sidwick, J. M., Greene, G., and Shearer, J. R. High-rate biological filtration: a comparative assessment. *Wat. Pollut. Control,* 1971, **70,** (4), 383

Discussion *by* G. Rincke, Germany

There are three points to which attention should be drawn:— 1. Limitations of the various parameters. 2. Effects of the filter packing materials used. 3. The fields of application of high rate filters.

1. The depth of filter required is dependent upon the extent of decomposition or oxidation of the organic matter and the concentration of organic matter in the wastewater and therefore the organic load B_R (in Kg BOD m^3/d) and the hydraulic load QF in m^3/m^2/h have to be properly established.

Using the low filter depth of 2.2 m one is obliged to use a low hydraulic rate i.e., 3-12 m^3/m^3/d as the volumetric load (see Table and Figure 5), and $0.27 - 1.08$ m^3/m^2/h in which region according to our experience a satisfactory volumetric distribution is not obtained.

In our previous work, reported at the 5th IAWPR Conference in San Francisco, it was shown that the optimum hydraulic maximum is 2–6 m^3/m^2/h. It is not possible to make a precise statement on the hydraulic load if this is increased by recirculation.

2. Packings. The authors refer to the low efficiency per unit area using Cloisonyle media. With this type of packing especially the specific surface area of $225m^2/m^3$ given by the manufacturers is not fully effective since the contact area of the tubes and the acute angular portions between the tubes cannot be fully utilised. In other vertical tubes of hexagonal or square cross section all the surface area available can be used.

3. Application. Under European conditions the application of these materials is as yet somewhat limited. For the partial purification of domestic sewage coarse media 40-80mm in diameter is adequate and there is no need to replace conventional media by more expensive plastic media. The more favourable cost-benefit given in reference 17 by Germain has not been realised in Europe.

However, the use of non-clogging filters for concentrated trade wastes e.g., from the food industry is very interesting and on economic grounds is often superior to other processes. Many examples of such plants exist in the United Kingdom.

It is desirable to check the applicability of formula 3 for ordinary trade effluents and to quantify the values for k_{15} and C.

Authors' replies to Discussion by G. Rincke

We have not obtained any conclusive evidence that surface loading rates as high as 2–6 m^3/m^2 h are necessary to bring about optimum performance. The maximum rate employed in our studies was 1.4 m^3/m^2 h, and, as clearly shown by the results, the efficiencies of the filters were not intrinsically better at this loading than at lower loadings. It is very probable that the optimum surface hydraulic loading for a given filter depends on the actual method of distribution and also on the specific surface area of the medium. A moving-arm distributer, as used in our studies, imposes a local rate of application much higher than the average rate of application for the entire surface of the filter. The loadings quoted by Dr. Rincke apply to static distributors and are certainly not valid for moving distributors which give effective wetting at relatively low rates.

It is agreed that a proportion of the surfaces of the "Cloisonyle" medium cannot be fully effective where they touch upon adjacent surfaces. This factor might reduce the effective specific surface area by about 10 per cent. However, we are of the opinion that the major factor causing the relatively low efficiency per unit area of this medium is the entirely vertical orientation of the tubes.

We would regard mineral media of 40–80 mm as rather too fine for high-rate filters treating strong sewage. In the UK, media of 70–150 mm size is preferred in order to reduce the risk of clogging. The costs of using mineral media for sewage treatment are certainly often more favourable than those involving plastics media, but in special cases plastics media are justified. Certainly, the use of plastics media in filters for partial treatment of strong industrial waste waters is frequently the most attractive choice on both economic and technical grounds. We hope to investigate the treatment of various trade wastes to determine if our mathematical model is generally applicable.

L. Holmberg, Denmark.

Plastic balls as a filter in biological trickling filters should be considered. Efficiencies of up to 22 times that of blast furnace slag have been obtained.

Reply

The suggestion by Mr. Holmberg that plastics balls should be used as media in biological filters is *not* a very attractive one on theoretical grounds. The sphere has the lowest surface area per unit volume of any shape and also the cost of manufacturing plastic spheres seems to be relatively high. It is difficult to find an explanation of the claim that spheres are much more efficient than slag, and further details of the comparison would be necessary before commenting further.

THE PROBLEMS OF WASTE WATER PURIFICATION IN THE CHEMICAL PHARMACEUTICAL INDUSTRY

K. MACK

Farbenfabriken Bayer AG, Werk Elberfeld,
Friedrich-Ebert-Strasse 217–319, Postfach 130105
56 Wuppertal-Elberfeld, F.R.G.

The composition of wastewater from an industrial plant is determined by the type of product manufactured and the methods used. The production programme of the pharmaceutical industry is generally characterized on the one hand, by the manufacture of relatively small quantities of active substances and on the other by a vast assortment of the most varied preparations. For the manufacture of preparations from animal or vegetable materials large quantitites of extractive agents are required. Chemical procedures consume large quantities of solvents to achieve the required degree of purity of the active substance. A further important factor is the wastewater from the biological production method, for example, the obtaining of antibiotics by fermentation. Basically, these wastewaters can be purified using biological methods. However, one or two prerequisites must first be fulfilled so that the biological purification process may run at optimal efficiency.

Nutrient requirements

The micro-organism, besides requiring organic carbon for the formation of cells also need nitrogen and phosphorus compounds. According to several authors [1, 2, 3] the biological breakdown should proceed at maximum efficiency at the ratio of $BOD_5 : N : P = 100 : 5 : 1$. For domestic wastewater this ratio is $BOD_5 : N : P = 100 : 25 : 4.5$.[4] One recognises that domestic waste contains an excess of nitrogen and phosphorus compounds which could be utilized to balance out the nutrient-deficient, industrial wastewater. The remaining trace elements such as sodium, potassium, magnesium, iron, manganese and sulphur required for cell formation are usually present in sufficient quantities in industrial wastewater.

Composition of wastewater

The properties of wastewater depend on the production of the factory and the applied method of manufacture. Problematic are the substances which, in certain concentrations, have a toxic effect on the micro-organisms required for the purification process. One must differentiate between toxicity to anaerobes and toxicity to aerobes. Most heavy metals inhibit the biological purification process at concentrations of $1-5$ mg/ltr. The toxicity to activated sludge increases in the following sequence: chrome VI, zinc, cadmium, chrome III, copper and nickel.[5] In the case of mercuric ions a concentration of even 0.1 mg/ltr. causes a significant inhibition of the breakdown by aerobes. Heavy metals are equally toxic to aerobes and anaerobes. A series of organic compounds behave differently, e.g. chlorinated hydrocarbons which are used very often and in large

239

quantities as solvents and extracting agents in the pharmaceutical industry. The damaging effects of these compounds on the activated sludge are relatively small. They are, however, extremely toxic to anaerobic sludge digestion.

TABLE 1

Anaerobic toxicity

	Concentration	% of inhibition of the methane gas after 48 hours
Chloroform	0.6 mg/ltr.	2
Chloroform	94.0 mg/ltr.	87
Methylene chloride	8.0 mg/ltr.	51
Ethylene chloride	8.0 mg/ltr.	68
Acetone	1,600.0 mg/ltr.	9
Ethanol	2,400.0 mg/ltr.	none
Methanol	1,200.0 mg/ltr.	12

A very important problem arises when wastewater containing such substances is passed into communal sewage treatment plants since the majority of these plants practise anaerobic sludge stabilization. These compounds are enriched in the primary sludge and, as a result, very quickly achieve critical concentrations in the digestion tanks.

Accordingly steps must be taken by firms to keep the concentrations of these compounds as low as possible in wastewater. This is the basic problem, i.e. to influence the properties of the wastewater by selecting the appropriate production technique and by preventing the escape of substances into the wastewater. This can be achieved as follows:

1. By selecting and developing the chemical reaction stages producing the final product according to the appropriate technical, wastewater criteria.
2. By correspondingly selecting the materials to be used.
3. Recovery and re-circulation of unconverted materials.
4. Separation of unconverted materials, e.g. in solvent separators, and then burning them
5. Preliminary chemical or physical treatment of wastewater before being subjected to biological purification (e.g. precipitation of heavy metals or stripping of solvents).

Due to the necessity of avoiding water pollution the manufacturing costs today present a very different picture from that of a few years ago. For example, in the last few years the recovery of solvents and raw materials has become a major priority. In future, before new products are manufactured not only their advantages but also their effects on the environment will have to be given more and more consideration.

The wastewater produced by the Wuppertal factory of Bayer A.G., is an example of the composition and properties of wastewater produced by a chemical – pharmaceutical factory and the associated problems of wastewater purification. Approximately 150 different active substances are manufactured in the Wuppertal factory. The following processes are used:

1. Pure chemical synthesis (e.g. Aspirin, sulphonamides, barbiturates, quat. ammonium compounds, thiophosphates).
2. Extraction and isolation from animal organs (e.g. Trasylol, Padutin, Campolon).
3. Fermentation (e.g. penicillins).
4. Combinations of 1. and 3. (e.g. semi-synthetic penicillins, such as Baycillin).

Table 2 shows the average quantities and properties of the wastewater produced.

TABLE 2

Quantities of waste water:		16,000 mtr^3/day
BOD$_5$:	1,500 mg/ltr. =	25 tons/day
COD :	2,700 mg/ltr. =	43 tons/day
Sodium chloride:	1.5 g/ltr.	
Sodium sulphate:	1.5 g/ltr.	
Nitrogen:	80.0 mg/ltr.	
Phosphorus:	9.0 mg/ltr.	
BOD$_5$: N : P = 167 : 8.9 : 1		

Acetone, ethanol, isopropanol and methanol are approximately 45% of the total BOD$_5$. No inhibitory effects on *Pseudomonas fluorescens* and *Escherichia coli* are produced. There is a separate cooling system in the factory whereby the water is cooled in cooling towers. Domestic sewage is fed into the municipal sewage system.

At the present time the factory wastewater is partially purified together with communal wastewater using biological methods. The quantity of factory waste is

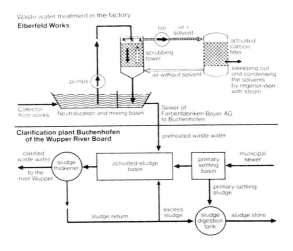

Fig. 1. Factory effluent of Farbenfabriken Bayer AG

approximately 20%, the BOD_5-load about 45%. Before feeding the wastewater into the city's sewage system it must first be established that chlorinated hydrocarbons contained in the factory wastewater will not inhibit the anaerobic sludge digestion. Thus after neutralizing the factory wastewater it is fed into a stripper unit in which the volatile solvents are, to a greater extent, blown off by introducing air into the unit against the current flow. The water to air ratio is 1 : 10. The air is passed through activated carbon filters and then returned into the stripper unit. The substance adsorbed by the activated carbon is displaced by steam and burnt after being separated from the aqueous phase of the condensate. In this way approximately 1 ton of solvent can be removed from the wastewater per day. Another safety measure used is the by-passing of the anaerobic stage and the feeding of the wastewater directly into the aeration tank of the aerobic stage. This necessitated the laying of a separate sewer for the waste water from the factory premises to the municipal sewage plant.

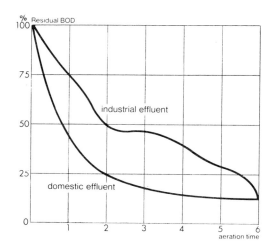

Fig. 2. Curve of BOD degradation by activated sludge according to Sawyer

It is a very important fact that the contents of domestic and industrial wastewater be broken down at different rates. Fig. 2 shows the course of the BOD_5 breakdown of industrial and communal wastewater according to Sawyer.[6]

In most cases, when industrial wastewater is fed into communal systems it cannot be expected that the standard purification will be up to present day demands, since communal purification plants are only constructed to take domestic sewage. The opinion is very often expressed that after appropriately laying out the sewage plant it is advantageous to purify industrial wastewater together with domestic sewage using biological methods. Many points favour this view providing the portion of industrial wastewater is small in relation to the municipal wastewater. Fluctuations in the quantities and concentrations and change in the properties of the industrial wastewater, e.g. pH value, temperature, nutrient supply, salt-content, could be balanced out by domestic wastewater, thus achieving a greater working reliability. However, while considering these factors others must also be taken into consideration, e.g. the biological breakdown rate of the components and the elimination of the odour of industrial wastewater. This could be favourably influenced by the use of a special process, e.g. Unox system. Biological breakdown trials were carried out in pilot plants with industrial wastewater alone and mixed with domestic sewage.[7] The results are shown in Table 3.

TABLE 3

Waste water	Industrial waste water		Communal sewage with indust. waste water 4 : 1		
Process	single	double stage	single	stage	double
Aeration period in hrs.	6	12	12	6	12
BOD$_5$ influent	1,700	1,700	300	365	348 mg/ltr.
BOD$_5$ effluent	112	60	35	41	20 mg/ltr.
Breakdown %	94	97	88	89	94
COD influent	2,540	2,540	617	762	616 mg/ltr.
COD effluent	405	305	231	271	177 mg/ltr.
Breakdown %	84	88	63	64	71

The treatment of the wastewater alone or when mixed with municipal wastewater from Wuppertal requires 2-stage treatment to achieve sufficient purity. The aeration period required for this purification method is 12 hours. For domestic waste alone an aeration period of 4–6 hours is sufficient to achieve the same degree of purity. Thus a tank of corresponding capacity would have to be installed for the 12 hour aeration period of the industrial wastewater which is only 20% of the total wastewater. Thus the question as to whether the wastewater from the chemical industry and in particular that from the pharmaceutical-chemical industry should be biologically purified together with communal wastewater cannot be answered with a simple 'yes' or 'no'. Each case must be examined on its own merits and the best solution found by carrying out breakdown tests and by studying local conditions.

During purification tests with the industrial wastewater it was interesting to note which compounds are easily broken down by biological methods and those which are not. The contents of the influents and the effluents of the first and second stages of a 2-stage pilot plant were gas chromatographically tested (head-space method).

The technical data are listed in Table 4.

TABLE 4

Pilot plant, type aero-acellator

	1st stage	2nd stage
Aeration volume	24 ltrs.	24 ltrs.
Aeration time	6 hrs.	6 hrs.
BOD_5 influent	1,600 mg/ltr.	120 mg/ltr.
BOD_5 effluent	120 mg/ltr.	40 mg/ltr.
BOD_5 breakdown	93%	97%
COD influent	3,400 mg/ltr.	540 mg/ltr.
COD effluent	540 mg/ltr.	300 mg/ltr.
COD breakdown	84%	88%
Volumetric Loading	6.8 kg $BOD_5/m^3 d$	0.5 kg $BOD_5/m^3 d$
Sludge Loading	0.8 kg $BOD_5/m^3 d$	0.15 kg $BOD_5/m^3 d$
dissolved oxygen	1.5 mg/ltr.	4.5 mg/ltr.

The following curves show the average daily concentration of some of the compounds found in the influent and effluent of the first and second stages over a period of several days.

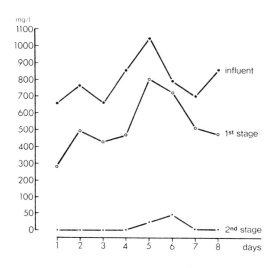

Fig. 3. Methanol decomposition

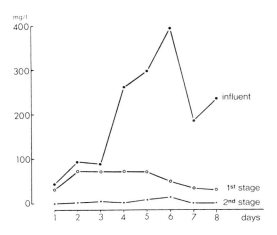

Fig. 4. Isopropanol decomposition

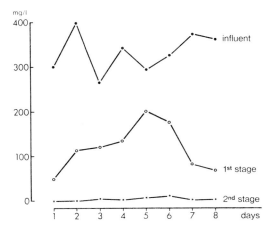

Fig. 5. Acetone decomposition

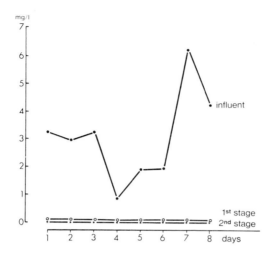

Fig. 6. Toluene decomposition

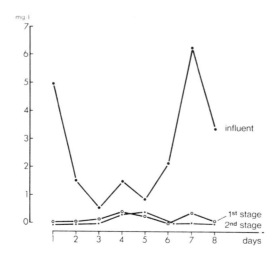

Fig. 7. Phenol decomposition

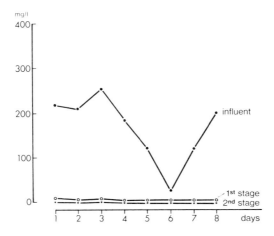

Fig. 8. Ethanol decomposition

As can be seen the breakdown rates vary from one substance to another. For example methanol and isopropanol are only partly broken down in the first stage. Relatively large quantities reach the second stage. In contrast toluene, phenol and ethanol are almost completely broken down in the first stage.

The biological breakdown is based on the ability of the micro-organisms to remove organic substances from wastewater and from these substances form new cell material. Many of the substances contained in the wastewater cannot be used directly for the development of cells but must first pass through one or more intermediate stages.

It is well known that the various microbial metabolic systems contribute greatly varying amounts to the conversion of matter. Thus, which intermediate stages develop and during which metabolic cycles these intermediate stages become active are factors which decide the rate at which the substance will be taken up by the micro-organisms. A good example is the biological breakdown of phenol by *Pseudomonas fluorescens*, over several intermediate stages, to acetic acid, pyruvic acid and 3-oxo-adipic acid.[8] Those compounds are broken down in the tricarbonic acid cycle.

The aerobic breakdown of primary alcohols having at least 2 atoms of carbon by certain species of pseudonomas to the appropriate aldehydes and fatty acids are then converted to acetic acids by β and ω oxidations as follows:

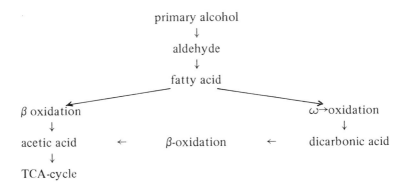

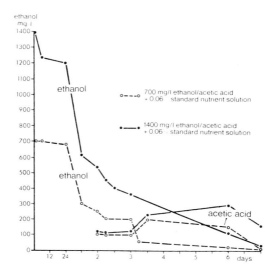

Fig. 9. Decomposition of ethanol in 0.06% standard nutrient solution with formation of acetic acid

Fig. 9 shows the breakdown of ethanol and the formation of acetic acid in relation to time. The test was carried out in 0.06% standard nutrient solution (Merck standard nutrient broth I). The effluent from a purification plant served as inoculum.

The secondary alcohols are initially oxidized to the corresponding ketone which is then converted, via its own enol form to the relevant fatty acid and formaldehyde. Then the fatty acid is converted to acetic acid as follows:

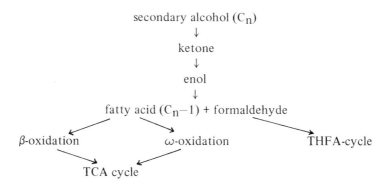

Fig. 10 shows the breakdown of isopropanol and the formation of acetone in relation to time also in 0.06% standard nutrient solution.

For many micro-organisms the tricarbonic cycle is responsible for the conversion of a substantial part of the substances. Only substances having two or more atoms of carbon can be processed in the tricarbonic acid cycle. Substances having only one atom of carbon can usually only be broken down in the tetra-hydro-folic acid cycle which plays a very minor role in the conversion of substances. It is therefore understandable that methanol, isopropanol and acetone, because of the splitting off of an atom of carbon and the formation of formaldehyde, require a longer period of time for the biological breakdown process than for example phenol, toluene and ethanol.

The knowledge of such breakdown procedures thus explains the varied breakdown rates for the different kinds of wastewater. Further, with this knowledge it should be possible, during the development of the manufacturing method, to influence the composition of the wastewater and thus influence its biological breakdown. This could be

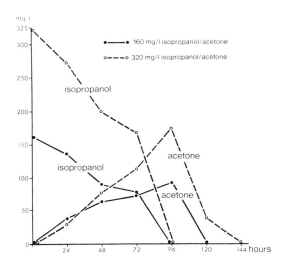

Fig. 10. Decomposition of isopropanol in 0.06% standard nutrient solution with formation of acetone

achieved by giving preference to substances which can be more rapidly broken down by biological means. Such factors will have to assume a more important role when choosing a product and the appropriate production method if the ever increasing demands placed on the purity of the wastewater and, in particular, the chemical oxygen demands are to be fulfilled without incurring astronomical costs.

REFERENCES

1. Nolte, E., Meyer, H.-J., Fromke, E. Jahrbuch vom Wasser VIII (1934), S.126.
2. Helmes, E. W., Frame, J. D., Greenberg, A. F., Sawyer, C. N. Sewage and Industrial Wastes 24 (1952), S. 496.
3. Sawyer, C. N. Biological Treatment of Sewage and Industrial Wastes, Reinhold Corporation, New York (1956).
4. Sierp, F. Zentralblatt Bacteriologie 155, (1950), S. 318.
5. Heukelekian, H., Gellmann, J. Sewage and Industrial Wastes 27 (1955), S. 70.
6. Sawyer, C. N., Kahn, P. E. Proc. 13th Industrial Wastes Conf. Purdue Univ. Lafayette, Engng. Bulletin 43 (1959), S. 341.
7. Bischoffsberger, W., Mack, K. Kommunal-Wirtschaft 9 (1968), S. 323.
8. Bacterial Metabolism, H. W. Doelle, Academic Press (1969), S. 376.

TREATMENT OF INORGANIC LIQUID WASTES

R. K. CHALMERS, B.Sc., M.Chem.A., F.R.I.C., M.Inst.W.P.C.
Senior Partner, Bostock Hill and Rigby, 37–39 Birchfield Road,
Birmingham B19 1SU, United Kingdom

Factors Determining the Degree of Treatment Required

While running swill waters may almost always be adjusted in quality by application of simple scientific procedures, it is comparatively uncommon to find cases where pre-treatment at the factory is not required before discharge to the sewers or to rivers. The degree of pre-treatment at a factory is determined by:

(a) The quantity and quality of effluent discharge permitted.

(b) The contaminants in the effluent.

(c) The extent of conservation measures adopted in the factory.

(d) The nature, volume and frequency of discharge of strong spent solutions.

(e) The degree of segregation of different types of effluents within the factory.

The effect of the standards imposed upon effluent quality is of prime importance. They vary widely with the location of the factory, the size of the sewage disposal works, and the requirements of the River Authority accepting the final discharge, whether this is passing through a sewage disposal works or directly to the river. In the United Kingdom standards for discharge to sewers vary from 1 mg/l total toxic metals to upwards of 70 mg/l total toxic metals. Standards for Cyanide (CN) range from Nil to 20 mg/l. For discharges to rivers standards range from 0.5 mg/l total toxic metals to 5.0 or more mg/l., and for cyanide from 0 to 0.1 mg/l. These standards are always examined, questioned, and if necessary contested. In some cases a distinction is made between discharges containing toxic metals in suspension and those in solution, and separate standards prescribed for these. It is rare to find a distinction made between the toxicity of different metals, although this is known to vary widely: by a factor of at least 100.

Minimizing Water Flows and Effluent Discharges

This is clearly of critical importance in:

(i) Reducing water costs.

(ii) Reducing size of treatment plants.

(iii) Reducing effluent disposal charges.

Methods for reducing water consumption in metal-finishing processes are well known and have been widely reported.[1,2,3] They include:

(a) Restriction of flows by constant flow orifices in pipelines.

(b) Conductivity controllers in rinse tanks.

(c) On/off pedals at tanks with hand-operated processes.

(d) Counter-current rinsing.

(e) Recycle or re-use of cooling waters.

(f) Use of fog, spray or intermittent rinses.

(g) Air agitation in rinse tanks.

(h) Chemical rinses after some processes.

251

At most new factories a critical examination of the need for water is made under the headings shown in Table 1.

TABLE 1

Examination of the Need for Water

---

(1)   Construct a Water 'Balance Sheet' for the factory. Relate input flows to output and to need.
Record water intake with all taps closed – in one factory amounted to 1 million gallons per week (4.5 m^3 x 10^3/week). Systematic reading of strategically placed input water meters is revealing and sometimes astonishing.

(2)   Exclude cooling and uncontaminated waters from direct discharge to drain. Initiate or extend re-use and recycle. Examine bleed-off rates.

(3)   Examine timing on automatic flushing systems, and spring loaded taps or pedal controls for domestic supplies – reduction at one factory 2 million gallons per week. (9.1 m^3 x 10^3 per week).

(4)   Compare day and night flows.
Higher night pressure in mains gave an unnecessary increase from 40,000 gallons per hour (182.1 x 10^3 per hour) to 50,000 gallons per hour (227.1 x 10^3 per hour) at one Birmingham factory.

(5)   Stop rinse flows when no work is present.
(a)   On/off pedals at tanks with hand-operated processes.
(b)   Mechanical linkages controlling water supply against plant cycles on automatic plants. Incorporate time-delay if required.
(c)   Conductivity controllers.

(6)   Examine efficiency of water use.
(a)   Avoid short-circuiting in rinse tanks.
(b)   Use air agitation whenever possible.
(c)   Counter-current rinsing.

(7)   Determine precisely the water flow needed to maintain the required quality in each rinse tank. Control the maximum flow by pipeline restrictors.

(8)   On all new process plants incorporate water and waste economy measures at the planning stage.

All of these economy measures are in use at one factory or another. Where they are combined the effect upon the use of water is dramatic.

Three examples of less usual, but valuable, approaches to water saving are:

(a) Reduction of flows to very low rates by the use of fog and spray rinses which are recirculated and discharged at intervals. By using small tanks of 500 gallons (2270 litres) capacity in series, and 2 or 3 rinses per process, the effluent discharge at one of the largest plating and aluminium anodic oxidation plants in the United Kingdom is reduced to 2800 gallons per hour (12.71×10^3 p.h.) from a total of six large process shops.[4]

(b) The sequential use of swill water for a number of different swill duties, with cross-connections between different process plants.[5]

(c) The use of vapour rinsing, using condensation on the work surface as the rinsing agent.

It is common, but not universal practice in the United Kingdom, to recycle cooling water either through lagoons or large holding tanks where space is available, or through evaporative coolers. There is extensive re-use of water in the steel industry.

In the metal-finishing industry re-use is commonly restricted to cooling water or to some sequential use of water between different processes. Many measures are, however, taken to minimize the use of water and of the contamination of swill water by use of drag-out tanks and similar saving methods. In general it is found that there need never be a problem with the quality of effluent swill waters. The water flow required to produce an effluent of any given quality can be produced at will by combinations of the methods listed above, with the use of static drag-out tanks. There are, however, problems, which must be considered individually in each case on the economic disposal (or recovery) of strong spent solutions, including retained drag-out solutions.

Industries installing new process plant in the metal-finishing industry sometimes examine critically the water requirement quoted by the manufacturers of the process plants. In competitive tendering price sometimes militates against the inclusion of water economy measures. The water requirement of a process plant before and after taking simple economy measures are shown in Table 2.

TABLE 2

Water Requirements of a Process Plant before and after
Simple Economy Measures

| | | Rinse Flow Rates | |
Process	Tank capacity (gallons)	as Tendered	as Modified
		gals/hr.	
Hot Soak Clean	225	105	100
Hot Cathodic Clean	175 ⎫	150	50
Hot Anodic Clean	175 ⎭		
Muriatic Acid (HCl)	175	270	60
Cyanide Copper Flash and Drag-out	175	270	180
Muriatic Acid	175	300	60
Nickel Plate and Drag-out	1250	270	60
Chrome Plate and Drag-out	530	360	60
Hot swill	–	45	15
		1770	585

Practical approaches to Pre-Treatment at Factories.

(a) A very large number of treatment plants exist for simple neutralization with or without sedimentation or removal of suspended solids. Comparatively few problems arise.

(b) Waste waters from the metal-finishing industry, particularly those containing chromates and cyanides, are treated by:

    (i)   Batch treatment with separation of flows.
    (ii)   Batch treatment without separation of flows.
    (iii) Flow-through treatment with separation of flows.
    (iv) Flow-through treatment without separation of flows.

There is a marked trend to flow-through types of treatment with strong spent solutions metered into flow being treated. Treatment without separation of flows, which is normally found to require the introduction of large balancing tanks, is unusual in the United Kingdom because of space limitations. Batch treatment is frequently restricted to relatively low flows. At large installations a careful assessment is normally made of the possibility and costs of adjusting the running flow of swill waters to reach the required acceptance standard subject only to neutralization and flow-recording. In addition some flows are usually found which require no treatment prior to discharge. Where full water economy measures are practised, however, it is sometimes found to be more economical to pass the whole of the reduced flow and the strong spent solutions through a common treatment plant. Examples of the practical applications of these approaches are as follows:

A. *Reduction of Process Shop Flows, Segregation of Uncontaminated Flows, and Treatment of the Whole Main Flows*

This approach to treatment was adopted for the motor car factory of Rolls-Royce Limited, Crewe, Cheshire following reduction of the plating shop flows from 16,000 gallons per hour (73.1×10^3/hr) to 3000 gallons per hour (13.61×10^3/hr). This was made possible by extensive re-organization of the plating shop. The flow diagram is shown in figure 1.

Flows to treatment are segregated into:

Cyanide-bearing flows
Chrome-bearing flows
General acid/alkali flows

In addition miscellaneous swill flows requiring only neutralization by-pass the main treatment plant.

There are a large number of outlying points in the factory where trade effluent arises from a wide variety of processes, and because of the distances involved strong spent solutions are retained in receiving tanks local to the process and transferred at a slow rate into the main transfer lines travelling ¼ mile to the treatment plant. Transfer pumps are interlocked so that no strong solutions can pass forward in the absence of the corresponding swill waters. The swill waters are pumped to high level and gravitated to the treatment plant along overhead P.V.C. pipelines. This method of transfer was found to be some £25,000 less costly than transfer by underground buried pipelines. At the treatment plant the cyanide flow is treated by alkaline chlorination under pH and Redox control, with a monitoring electrode downstream of the treatment tank to recycle any flows showing a cyanide content.

Chrome flows are treated by SO_2 dosing at controlled pH, under Redox control, with a similar recycle system. Treated chrome and cyanide solutions mix with the acid alkali

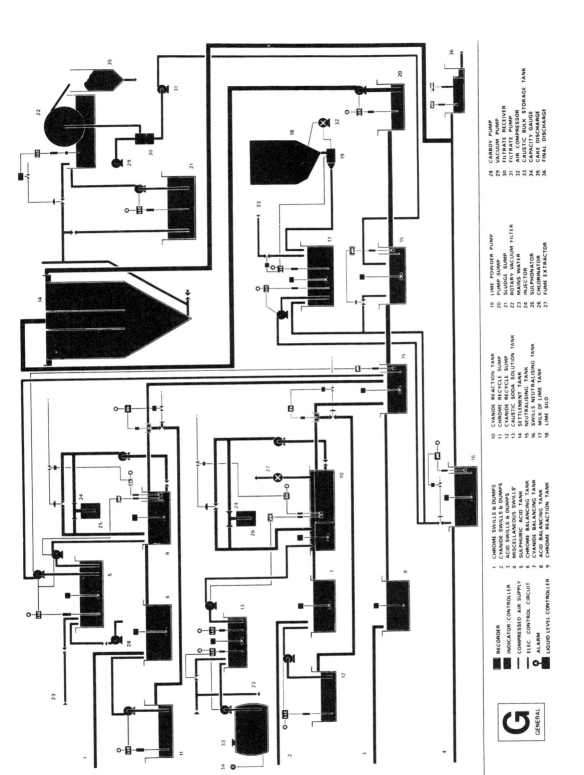

G GENERAL

- RECORDER
- INDICATOR/CONTROLLER
- COMPRESSED AIR SUPPLY
- ELEC CONTROL CIRCUIT
- ALARM
- LIQUID LEVEL CONTROLLER

1. CHROME SWILLS & DUMPS
2. CYANIDE SWILLS & DUMPS
3. ACID SWILLS & DUMPS
4. MISCELLANEOUS SWILLS
5. SULPHURIC ACID TANK
6. CHROME BALANCING TANK
7. CYANIDE BALANCING TANK
8. ACID BALANCING TANK
9. CHROME REACTION TANK

10. CYANIDE REACTION TANK
11. CHROME RECYCLE SUMP
12. CYANIDE RECYCLE SUMP
13. CAUSTIC SODA SOLUTION TANK
14. SETTLEMENT TANK
15. NEUTRALISING TANK
16. SWILLS NEUTRALISING TANK
17. MILK OF LIME TANK
18. LIME SILO

19. LIME POWDER PUMP
20. PUMP SUMP
21. SLUDGE SUMP
22. ROTARY VACUUM FILTER
23. MAINS WATER
24. INJECTOR
25. SULPHONATOR
26. CHLORINATOR
27. FUME EXTRACTOR

28. CARBOY PUMP
29. VACUUM PUMP
30. FILTRATE RECEIVER
31. FILTRATE PUMP
32. AIR COMPRESSOR
33. CAUSTIC BULK STORAGE TANK
34. CAPACITY GAUGE
35. CAKE DISCHARGE
36. FINAL DISCHARGE

waters for lime neutralization and settlement in a 4 hour retention above ground settlement tank. Sludge is filtered on a rotary vacuum filter. The treatment plant is housed in a building that was formerly a coal bunker house, with a strong steel frame, and was readily adapted for the purpose.

B. *Adjustment of Swills Quality and Treatment of Strong Solutions only*

This treatment is adopted at Joseph Lucas Limited, Great King Street, Birmingham and at the Rolls-Royce, Glasgow, aircraft engine factory. Treatment at Glasgow is of unusual interest because the solutions for treatment have been kept as strong as possible, consistent with the treated liquors remaining pumpable. The flow diagram is shown in figure 2.

All wastes other than cyanide wastes are treated at a central treatment plant, apart from substantially uncontaminated flows. Because the cyanide wastes are low in amount, they are treated locally at their points of origin. In this treatment a separation is made between cyanide solutions carrying toxic metals and alkali cyanide solutions. This is because the transfer rate necessary for the metal-bearing cyanides would contribute excess metals to this flow which by-passes settlement.

In developing this treatment it was found that:

(a) Chrome solutions could be treated at a strength which produced a sludge needing no consolidation, and that could by-pass settlement, prior to filtration.

(b) Strong solutions, particularly from chrome plating and electro-polishing required addition of 3 volumes of swill waters as a carrier flow.

(c) Lime was preferable to sodium hydroxide as a general neutralizing agent.

(d) Wet sludge production could amount to 100% to 400% of the original waste solution treated, requiring sludge filtration for economic disposal.

(e) Full 'fail-safe' instrumentation was advisable, with monitoring systems to recycle any treated cyanide, chrome, or acid not reaching the limits required.

(f) Near points of origin of strong solutions, polypropylene was the pipeline material of choice.

(g) In very strong solution treatment control electrodes need careful siting and cleaning.

(h) At full plant load, reagent supply needs special attention with regard to the temperature of gaseous reagents and distribution of neutralizing agents.

Hydrofluoric acid solutions are separated for slow rate transfer to the spent acid flow to treatment in order to minimize effects upon pumps and pipelines.

Strong spent solutions are collected locally to their points of origin, and never discharged without automatic addition of diluting water. They are transferred forward to strong solution retaining tanks at the treatment plant area.

Emulsified oils are separated by treatment in batch tanks with sulphuric acid and aluminium sulphate.

C. *Treatment with Demineralization of Selected Flows to Minimize Discharge to the Sewer*

At the Cannock factory of Joseph Lucas Limited trade effluent is produced from two large automatic copper – nickel – chrome rack plating plants, a rack zinc plating plant, and a phosphating plant.[6]

The essential effluent plant design criteria were that the total effluent production was to be 20,000 gallons per hour (91 l. x 10^3/hr) the maximum permissible discharge to the sewage works was to be 14,000 gallons per hour (64 l. x 10^3/hr) and that the factory

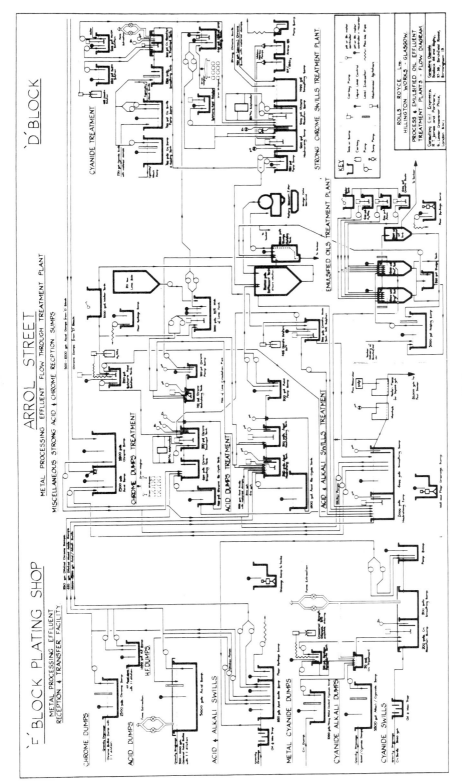

Fig. 2

requirement for demineralized water would be 8600 gallons per hour (39 l. x 10^3/hr).

A High Court injunction restrained the discharge from the sewage works of any effluent materially affecting the quality of the receiving stream. This required the production of a trade effluent of exceptionally high quality. The initial flows consisted of:

> 7700 gallons/hour (35 l. x 10^3/hour) alkaline effluent
> 5000 gallons/hour (25 l. x 10^3/hour) acid effluent
> 4000 gallons/hour (18 l. x 10^3/hour) cyanide liquor

All of these flows required treatment to meet stringent Local Authority requirements, including limits for chromate (as Cr) of 2 mg/l and total toxic metals of 10 mg/l.

For treatment seven separate pump pipelines were provided for:

(a) Cyanide-bearing running swills.
(b) General acid–alkaline running swills.
(c) Chrome-bearing swills plus some relatively uncontaminated swills using deionized water to demineralization.
(d) Strong cyanide dump liquors.
(e) Strong acid dump liquors.
(f) Strong caustic dump liquors.
(g) Strong chromic acids from one point.

Strong liquors are delivered into holding tanks for slow transfer into the running flows for treatment. Control of treatment is essentially automatic, with redox control of cyanide chlorination and of chromate reduction, and fully automatic ion-exchange cycles.

Treatment of the flow to be demineralized is the reverse of the normal cycle, with anion exchange followed by cation exchange. Pre-treatment of this flow is by flocculation, sand filtration, chlorination and activated-carbon filtration. Split elution is used on the regeneration cycles, the discarded first half of each regenerant flow being treated as a strong dump solution. The pre-treatment liquors from cyanide treatment and chromate treatment join the main flow of effluent for pH controlled lime neutralization and then pass into an 88,000 gallon (400 l. x 10^3) circular sedimentation tank. Sludge is separated, consolidated and filtered on a rotary vacuum filter. The treatment plant is running under its design capacity, in-plant economy measures introduced after the construction of the production plant and the effluent plant, which proceeded simultaneously, having produced considerable saving.

Typical results from the effluent discharged are:

pH value	9.0
Cyanide (as CN)	0 to 0.1 mg/l
Sulphate (as SO_4)	1050 mg/l
Suspended Matter	Less than 100 mg/l
Oil	Trace
Cadmium, chromates, lead	Nil
Copper	2 mg/l
Iron	0.2 mg/l
Nickel	0.8 mg/l
Zinc	2 mg/l

Final 'polishing' treatment of the effluent takes place at the sewage works. The effluent is delivered to the works through a separate pipeline and is then mixed with an equal volume of sewage and treated on two separate biological filters.

Sludge filtration was introduced after the treatment plant had been operating for four years. At the same time sodium hydroxide was replaced by lime as the main neutralizing agent. This was on grounds of cost, much improved filterability of the sludge and the need to limit sulphate in the discharge to the sewer. The savings resulting from these two modifications amounted to £6500 per annum.

At this factory an unusual approach to rinsing after nickel plating has eliminated the flow of effluent from this source. A flow of 10 gallons (45 litres) per hour of demineralized water is fed into the final tank of a triple counter-current rinse system. 10 g.p.h. is less than the evaporation rate from the nickel plating vat, and the flow of water delivers from the counter-current rinse system directly into the process vat. By the use of demineralized water no foreign constituents are added to the plating bath, and carbonates and normal breakdown products are removed by the regular routine plating solution purifications.

D. *Re-use of Swill Water after Integrated Waste Treatment*

Where space in the production lines permit, and particularly where the treatment problems are not dominated by large volumes of spent strong solutions, treatment is applied in some cases by chemical rinses followed by normal rinses discharging to waste. The chemical rinses operate with high concentrations of chlorine for cyanide oxidation or of bisulphite, hydrosulphite or hydrazine for chromate reduction and are pumped round at high rate to in-line settlement tanks. The following swill is consequently very lightly contaminated and in many cases suitable for re-use. This Lancy integrated waste treatment[7] is finding increased application in cases where re-use of water rather than reduction of flows is required, and in a number of new installations.

Re-circulation of water after chemical treatment, without specific separation to give separate treatments for individual processes, is also possible.[8]

E. *Treatment with Recovery of Metals*

This is practised in a number of plants:

(a)   After some integrated treatments, as in D.
(b)   Ion-exchange for silver, gold and chromium.
(c)   By electrolysis for copper and tin.

F. *Small-Scale Waste Treatments*

A very large number of small process plants in the metal-finishing industry have disproportionately large effluent problems. In many cases the problems are reduced by the use of static drag-out tanks after significant processes – often disused tanks in the factory can be adapted for this use – and discharge of the running flow of effluent subject only to neutralization. The retained strong spent solutions are then either treated in a batch treatment process of comparatively small size, or in some cases are removed from factory sites by tanker for disposal.

Disposal of wastes to the land is controlled under the Water Resources Act, 1963 and by Local Planning Authorities. One case exists of a disused coalmine with a unique geological formation making it similar to a clay bottle for which a company has the necessary authority for disposal of toxic wastes. The cost of tanker disposal is high, and sometimes special measures are taken to reduce the volume of wastes discarded.

Evaporation has been successfully applied to the reduction of volume of drag-out solutions, particularly after zinc plating processes, and concentration by ion-exchange is also possible.

Water Re-use

At least one international Company has established a Group policy that no liquid trade effluents shall be discharged from its manufacturing plants, and all waste waters shall be recycled.

The conditions necessary for re-use of effluents are:

(i)   The quality must be acceptable for the processes to which the water is returned. It is usually advisable to divide the processes into those where the water quality is critical and those with non-critical requirements. It may be possible to return water to processes from which it was not derived — to boiler feed, cooling tower make-up, or to general service water for floor cleaning and similar crude uses.

(ii)   The water must generally be rendered non-corrosive and non-scaling.

(iii)   A minimum of dissolved solids should be added in the course of treatment.

(iv)   The cost of treatment and return of water should normally be less than the cost of purchase of raw water and disposal of effluent.

(v)   There must be a real need for the returned water. It is very much cheaper to reduce water consumption wherever possible than to treat unnecessarily inflated volumes for re-use.

At one motor car factory water consumption was reduced from 550,000 gals/day (2.5×10^3 m^3/day) to 285,000 gals/day (1.0×10^3 m^3/day) by recycling cooling waters and eliminating unnecessary flows. The saving in water purchase costs alone amounted to £11,000 per annum.

There is a wide choice of possible approaches to treatment of effluents for re-use, and all of them need to be considered: every case is different. They include:

(a) Chemicals or physical treatment and recirculation of segregated rinse flows individually.

(b) Demineralization and recirculation of segregated rinse flows individually.

(c) Demineralization of all combined rinse flows.

(d) Demineralization, or metal removal by selective ion-exchange and recirculation of dilute rinses; neutralization, settlement, pH adjustment and selective weak cation exchange of concentrated rinses.

(e) Conventional pre-treatment, neutralization, and settlement, followed by filtration and return of water to non-critical rinses.

(f) Rinse water recirculation after pre-treatment of any cyanide and chromate, followed by base-exchange softening.

(g) Sequential use of water on the production plants.

(h) Electrolytic treatments.

(i) Evaporation processes.

(j) Reverse osmosis:

    (a)   On individual waste water streams at source.

    (b)   On combined waste waters before chemical treatment.

    (c)   On combined waste waters after chemical treatment and settlement.

Four case histories illustrate the differing solutions reached in different circumstances:—

(A) 7000 gallons per hour (32 m^3/hr) of metal finishing effluent for 8 hours per day. Cost of water purchase and disposal as effluent is 34p rising to 68p/1000 gallons.

	Approach	Water Saving	Annual Reduction in Water Charges	Treatment Plant Costs	Running Costs
(1)	Demineralization of all rinses	90%	£8,550	£50,000	£0.35/1000 gallons (4.55 m^3)
(2)	Chemical treatment, filtration and return to non-critical rinses	28.5%	£2,720	£35,000	£0.20/1000 gallons (4.55 m^3)
(3)	Swills recirculation after pre-treatment and base-exchange softening	57%	£5,540	£40,000	£0.15/1000 gallons (4.55 m^3)
(4)	Sequential use of water on the production plant	28.5%	£2,720	Negligible	Nil
(5)	Selective ion-exchange on suitable rinse waters	66%	£6,300	£35,000 to £40,000	£0.12/1000 gallons (4.55 m^3) approx.

Recommendations:

Apply sequential use of water, saving 2000 gals/hr. at virtually zero cost. Apply the residue after conventional precipitation treatment and filtration to the revised non-critical rinses, saving 1000 gals/hr., at zero capital cost. (Reduction in size of treatment plant = cost of sand filters.) Follow by selective further treatment of the sand filtered water, applied only to the restricted flows that actually need very high quality water.

(B) 30,000 gals/hr. (136 m^3/hr.) of metal-finishing effluent from data processing equipment manufacture.

        Strong spent solutions excluded from treatment.

        Rinse waters segregated into strong and weak streams.

        Raw water costs £0.22/1000 gals. (4.55 m^3)

	Approach	Water Saving	Annual Reduction in Water Charges	Treatment Plant Costs
(1)	Demineralization of Total Flow	90%	£3,600	£90,000 Operating Costs £25,000 per annum
(2)	Separation of strong and weak rinses; settlement for strong rinses only, to reduce size of plant; re-combined flow to final polishing treatment by selective ion-exchange or paper pad press filtration.	90%	£3,600	£60,000

Recommendation:

Apply method (2).

(C) 5000 gals/hr (22.7 m³/hr) of effluent from data processing equipment manufacture.

Water costs: £0.075/1000 gallons (4.55 m³)
No trade effluent acceptance charges.
Existing effluent pre-treatment plant prior to discharge to sewer.
Required: maximum economic re-use of effluent.

Approach	Water Saving	Additional Treatment Plant Cost	Running Cost
(1) Direct re-use of 3250 gals/hr (14.8 m³/hr) after cooling	65%	£10,000 Plant £50,000 Total	£0.075/1000 gals (4.55 m³)
(2) Re-use of filtered treated effluent for non-critical processes after sand filtration 650 gals/hr (3 m³/hr)	13%		
(3) Further treatment of remaining effluent for total recycle.	95%+	£150,000 to £200,000	—

Recommendation:

No economic advantage until water costs rise or effluent acceptance charges introduced.
Then apply method (1), followed by (2).
Re-assess approach (3).

(D) 10,000 gals/hr (45.46 m³/hr) of effluent from television tube manufacture. Cost of water purchased and effluent discharge 25.5p/1000 gallons.

Approach	Water Saving	Annual Reduction in Water Charges	Treatment Plant Costs	Running Costs
Re-use of treated effluent after chlorination, activated carbon treatment and dealkalization, by feed to evaporative cooling tower as make-up	20% average 66% maximum	£4,200	£10,000	£0.09 per 1000 gallons

Production of very high quality effluent:

In some cases of direct discharge to a river, or where there are special circumstances requiring very strict limits on discharges to sewers, effluents of an exceptional quality have to be produced. The high quality required can be obtained, usually at very considerable expense, by the application of final polishing treatment to remove the remaining traces of metals after more conventional pre-treatment. Typical results from polishing treatments are as follows:

Results of polishing treatments

Results in milligrams per litre.

	1. 10,000 gal/h (45.5 m^3) from conventional CN, Cr and neutralization treatment followed by sand pressure filtration.	2. 3,000 gal/h (13.6 m^3) from conventional CN, Cr and neutralization treatment with 10 h settlement retention, followed by paper filtration in filter press.	3. Experimental polishing of Cu, Ni, Zn effluent on weakly acidic cation exchanger.
		Metals present in treated effluent	
Cadmium, Cd	0.7	0.8	—
Chromium, Cr	1.4		—
Copper, Cu	2.0	0.04	—
Nickel, Ni	2.6	0.1	0.1
Zinc, Zn	0.6	—	0.03

Typical programme of investigations, recommendations and action in the development of a trade effluent treatment installation.

1. Establishment of water balance for the factory, relating water intake to water output.
2. Establishment of flows of liquors at each point of use, and of possible methods of treatment.
3. Determination of contaminants discharged at each point of origin: (a) in swills; (b) as strong 'dumped' solutions. Construction of a dumping schedule for all regular strong discharges.
4. Determination of the results of combining flows from various sections of the factory.
5. Survey of existing drainage routes: levels, capacities, and present state.
6. Segregation of the drains: e.g. plating shop; individual trade effluent sources; surface water; domestic discharge.
7. Determination of practicability and costs of the various possible routes for the drains.
8. Determination of the possibilities and effects of water or effluent recirculation or re-use.

9. Comparison of benefits and costs between local product recovery and recovery from one central plant collecting from all sources.
10. Physical nature and outline arrangement of the proposed effluent treatment plant.
11. Outline flow diagram for the treatment plant(s) proposed.
12. Outline specification for mechanical plant and civil plant.
13. Report to client, with recommendations; discussion and agreement on the treatment system to be installed.
14. Discussion and negotiation if necessary with the Local Authority and River Authority.
15. Detailed specification for the mechanical plant and civil construction.
16. Contract drawings.
17. Contract documents.
18. Invitation of competitive Tenders, if required.
19. Analysis of Tenders and report, including capital and running costs and assessment of plant reliability.
20. Placing of contract:
     Building works: foundations and drainage, plus building if required.
     Mechanical service, including pipework.
     Electrical services, including instrumentation and control equipment.
     Treatment plant(s)
21. Preparation of over-all construction programme.
22. Site meetings and progress meetings.
23. Works tests on plant to be supplied — as necessary.
24. Site tests on installed plant.
25. Record drawings of plant as installed.
26. Supervising supply of adequate operating and maintenance instructions.
27. Staff instruction as required.
28. Plant commissioning, and commissioning tests.
29. Advice as necessary, and correction of any defects during the maintenance period of the plant.

REFERENCES

1. Chalmers R. K., Minimising trade effluent discharges. Wat. Pollut. Control, 66 1967 (1) 49.
2. Harris E. P., 'A survey of nickel and chromium recovery in the electroplating industry' 1960. London DSIR.
3. Kushner J. B., Plating, 1949 36, 798 and 915.
4. Pianoforte Supplies Ltd. Roade, Northants. Private communications.
5. Chalmers R. K., Pre-treatment of Toxic Wastes. Wat. Pollut. Control. 68 1970 (3) 281.
6. Pinner R., Lancy integrated waste treatment, Technical literature.
7. Chalmers R. K., Trade effluent treatment at the Cannock factory of Joseph Lucas Ltd. J. Proc. Inst. Sew. Purif. 1965 (4) 357.
8. Mattock G., Water re-use from industrial effluents by the method of controlled recirculation. Chemistry and Industry. 1970 pp. 46–53.

Discussion *by* Howard Edde *on*

"Paper Pulp Wastewater Treatment"

Ekono Consulting Engineers
Helsinki, Finland and Seattle, Washington, USA.

Introduction

The pulp and paper industry has long been established as a major industry in the world's more industrialized nations and is rapidly increasing its significance in the developing nations as well.

This paper will direct attention toward the recent technological developments in the pulp and paper industry and the impact these advances have made on the treatment of waste waters from this industry.

There are several different important production processes which can produce paper pulp, the most significant of which are sulfate, sulfite, and mechanical pulping each having a number of modifications of importance to the industry.

The pollution load from these processes are closely tied to product yield, internal chemical recovery, product quality requirements and the economic profit incentives associated with the process. The yield ranges of the major wood pulping processes are indicated below:

Yield Ranges of Wood Pulping Processes

PROCESS	Unbleached Range %	Yield Mean %
Groundwood	90–95	93
Chemi-Mechanical, Chemigroundwood	85–90	87
Semichemical	60–85	75
Chemical	53–52	48

As the unclaimed portion of the wood becomes dissolved or suspended in the spent pulping liquors, low pulping and bleaching yields are generally associated with high pollution potential. These pulping liquors are usually concentrated by evaporation and burned for chemicals and heat recovery. The extent of this operation has in the past usually been based upon economical considerations alone. Thus, high cost of chemicals in the sulfate industry has justified 95-98% liquor recovery while the sulfite industry has practiced from zero recovery up to 85% dependent upon cooking base. mill size and fuel price.

Approach to the Problem

In the past the incentive for the manufacturer to reduce his pollutants discharges by internal technological measures was largely limited to his own cost economics of production. However, rarely in the past did government imposed environmental control regulations recognize internal measures as a valid action toward pollution abatement.

Thus, throughout much of the world the lack of understanding and communication between industry people, with their interest in producing a product, and regulatory authorities, whose objective has been pollution control without regard to the wisdom of *where* capital investments on control measures would yield the maximum return has in many cases resulted in less than optimum solutions.

The Internal-External Treatment Cost Optimization

The need for a systematic approach to compare alternatives and evaluate and store operating costs led the authors organization to develop a Systems Design and Optimization Program for the Chemical Recovery Cycle and Effluent Treatment (SYDOP-CR-ET).

The main objective of the chemical recovery (CR) portion of the program is to generate investment and operating cost data for any feasible combination of process units, equipment, and operating parameters. In the course of this, the program also develops materials and heat balance information, including amounts and characteristics of solid, liquid, and gaseous effluents. The details of the program have been previously published.[1]

The effluent treatment (ET) portion of the program yield size requirements and investment and operation costs for unit processes commonly employed in external treatment of pulp and paper mill effluents. The input data are generated by the SYDOP-CR routines or additional information such that all effluent sources are described in quantitative and qualitative terms. As local conditions usually affect the optimum effluent treatment solution, data describing treatment requirements and land cost conditions are required. The cost models were derived from our firm's files and the most recently published and widely accepted data.

A digital computer is used for the simulation, and the CR-ET optimization is accomplished by varying imput possibilities and specifying subroutes for the respective portions of the program. A large

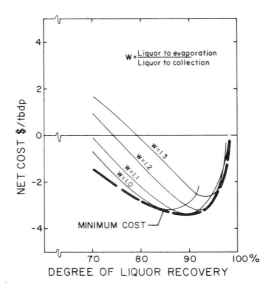

Fig. 1 Net cost of liquor recovery.

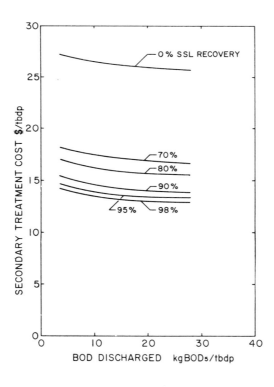

Fig. 2 Treatment cost vs BOD$_5$ discharge loads for various degrees of SSL recovery.

number of treatment combinations can be evaluated and the output analyzed by the user to select the most satisfactory combination to suit local conditions. The print out consists of unit size requirements and capital and total treatment costs presented in a tabular form. The final decision marking in process selection is, of course, retained by the user.

An Example of the Sulfite Mill Optimization Problem[1]

A case example is presented to demonstrate the cost optimization (procedure) between spent sulfite liquor recovery degree and external effluent treatment for a 240 metric ton per day pulp mill.

A number of internal liquor recovery alternatives exist to improve production technology while minimizing pollutant discharges. Figure 1 illustrates optimum inplant equipment selection for various liquor recovery degrees when evaluating washing filters and four effect evaporators with W=1.0, 1.1, 1.2 and 1.3 (see Figure for explanation of W). These data show that the minimum SSL recovery cost occurs at different degrees of recovery for the W values. This is but one example of the method for evaluating the effect of inplant changes.

The ET Computer program is similarly used to calculate secondary treatment cost for different liquor recoveries and secondary treatment efficiencies.

The allowable (BOD₅) discharges were set at 5 Kg/ton of pulp for these calculations. The effluent load used for the calculations in the computer program are as follows:

	Effluent Amounts & Loads		
Wastewater from	Flow m³ t bdp	BOD load kg/t bdp	SS loss kg/t bdp
Woodroom and misc. losses	15	5	
Bleach plant	80	20	
Screening room (1)	80	275	40
Secondary condensate (2)	—	38	

1. At 0% liquor recovery.
2. Based on 100% liquor recovery.

Capital charges were based on ten years at 6% interest rate. The program checks buffering capacity of the activated sludge process vs acidity, and neutralization was found to be required for bleaching effluents and secondary condensates. Figure 2 shows a plot of treatment cost vs unit BOD₅ discharge loads for various SSL recovery degrees.

Figure 3 shows the cost picture for internal-external treatment investments optimization. It can be seen that the minimum total cost for the required final effluent discharge limitations occurs at a higher liquor recovery degree than the optimum point when considering internal liquor recovery optimization alone.

An Example of a Sulfate Mill Optimization Program[2]

As with the sulfite industry, perhaps the most important new development in the sulfate industry are directed towards reducing pollutant discharges at their point of origin in the production operation.

Two classes of these modifications will be discussed, first, involving only process modification which add to production cost but yield lesser pollutant losses, and second, an advanced design mill is considered as a new baseline in which the advanced mill incorporates new technology capable of both reducing production cost and lessening pollutant losses.

The internal measures considered include:

b closing brown stock screening
c oxygen bleaching
d increasing dilution factor
e condensate stripping
f spills collection
h dry debarking

Figure 4 shows decreases in overall total treatment cost considering internal investment cost and corresponding cost for the specified level of required external treatment when subsequent internal treatment measures are adopted in the order of b, be, bef, and befd, respectively. These measures add to the cost of production and thus are implemented only by the incentative of environmental control requirements.

The combination of internal-external measures may be different for different objectives. For example, in Figure 4 to meet a discharge limit of 18 Kg BOD₅/t the internal measures would be befd

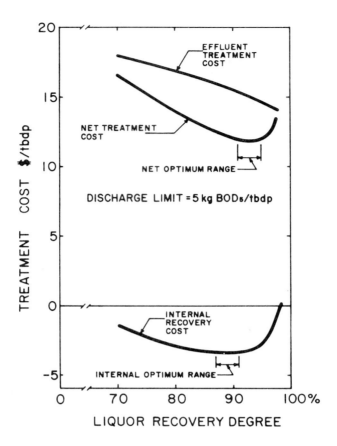

Fig. 3 Case example optimization results for internal-external treatment cost investment.

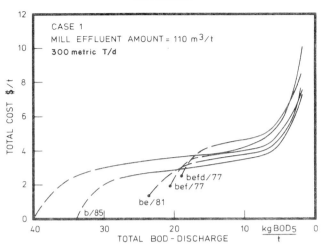

Fig. 4 Effluent BOD-reduction costs in a bleached sulphate mill by combination of internal external measures. (Total cost = 15% investment charge and operating costs).
Legend: Internal measure b/85 Effluent m³/t
        b = closing brown stock screening
        d = increasing dilution factor in washing
        e = condensate stripping
        f = chemicals spills collection

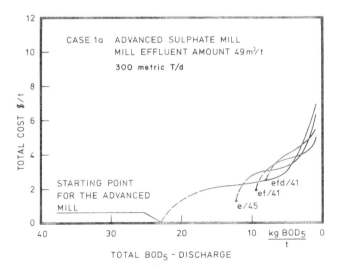

Fig. 5 Effluent BOD-reduction costs in an advances bleached sulphate mill by combination of internal and external measures (total cost = 15% investment charge and operating costs)
Legend: Internal measure b/85 Effluent m³/t
        b = closing brown stock screening
        d = increasing dilution factor in washing
        e = condensate stripping
        f = chemicals spills collection

with no external treatment whereas a discharge limit of 8 Kg BOD_5/t could best be met with internal modification b only followed by external treatment to the specified discharge limit in an aerated lagoon. Thus, in the example shown the internal modifications efd would not contribute to the overall least cost solution whenever the final effluent discharge is limited to approximately less than 15 Kg BOD_5/t.

Preliminary cost calculations indicate that internal process modifications c, and h will result in a net lowering of pulp production cost while lessening pollution. For this reason, these modifications can be expected to find near universal industry acceptance in future new mill situations and represent the basis for an advanced modern sulphate pulp mill.

Figure 5 shows the net total internal and external effluent treatment cost for the advanced modern sulphate pulp mill to meet different discharge level requirements. The trend toward increasingly stringent final effluent requirements, such as those presently being considered in the United States, point to the need for internal-external waste treatment combinations capable of providing minimum cost solution for bleached sulphate pulp mill final effluents of approximately less than 8 Kg BOD_5/t. Comparison of total treatment cost data for these very stringent discharge limits, as presented in Figures 4 and 5 for the modern conventional sulphate pulp mill and advanced modern sulphate pulp mill, respectively, indicate that rapid adoption of the advanced production facility will be increasingly attractive when future stringent discharge regulations become adopted as law and are enforced.

Discussion

This paper has mainly dealt with the influence of environmental considerations and the cost thereof as related to technological developments and problems arising from the manufacture of paper pulp. It has emphasized the need for consideration of internal control measures when developing solutions to the overall waste water treatment requirements.

For most receiving waters of measurable assimilative capacity it has become accepted by most parties that some form of "fail safe" environmental controls should exist externally from the production operation. The greater the equalization capabilities of these treatment facilities, the greater their value in protecting the recipient. Many internal pollution control measures cannot provide the required fail safe assurance. This has been an important factor in regulatory agencies requirement of external treatment. Furthermore, external treatment is more flexible to expansion with mill production increases or changes in receiving water quality requirements, i.e., it permits the designer to plan his design against a moving target. Thus, it is that continued reliance will be placed on the conventional and advanced waste water treatment facilities in solving most of the pulp industries water pollution abatement problems.

Unfortunately, regulatory agencies have in the past been unable to remain current with industrial technological developments and have not appreciated the benefits of internal conservation measures. Regulatory agents are themselves often pressured by the ecological conscious public opinion to document the environmental protection measures taken in the region of their regulatory jurisdiction. It is more reassuring to the general public opinion if reference can be made to a large new external effluent treatment plant, even though it perhaps requires large land areas including storage space for difficult to handle sludges. A compact internal conservation measure does often not convey the same measure of protective assurance to the uninformed public.

The purpose of this paper has been to present a tested and proved method of optimization which will aid in the decision of where investment in pollution control measures can be accomplished at least cost to meet a specific set of regulatory standards.

The calculations presented indicate that in many cases a combination of internal and external treatment will yield the greatest returns on the investment. This also has the advantage that less inorganic chemicals and less lignins are discharged into the receiving stream. Also, a recipient water in one location may accept the various characteristics in relatively different proportions than a recipient in another location. Effluents have varied characteristics (BOC, color, toxicify, pH, etc.). Different alternative treatment units and sequences of units may not give the same end results for all characteristics and it may become necessary to give different weights to reduction of some properties than others. In these cases treatment to remove a quality parameter other than BOD may control cost.

REFERENCES

1. EDDE, H., *et al* "Pulp Mill Internal–External Pollution Control Economic Modelling for Receiving Water Protection," Proceeding of the International Symposium on Modelling Techniques in Water Resource Systems, pp. 289-308, May 9-12, 1972, Ottawa, Canada.
2. Study of Pulp and Paper Industry's Effluent Treatment, Report prepared by EKONO Consulting Engineers, for the Food and Agriculture Organization of the United Nations Advisory Committee on Pulp and Paper, 13th Session, Rome, Italy, May 15-16, 1972.

"Petrochemical Water Pollution Control"

Discussion *by* Roy F. Weston, D.A.A.E.E.

Diplomate American Academy of Environmental Engineers;
Chairman of the Board of Roy F. Weston, Inc. Environmental Scientists and Engineers,
Lewis Lane, West Chester, Pennsylvania 19380

I have been requested to present a brief discussion of current trends and future developments in the control of water pollution by the Petrochemical Industry.

The trends and developments in the Petrochemical Industry water pollution control will be much more dependent on regulatory policies and standards than on technology per se. Technological trends and developments will properly relate to regulatory policies and standards.

Since I believe that regulatory policies and standards are the dominant forces in establishing technological trends and developments, I have chosen to spend most of my time discussing matters influencing regulatory policy.

Simply and directly it can be stated that the trends and developments in pollution control are and will be a direct function of society's environmental quality objectives and its willingness to utilize available resources to accomplish those objectives.

Some discussion of the factors behind this statement are in order.

What I have to say is obvious to all and has been said before at this conference. I repeat these statements here because they should be repeated again and again to be sure that we professionals in the field retain the proper perspective.

Man, in his quest for a better life, has increased his numbers and has developed a complex system of commerce and industry. In the process of living and making a living, he generates waste by-products that he customarily discharges into the environment (into the public domain). The more complex and affluent his society becomes, the more by-product waste he generates per unit of population.

The discharge of by-product wastes into the environment can have an adverse effect. Their impact on the environment ranges from minor influences on local ecological balances, to local nuisances, to destruction of valuable food resources, to catastrophic episodes of toxicity, disease, and death. Thus, the system man has built to enhance his quality of life has the potential of seriously degrading his quality of life.

As a society becomes more affluent, its members show an increasing desire to enjoy the luxury of an unimpaired natural environment. Thus, it is obvious that more stringent pollution control requirements will be the future way of life, whether the objective is born of necessity to maintain a minimum acceptable quality of life or whether the objective is the luxury of an unimpaired natural environment.

The trends and developments in regulatory policy in the United States are worthy of note.

History is recycling.

The first water pollution control laws in the United States permitted no pollution. The administration of the letter of these laws became impractical, because neither industry nor the public was prepared for such a concept at that time (35 or more years ago), and because it was obvious that water courses could accept some pollution without adversely interfering with other beneficial uses. Therefore, pollution control for its own sake was abandoned in favor of the rational concept that it was in the public best interest to utilize a portion of the capacity of the environment to accept pollution so long as the environmental quality was adequate to protect assigned beneficial water uses. This concept of water quality protection was adopted as Federal law.

The design stream flow for the application of this policy has been selected as the average 7-day minimum flow that occurs at the average frequency of once every 10 years. That is, the design flow is equalled or exceeded 99.9 percent of the time.

Immediately, there were those who contended that water quality criteria and stream standards were the vehicle for legalizing pollution and for allowing the degradation of all public waters to a barely acceptable level. Consequently, the concept of non-degradation of clean waters and the concept of some minimal level of pollution control for all situations became a policy.

Later, with the speed of pollution abatement implementation, the impatience of the American public and its public administrators led to the partial, if not complete, abandonment of the water-quality-criteria water-quality-standard concept which attempted to establish effluent standards based on available technology. The application of this concept would require a predetermined minimal degree of pollution control regardless of the environmental requirements. However, in case the minimal degree of pollution control did not comply with the environmental requirements, a higher degree of control would be required to ensure compliance. In the process of establishing industrial effluent criteria, it has been recognized that in-plant pollution control, as well as effluent treatment, is a valid technique for attaining a desired effluent quality objective.

Recently, legislation has been proposed that would require, in time, a zero industrial wastewater discharge, i.e., zero pollution. Thus, the cycle has been completed.

Although the purpose of the recent administrative moves was to accelerate pollution control, the practical result has been the delay of the implementation of many control programs.

Other developments in the United States and some other parts of the world are worthy of note. So-called "environmentalists" are advocating population control and control of industrial development on the basis that such control is essential to maintaining a satisfactory environmental quality. The emotionalism created by the doomsday proclamations of purported experts and the image created by industry that it cannot or will not prevent pollution nuisance have led local residents of affluent societies to stop the location of new industry in their neighborhoods. The consequences of these actions, particularly relative to the power industry, have helped industry to recognize that it must conduct sophisticated detailed environmental baseline and monitoring studies so that reliable scientific information is available for evaluating pollution impact.

The recent U.S. Environmental Quality Act has made the preparation of environmental impact statements a way of life.

Historically the pollution control profession, as well as industry, has solved control problems a step at a time; attacking the most obvious before concerning itself about the more subtle aspects of the problem. Current environmental impact evaluations require consideration of all aspects of the problem at one time.

In evaluating the overall impact of pollution control, the economic impact on industry has come under scrutiny. As pollution control requirements become more stringent, the economic impact on industry becomes more significant. This is so, particularly if pollution control standards are not applied uniformly relative to place, degree, and time. Since practical considerations have prevented such uniform application of standards, economic inequities have existed and now exist within nations and between nations. Consequently, those who control pollution for the public good are competitively penalized. Therefore, more and more professionals and some nations appear to have come to the conclusion that the desired pollution control will not be attained until policies are adopted that will make it pay to abate rather than create pollution. If such policies are adopted and soundly applied at the same time, the increased costs of pollution control can be passed on to the using public, as they should be, instead of coming out of individual corporate profits.

Concern over economic impact leads to concern for cost effectiveness. Consequently, it has been appropriate to look at new approaches for solving all problems.

Apparent trends may be summarized as follows:
1. Environmental quality protection will continue to be the primary criterion for pollution control. 2. Some predetermined degree of pollution control, labelled "good practice" for "best available treatment," will be established as the minimum degree of pollution control that must be provided irrespective of the environmental requirements. 3. Industrial pollution control technology will recognize the use of in-plant pollution prevention techniques, as well as effluent treatment, for attaining effluent quality objectives. 4. The control of pollution "incidences" (i.e., abnormal discharges due to accidents or to equipment, process, or manpower failure) and the day-to-day reliability of pollution control systems will be much more stringent. 5. Continuous effluent and stream monitoring will become a standard requirement. 6. Increasingly, industry will be under economic pressure and public opinion and regulatory agency pressure to approach or achieve zero discharge. 7. Comprehensive scientific team investigations requiring diverse physical, chemical, biological, and social science disciplines will be required for evaluating the impact of existing and proposed pollution-producing facilities. 8. The economic impact of pollution control on both the private and public sectors will receive more attention. Least-cost solutions to system and regional problems will be required. 9. The imposition of an environment use charge is within the realm of possibility. (Such a charge is already in use in some countries at this time). 10. Legal technicalities will have a greater impact on the details of scientific and engineering solutions.

These trends present a tremendous technological challenge. The pollution control profession will be expected to achieve the objectives of society at the cost society is willing to pay or can afford to pay. The challenge is independent of the political or economic system under which the society operates. In any case, the control is determined by that portion of its capital and labor which a society is willing to divert to the necessity or luxury of environmental quality control.

The Petrochemical Industry is a product of man's quest for a better way of life. Its activities generate a variety of water pollution problems that are as complex and as diversified as those of any other industry. The industry's wastewaters may create problems of oxygen depletion, pH change, taste and odor production, color addition, oil slick development, metallic ion and/or organic compound toxicity, aquatic plant nutrient addition, salinity increases, and temperature rise.

In endeavouring to solve its problems, the industry was one of the first to think of pollution control in terms of its two basic alternatives: prevention and cure.

A more classical way of looking at the industry's alternatives would be that of considering three levels of sophistication of pollution control technology. These are:

FIGURE 1
PETROCHEMICAL PLANT
VARIABILITY OF TOTAL PLANT FLOW
MONTH OF FEBRUARY 1970

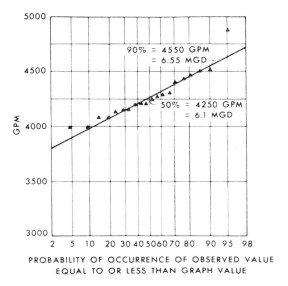

Fig. 1 Petrochemical Plant — variablility of total plant flow month of February 1970.

FIGURE 2
PETROCHEMICAL PLANT
VARIABILITY OF WASTEWATER BOD_5

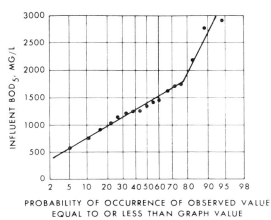

Fig. 2 Petrochemical Plant — Variability of wastewater BOD_5.

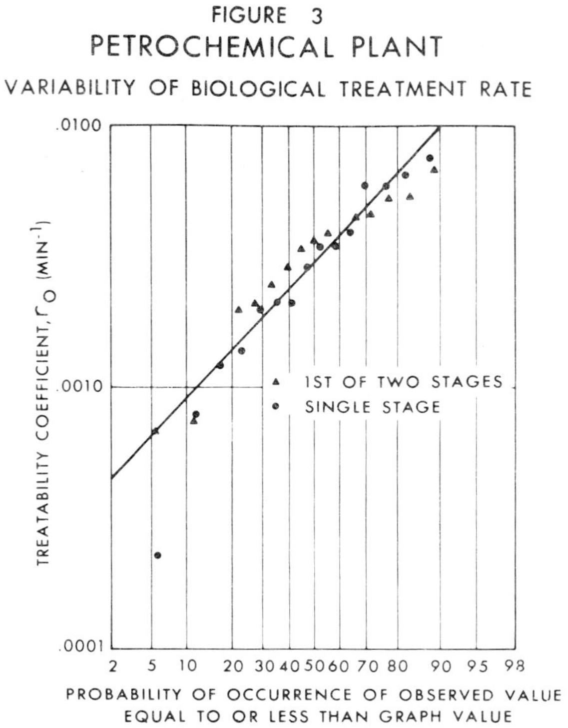

FIGURE 3

PETROCHEMICAL PLANT

VARIABILITY OF BIOLOGICAL TREATMENT RATE

Fig. 3 Petrochemical Plant – Variability of Biological treatment rate

1. Effluent Treatment concerned solely with the treatment system for municipal and industrial effluents.

2. Pollution Abatement concerned with the entire pollution generation system as well as the effluent treatment system and, therefore, includes all effluent pollution-reduction measures.

3. Environmental Protection concerned with the total environmental system with all the measure (including pollution abatement) that may be utilized to protect the system.

A typical example of the kind of problems involved in treating a petrochemical plant wastewater is provided by the data presented in Figures 1, 2 and 3. Figure 1 indicates a flow variability of about 15 percent, Figure 2 shows an organic loading variability of over 350 percent, and Figure 3 summarizes a rate of treatment variability of about 1,100 percent between the 10- and 90-percent probabilities of occurrence. This wastewater had to be treated to an average BOD5 which required a median removal efficiency of 96.3 percent; the maximum BOD5 could not exceed 100 mg/L. Experimental treatment data plus a mathematical model made it possible to design a two-stage activated sludge system that reliably meets the effluent requirements.

In two other instances, activated sludge could not be used successfully even though aerated lagoons could provide acceptable BOD5 removal. However, in other cases, the more stringent requirements for treated effluent concentrations of BOD5, COD, TOC, nutrients, metals, oils, and color have made conventional treatment techniques unfeasible both technically and economically.

Thus, it is apparent that careful investigation and evaluation of the entire waste generation, conveyance, treatment, and disposal system is essential to the development of the least-cost solution. In-plant studies may demonstrate economies by simple actions, e.g., separate collection and equalization, at-the-source treatment of troublesome discharges, or by more sophisticated actions such as process equipment changes, process reagent changes, or basic changes in the manufacturing processing. In the future, there will need be much closer liaison between the process developer, the process designer, and the environmental engineer than there has been in the past.

As our water pollution control problems become more complex, more serious, and more costly, advancement to the environmental protection level of technology will be required. I expect to see the day when a stretch of stream, or possibly an entire river basin, will be managed as a system, with monitoring and control facilities designed so that water quality within the system can be modified as needed to control to a desired objective.

As a practical and practicing environmentalist, I am one of those who firmly believes that man has the wisdom, if he has the will, to attain the physical comforts, conveniences, and luxuries of an industrial society and at the same time maintain the amenities of a healthful and pleasant environment.

R. Lepropre, Belgium

In waste water from paper mills, metallurgical operations and petrochemical plants, suspended matter or hydrocarbons or oils may be present. Our experience, using flotation with these effluents is promising, both with suspended matter removal and COD reduction. Have the authors any similar experience?

N.A. Lever, South Africa

In South Africa control authorities have recently begun to recognise the need to control discharges on a mass basis, i.e. pounds per ton of product and pounds per day rather than on concentration alone.

Matthew Gould, USA

There is a need to design wastewater systems to allow for human errors and mechanical upsets. Effluent requirements should be expressed in terms of pounds/unit of production rather than ppm in order to encourage wastewater reuse. Although I am against effluent taxes for discharge to surface waters I support equitable user charges for discharges into municipal or public systems.

J. Bernard, France

What is the suspended solids content of the treated waste in the cases of mixed sewage treatment (page 5 – three last columns of the Table 3) and what are the corresponding upward velocities and detention lines in the clarifiers at the different stages?

SAFETY CONSIDERATIONS FOR DETERGENTS

N. R. ARTMAN

Miami Valley Laboratories

The Proctor Gamble Company, Cincinnati, Ohio 45239

Questions increasingly arise about product safety and the effects of products on the environment. These are proper questions, and the detergent industry, like all other industries, must respond to them. We have a particularly high responsibilty in the area of water quality, for most of our products eventually find their way to waste waters. This paper describes how that responsibility is met by one of the large detergent manufacturers.

We see water quality questions as ultimately relating to questions of human safety, and, as manufacturers of food and personal care products, as well as of detergents, we naturally think first about personal safety. Then, in our investigations of environmental safety, we interpret our data in the light of prior knowledge about the human safety of our products. Much of what follows will describe how we investigate a new substance or formulation proposed for one of our products.

PRINCIPLES FOR SAFETY EVALUATIONS

Decisions concerning product safety are as important as decisions concerning product profitability, and the self-interest that necessitates our making good business decisions also requires that we follow very high standards in our concerns for human and environmental safety.

Our goal is for no product of our company, under any conditions of use or misuse, to cause harm to people or the environment. Our concerns include all components of our products, the impurities that will accompany them in commercial supplies, the interaction products of the components with other substances they may encounter, and all their breakdown products. We must consider biological activity, persistence, and accumulation.

Details of our concern for human safety are the same in all countries. Race and nation do not modify a person's susceptibility to chemical injury. But our responses to environmental concerns must take local situations into account. A material that is objectionable for use in one area may be perfectly satisfactory in another, owing to differences in population density, climate, sewage handling systems, etc.

We recognize that there is no such thing as absolute safety. Mankind has never lived under a zero-risk policy, and cannot do so at present. Rather our decisions relating to product safety have to be based on evaluation of the benefit/risk ratio associated with use of the product. For each new material that we consider using, we must answer the question, 'Are the benefits to be gained from the proposed use enough greater than the perceived risks to justify our going ahead?'.

There is no doubt of our Company's ability to assess the benefits that a new product will offer; at least, if we fail, we hurt only our stockholders. We feel competent also at judging risks, but no group of limited size can stay abreast of all the developments in contemporary knowledge and thinking. Therefore we use disinterested consultants, experts, generally from the academic community, and we rely on them for input into our judgment-making processes. And ultimately our judgments are scrutinized by government regulatory agencies in many countries, and by individuals who speak out in what they perceive to be the public interest.

SAFETY GUIDELINES

Although we have guidelines to follow in designing safety testing programs, and a battery of test procedures, there is no standard step-by-step procedure for finding out whether a material will be safe. For every new material, we devise a testing program based on our judgment and that of our consultants. The design of the program takes into account the properties of the material, its intended use, and any prior knowledge of its safety. The programs usually include studies of acute, subacute, and chronic effects in several kinds of organisms, and then look for such subtle effects as mutagenicity and teratogenicity. Human safety tests and environmental safety tests are closely intertwined. Not every substance needs to be subjected to every test, but in designing a program for a new material we need, as a minimum, to have considered each of the following.

Chemical Properties. Detailed knowledge is needed about the chemistry of the substance being tested — its oxidation states, its interaction products, possible impurities, and decomposition products. Sensitive and specific analytical methods must be available, and it is very helpful if the substance can be labelled with radio-isotopes.

Pharmacology. Absorbability through the gut, skin, and lungs, metabolizability by plants and animals, metabolic products, target organs, mechanism of action, and half-life in the organism have to be considered.

Toxicology. Acute toxicity—oral, percutaneous, and inhalation—may be measured in several species. Corrosiveness to oral mucosa, eye irritancy, skin irritancy, and allergenicity are prime concerns. Subacute toxicity tests provide information needed for the design of chronic studies. Long-term assessments of toxicity, teratogenicity, mutagenicity, and carcinogenicity are measured with large numbers of animals, usually of two or more species. Such experiments establish levels of the material that have no detectable adverse effects on test animals. These levels are divided by safety factors to give levels that must not be exceeded for human exposure.

Consumer Testing. Proposed new products, after thorough safety testing in animals, are subjected to human tests. At first, the products are exposed to small panels of volunteers, who have given their informed consent. The subjects are kept under strict professional surveillance during these clinical tests. Exaggerated conditions of exposure are generally used in these tests. Later, larger numbers of subjects can be exposed to the material, until we come to the large-scale tests involving hundreds or thousands of people. Here we look primarily for performance benefits, but any unforeseen properties that might affect the users will also be detected at this stage.

Biodegradability. Routine biodegradability screening tests of new substances involve seeding aqueous solutions of the test substance with micro-organisms and observing the rate of disappearance of BOD or the rate of formation of carbon dioxide. These tests are simple enough to be used for simulating the variety of conditions under which the substance may appear in nature. Direct measurements are also made of the rate at which the substance disappears after it has been added to a specimen of water collected from a stream, lake, or estuary. Acclimation studies are also important parts of the biodegradability testing program. Owing to the analytical difficulties at very low concentrations, it is sometimes convenient to do these tests using radio-isotope labelled compounds.

Environmental Toxicity. Concerns include acute, subacute, and chronic toxicity to micro-organisms, algae, plants, fish and other animals in the food chain. Possibilities for accumulation and for interactions with other environmental factors are also important. Standard test methods are available for answering some of these concerns, but for others methods are still being developed. In some cases, the state of the art is such that each investigation assumes the magnitude and complexity of a research project.

Treatability. A series of tests determines whether a substance will affect the treatability of sewages and waters containing it, and whether the substance will be removed from waters containing it by conventional treatment processes. Designs for these tests are straightforward; the test material is added to a sewage or simulated, sewage, and the mixture is taken through lab-scale simulations of settling, activated sludge treatment, trickling filter, sludge digestion, septic tank, adsorption, chemical precipitation and coagulation. The results of these tests are suggestive, not conclusive, and need to be followed by actual field tests. An ideal field test is carried out at a municipal sewage works having two identical systems side by side, where we can add our material to the influent of one system and use the other as control. Removal of the substance and effects on the processes can then be observed under natural conditions.

For septic tank field tests we recruit panels of home septic tank owners who agree to work closely with us. To some panelists we furnish a product containing the test material. Other panelists are given control products. Then, for periods up to one year, we measure water and product consumption in each household, as well as the level of product in the vault and in the effluent of each septic tank. Several analyses indicative of the condition of the septic tank operation are made at intervals. Only by such extensive and practical test procedures can we be sure that a new material destined for one of our products will not cause problems.

Projection of Levels. Safety, in the final analysis, represents a balance between toxicity and level. What substance is not harmful at sufficiently high levels—and 'harmless at sufficiently low levels? Our biological testing programs tell us that a given substance will produce certain effects at certain levels. We must compare those levels with the levels that will occur when the substance is used in our products. Hence we must make predictions about the quantities of material we (and others) will use, and about its distribution in time and space, and about its ultimate fate in the world.

From marketing and formula data we can predict with good reliability the total quantity of a substance that will be used in a given period of time, and thus the quantity of it that will appear in the average home. Most detergent products go down the sewer, and adequate data are available describing the average *per capita* production of sewage. For Europe and the US we know the disposition of that sewage — the fractions of it that are treated in each of several ways, and the fraction of it that flows untreated into waterways. When this information is combined with our projections of levels in usage, and our knowledge of treatability, we can predict the levels at which the substance will appear in raw sewage, the fractions of it that will be removed by treatment, and the quantities that will appear in effluents as they enter waterways. Having already studied the disappearance of the material from natural waters, and having also some information about the dilution of sewage effluents by receiving waters, we can predict whether, and at what level, the substance may enter the intake of a downstream drinking water system Our treatability tests give information about the removal of the substance during water treatment. Therefore we can predict the level at which the substance may appear in drinking water. This level is to be compared with the levels that we already know from our safety testing program to be safe or harmful.

These projections of level are tied to reality by our monitoring program. We regularly analyze water samples from many sources for substances of present or potential interest to us. Analytical values on water samples taken before and after introduction of products that know about enable us to verify our earlier predictions and make subsequent ones more accurately. And this monitoring program assures us that the materials already present in our products do not enter the environment at levels of questionable safety.

Special Concerns. An essential part of our safety testing program is to look for items of concern that are peculiar to the substance of interest. About these we cannot generalize; we can only put our best minds to work on the question, 'What possible unforeseen harm might come about through the use of this substance or by its interacting with other agents in the environment?'.

HISTORY OF NTA

To make the items mentioned above more meaningful, we shall examine how they have been applied in a real case, that of NTA. The NTA story exemplifies the kind of safety testing program that we now consider necessary.

Chemistry. NTA is the trisodium salt of nitrilotriacetic acid. It is a colorless, readily soluble salt. It is a powerful chelator of metals, comparable to citric acid, or the humic acids of soil. It is easily oxidized to carbon dioxide, water, and nitrate. Highly sensitive and specific analytical methods have been developed to measure it.

Toxicity. Acute toxicity tests with NTA give LD_{50} values ranging from 1 g/kg in monkeys to more than 5 g/kg in dogs. NTA induces the protective mechanism of emesis. In subacute toxicity tests with rats there was kidney damage when NTA was fed as 20,000 ppm in the diet, but not at levels of 5000 ppm or less. A two-year feeding study showed minimal effects on the kidneys at the 5000 ppm feeding level and no effects at lower levels.

Since the maximum levels anticipated for human exposure are of the order of 25 parts per billion (thousand million), very reassuring safety factors are provided.

Metabolism. Experiments with isotope-labeled NTA show that, in the rat, NTA is absorbed from the gastrointestinal tract. It is quickly excreted in the urine; less than 1% of the ingested dose appears as respiratory CO_2, and this we attribute to the metabolism of radioactive impurities present in the sample studied. Even after the rats' intestinal microflora had been acclimated by preliminary feeding with NTA, no biotransformation products could be detected.

Dogs, like rats, absorb NTA from the gastrointestinal tract, but rabbits and monkeys do not. Human studies show that NTA is largely unabsorbed, and that the fraction that is absorbed promptly appears in the urine, unchanged. Traces of NTA appear in bone during rat feeding studies, but the quantity does not increase as the experiment progresses. There is no measurable effect on bone growth or strength. The effects of NTA on zinc, calcium, and sodium metabolism have been studied, and results are quite satisfactory. Prolonged feeding of high levels of NTA had no effect on liver enzyme functions.

Reproductive Effects. Experiments with rats and rabbits gave no indication that NTA is embryotoxic or teratogenic, even at levels as high as 5000 ppm. Several experiments with mice and lower organisms gave no evidence of mutagenic activity.

Carcinogenicity. The possible carcinogenicity of NTA was investigated in a two-year feeding study with rats. No evidence was seen to suggest that NTA might stimulate tumor formation. If NTA had been a carcinogen, there is a very high probability that it would have been detected in this experiment. We, together with our consultants, after considering all available information, have concluded that we have reasonable and adequate assurance of NTA's lack of carcinogenic hazard. But at the time this is written (November, 1971) the US Surgeon-General has stated that NTA has not been proven not to be a weak carcinogen. Our efforts to answer this concern are limited by some laws of nature. It is not merely that we are confronted with the impossibility of proving a negative. The only way to detect weak carcinogens is to feed them at high enough levels so that their weak action is exaggerated and becomes observable, preferably as a dose-related response. We fed NTA at levels up to one-half percent of the diet, 100,000 times the expected human dose, and produced no evidence of carcinogenicity. High doses, which conceivably might have revealed an extremely weak carcinogenic activity, could not be used because other effects of NTA would have interfered. Our satisfaction with the present evidence of NTA's safety is great enough that we are continuing to use it in our products in countries where it is permitted.

Metal Chelates of NTA. Because NTA is a powerful chelator, questions arose early about the products formed by the interaction of NTA with heavy metals in the environment. Therefore a number of NTA-heavy metal complexes have been investigated for their possible effects as mutagens, teratogens, and carcinogens. Results of the latest experiments are very reassuring, and suggest that NTA may have some ability to protect animals from the harmful effects of heavy metals.

Significance of Findings. I have given much space to non-water-pollution aspects of the NTA story. This has been done to make clear that data on levels of a material in streams and on its survival through water and sewage treatment works mean very little unless we know what effects the material will have on people. Human safety testing is an inherent part of our environmental safety testing program.

Biodegradability. A number of experiments have shown NTA to be intrinsically biodegradable in aerobic systems. Both activated sludge and trickling filter units, once acclimated, consistently give greater than 90% removal of NTA from incoming sewage, and frequently this value approaches 100%. The products of degradation are water, carbon dioxide, and ammonia or nitrate. The chemical intermediates of the biodegradation pathway have not been identified, because they have no more than transient existence. The effects of temperature, loading, acclimation and oxygen level have been studied. NTA apparently is not degraded under strictly anaerobic conditions, but such conditions are rare in the world. The oxygen tension in septic tanks is great enough to effect substantial (60% in one experiment) removal of NTA from sewage passing through the vaults. Further degradation occurs in the leaching fields of the septic tanks. NTA added at levels of 5–50 mg/l to unacclimated river water is degraded after 8–12 days; when the water has previously been acclimated, the disappearance times are 2–6 days.

Most NTA-heavy metal chelates degrade as rapidly as NTA itself. Degradation of the mercury and cadmium chelates is a little slower. The presence of NTA in a sewage influent stream containing heavy metals does not cause elevated levels of those metals in the effluent.

Projected Levels of NTA. Predictions have been made of the levels at which NTA may appear in the environment if it is widely used. Assuming 50% replacement of all detergent phosphates with NTA, no biodegradation of NTA, and 10% contamination of drinking water with sewage, then the drinking water could contain as much as 4 ppm NTA.

Field tests show this prediction is much too high. One test was made in a densely populated area where each home has its own well and septic tank. The water table is shallow, and the soil is sandy. This area was chosen as the 'worst possible case'. Detergents containing NTA were being sold in the area at such levels that 6–10% of the wells should have yielded water containing 2000 ppb NTA. But analyses of samples from 279 wells showed one with 125 ppb, 5 with levels of 25 to 76 ppb, and the rest with none detectable at the 25 ppb limit of analytical sensitivity. At the same time, in the Great Lakes region of the US, it was estimated that, in the absence of biodegradation, NTA might appear in municipal waste waters at levels of 3–5 ppm, and possibly in receiving streams. Analyses of receiving streams showed 3 samples containing 39 to 190 ppb, and 32 samples containing none detectable. It is clear that most of the NTA was being destroyed before it reached the points where samples were taken.

It now appears that, should NTA replace phosphate in all detergent products, its level in receiving waters would average no more than 25 ppb. A person consuming such water would receive NTA doses of the order of 1 μ g/kg/day. In our animal toxicity tests, doses of 20 mg/kg/day had no harmful effects. Thus a safety factor of at least 20,000 is provided, which amply allows for upward deviations from the average levels and for variations in biochemical response. The safety factor becomes even greater when we recall that NTA was well absorbed by the test animals, but is poorly absorbed by man.

This projected safety factor is valid only if predictions about the levels of NTA in the environment are correct. Therefore, if NTA is to be widely used, there must be a continuation of our environmental monitoring program to find out the levels that occur over a longer time period.

HISTORY OF ABS

The history of ABS (alkylbenzene sulfonate) illustrates some of the broader points of safety testing. Before ABS was introduced as a household cleaning agent it was tested to establish its safety to humans. On the basis of these tests, ABS was recognized as remarkably safe, and subsequent events have fully justified that early optimism. Relatively little consideration, by today's standards, was given to its potential effects on the environment beyond the obvious recognition that, having a low inherent toxicity, it would not be harmful to living creatures at the anticipated concentrations. During the last 25 years the world has become much more conscious of the need and desirability to maintain and restore the cleanliness and beauty of our surroundings, just at a period when population growth and affluence have greatly increased the difficulty of that task. As a result, we are now asking questions that were not asked in the 1940's, especially questions about the effects of technology on the environment.

As is now well known, ABS caused foam on sewage treatment works and waterways in certain parts of the world. Elsewhere, little or no difficulty has arisen, and our company's response has been different, depending on the circumstances. Where the problem arose, it was relieved by the introduction of linear alkylbenzene sulfonate (LAS). In other places, ABS is still in use. In some places, laws were passed regulating the use of ABS, but in many places, including the US, the problem was met by the industry's agreement to replace ABS with LAS at the earliest practicable time. This kind of arrangement is the most satisfactory, both to industry and to the public. No law could have brought about the change had not LAS become available from the chemical industry when it did.

The question remains, 'What are we putting into detergents today which, 20 years hence, in the brilliant light of hindsight, will be seen to have been undesirable?' The answer is that no one can ever be sure that unanticipated effects will not result from his actions; that is within the definition of the word 'unanticipated'. But the present expanded scope of our environmental safety activities, and the degree to which our attention is directed toward environmental concerns, have certainly decreased the likelihood of unanticipated events.

We are limited by the contemporary state of knowledge and thinking, but the best available minds, both inside and outside our company, are helping us anticipate difficulties before they arise.

FORCES SHAPING OUR POLICIES

Three kinds of forces influence our product safety decisions. First is our continuing need to sell products that will universally be recognized as good, effective, and safe. This is simply good business.

The second force is public opinion. Public opinion is easily swayed these days, and sometimes we are hard pressed to keep up with its shifts. We have a responsibility to influence public opinion ourselves, into directions that are consistent with our knowledge of what is right.

The third force acting on us is the force of law. This is related to the public opinion force, for the appearance of laws and regulations ultimately responds to *vox populi*. We have two responsibilities towards the law. Our first responsibility is to obey the laws that have been passed and to live within the regulations that have been issued. Our second responsibility is to help shape future laws and regulations by making available our body of expert knowledge as input to the law-making process.

We do business in many different countries, and we must always be prepared to respond to differences in their laws, regulations, and environmental situations. This responsibility is borne by the people who have the primary responsibility for our business in each country. They are the ones who stay aware of local situations and call on all the technical resources of the company to ensure that we act at all times in conformance with our stated policies.

Procter & Gamble standards are sometimes higher than government standards. For example, we are very concerned about the corrosiveness of some detergents to mucosal tissue. In the US, no laws prohibit the sale of highly alkaline products that are damaging when ingested or when spattered into the eye. It would be contrary to our policies (provided we have a choice to sell a material so strongly alkaline that it would be hazardous to have in the home.

But there are other cases where laws force us to do things that we would rather not. For example, some communities (in November 1971) are in the process of enacting laws regulating the content of phosphates in household detergents. If these laws go into effect it may be necessary for our company to provide different formulations and different labels for different localities within the US. We don't like this, partly because it will mean extra expense for us and our customers, and partly because it will lead to consumer dissatisfaction with detergent performance. But our chief concern is that such laws represent an inefficient and ineffective solution to the eutrophication problem. These laws might have some justification if they were being considered only in areas where the phosphorus content of local sewage is known to contribute to eutrophication, and where reduction of phosphates in detergents would significantly lower the phosphorus content of sewage, but that is not the case. The introduction of such legislation seems sometimes to reflect other forces than a genuine and informed concern for environmental quality.

RECAPITULATION

Our company, and, we presume, most responsible companies are honestly concerned about the safety of their products. We have extensive programs of testing and predicting product safety. The strength of our safety testing program and the strength of our market experience over the years reassure us that we, as a company, have carried out our safety policies well. Our predictions are not infallible, but our methods of making them are always being improved. We are sensitive to public needs and desires. We do not accept the zero-risk philosophy. Instead, we believe that the benefits and risks of each product must be weighed against each other. We solicit and accept expert guidance from outside our company in evaluating benefits and risks.

Discussion Synthetic Detergent Constituents and Eutrophication
by E.J. Hudson

I think Dr. Artman underestimates the potential problem of an increased carry-over of heavy metals during sewage treatment due to the use of organic chelating agents in the quantities required for detergent use.

I would, however, like to discuss the basic principles of changes which may be required in synthetic detergents through environmental considerations, using eutrophication by way of illustration. The role of synthetic detergents in eutrophication is a unique situation which is unlike either of the cases cited in the two papers which have been presented.

When the contribution made by synthetic detergent constituents to eutrophication is considered, one is concerned only with the phosphate builder which provides a source of one of the chief plant nutrients. But here one can see the difference between this problem of water quality and others which involve synthetic detergents. To put the effects of synthetic detergent constituents into proper perspective, the phosphate builder is one of several sources of the abundantly available nutrient element, phosphorus, which in itself is only one of the many nutrients required to stimulate algal productivity in natural waters.

In order to culture algae it is necessary to provide at least 15 nutrient elements either singly, or in combination, with such compounds as chelating agents and vitamins in the case of certain heavy metals. Quantitatively the three most important nutrient elements are carbon, nitrogen and phosphorus which, from typical dry-weight analysis of algae, are required in the ratio of 106:16:1. In addition to the specific nutrient requirements of algae their growth is very dependent upon light intensity, temperature and climatic conditions in general.

Hence one would expect to find, and this is very true in practice, that eutrophication is not a universal problem. Since so many factors are involved simultaneously and any one of these may be critical in controlling algal growth, the effect of eutrophication on water quality is very localised. For instance, it has been estimated that only some 15% of the US population is affected by eutrophication as distinct from general problems of water pollution and a survey by the Environmental Protection Agency showed that out of around 4,000 lakes in the US, only 1% exhibited eutrophication. In the United Kingdom, eutrophication is not a problem except in certain water storage reservoirs where heavy algal growths have been witnessed for the past 50 or 60 years and effectively controlled by good reservoir management.

The chief sources of the particular nutrient phosphorus can be listed as: Domestic effluents, rural effluents and run-off — cultivated land, forestry regions and farm effluents, urban run-off, direct rainfall, aquatic flora.

The phosphate residues from domestic detergents appear in domestic effluents and comprise, usually, between about 30% and 50% of the phosphorus in the effluent, the remainder arising from human excreta. The contribution of phosphorus from domestic effluents amounts generally to about 50% of the total from all the sources. Hence, synthetic detergents can be said to provide around 15-25% of all the phosphorus appearing in natural waters.

Thus the role of synthetic detergents in eutrophication is not one of producing a specific pollutant which can be changed by reformulation to prevent the problem with which it is associated. The tripolyphosphate used in detergents is a harmless compound which has been shown to be rapidly hydrolysed, in sewers and during sewage treatment, to the ortho-phosphate form in which phosphorus is naturally present in the environment and which is essential to all life.

Why then has so much discussion evolved around the replacement of phosphate builders in synthetic detergents?

Firstly, of the major nutrients required by algae, phosphorus is likely to be the most easily controlled by controlling inputs to natural waters to bring it close to a limiting concentration. This is because carbon dioxide is available in the atmosphere and most objectionable algae, the blue-greens, can fix atmospheric nitrogen.

Secondly, it is argued that by eliminating phosphate builders from detergents it is possible to remove totally one source of phosphorus which will lessen the total input to natural waters considerably. In view of the figures presented above, I would argue that the extent of the reduction by this means alone would be insufficient to have any effect, since the reduction in algal productivity of a body of water is not proportional to the reduction in phosphorus input at the concentrations normally encountered in eutrophic bodies of water. That is to say, at concentrations which are already well above the threshold level. The possible ramifications of the introduction of a replacement for phosphate builders are then all-important as shown by Dr. Artman's paper.

Thirdly, it is possible to remove phosphorus easily and relatively cheaply from sewage effluents by existing technology using a chemical flocculation process. This provides an effective means of reducing phosphorus inputs to natural waters from domestic effluents and intensive farming effluents. The question of whether or not detergent phosphates should be replaced when improved sewage treatment processes are installed then becomes one of cost.

In connection with the concept of benefit/risk calculations introduced by Dr. Artman, the most important calculation which should be carried out is not that for a particular substitute for detergent phosphates. The calculation which should be made is that of the benefit/risk ratio, including the respective costs, of adopting an overall policy directed towards improvement of water quality where eutrophication occurs against the risk of introducing substitutes for detergent phosphates as a partial policy.

The first parameter which must be evaluated is the actual cost to the public at large of eutrophication. Against this must be set the costs of the various corrective measures proposed and the benefits which will accrue from these and their attendant risks.

No figures appear to be available for the cost of eutrophication in those areas where it is considered to be a significant problem. What is available is the cost of improved sewage treatment to remove phosphorus chemically which, from operating experience has been shown to be of the order of $1 to $2 per person per year. The benefits of such processes are far more than just removal of phosphorus, including increased efficiency of BOD removal, removal of viruses and parasite eggs and removal of metals with the further improvement of increased hydraulic throughput.

Since eutrophication is a localised problem, the introduction of improved sewage treatment will only be required in those areas. If such treatment plants are installed then a further calculation is required to examine the cost saving which could be brought about by eliminating detergent phosphates from crude sewage. Experience has shown that a 50% reduction of phosphorus in crude sewage results in only about 20-30% saving in the cost of treatment chemicals. Capital costs, of course, remain the same since the phosphorus from human wastes still has to be removed. Furthermore, introduction of an organic chelating agent to replace detergent phosphates may well nullify this small cost saving since, unless it is completely biodegraded in the sewage treatment process prior to the chemical flocculation state, it will cause an increased use in chemicals due to its chelating power.

If synthetic detergents continue to use phosphate builders, then the adoption of improved sewage treatment involves no risk other than that the reduction in phosphorus input achieved may still not be sufficient in some circumstances to reduce eutrophication. If a replacement for phosphate builders is used in addition to improved sewage treatment, then all the risks outlined by Dr. Artman have to be evaluated against the small cost saving of chemicals used in the treatment process.

If the technique of benefit/risk calculations is adopted to evaluate the possible approaches to improving water quality in eutrophic bodies of water, then it would appear that the maximum benefit at the least risk will be obtained by upgrading sewage treatment processes in those areas where eutrophication is regarded as a serious problem whilst continuing to use phosphate builders in detergents.

Reply

I certainly agree that phosphate removal by tertiary sewage treatment is the right thing to do to ameliorate certain eutrophication problems. I also would agree and emphasize that eutrophication is a local problem, affecting only certain areas. In the U.S., of course, certain localities have prohibited the use of phosphates in detergents. I think such regulations are mistaken, but where they exist, our Company has to obey them; in these situations we have little choice other than trying to devise alternative detergent formulations that will be safe, effective, and permissible.

J.A.G. Taylor, UK

Studies at the Water Pollution Research Laboratory (UK) with N.T.A. added to pilot-units operating on domestic sewage showed that at $6-8°C$ no degradation occurred. At $15-20°C$ acclimatization occurred relatively rapidly and satisfactory ($>90\%$) N.T.A. removal was achieved. However, when the temperature was again lowered, then below $10°C$ satisfactory removal was no longer achieved and generally fluctuated around $60 - 80\%$. Work by Dr. N. Harkness of the Upper Tame Main Drainage Authority, Birmingham, UK, shows that using sewage containing metals derived from industrial effluents zero degradation of N.T.A. occurred at $15°C$. Harkness also found the levels of four metals analysed were higher in the discharge from the N.T.A. containing unit by 20–50%; 70% of the metal normally retained by the sludge was extracted by the N.T.A, and appeared in the effluent. See R. Wilsson, Water Research, 1971, Vol. 5 pp 51–60 for further effects of NTA on metals removal. Thus, N.T.A. degrades inadequately in domestic sewage at temperatures found during the winter in sewage works in most Northern countries; it does not degrade at summer temperatures in the presence of metals at levels to be found in practice in many industrialised areas, and it transports additional quantities of heavy metals into the environment.

Reply

The comments about the non-biodegradability of N.T.A. at low temperatures are interesting and puzzling. We have evidence from laboratory units, pilot units, and field studies, showing that N.T.A. does degrade even at low temperatures. The rate of degradation does diminish as temperatures fall, just like the rate of any other chemical or biochemical reaction, but it is clear that there are organisms that attack N.T.A. at low temperatures.

Regarding the mobilization of heavy metals by N.T.A., any realistic assessment of this situation requires an appreciation of the complexities of multi-component equilibria and of the complexities of natural waters. N.T.A. does chelate nickel. But what are the extent and the effect of this chelation in the presence of other chelatable metals, especially calcium and iron, and in the presence of other, naturally occurring, chelants? Most laboratory work has been done at high concentrations of N.T.A. or heavy metal or both; the use of high levels simplifies the analytical chore, but divorces the situation from reality. No laboratory experiment, using a simplified system, is going to give us information that can be directly extrapolated to a natural water-way. Only field studies and monitoring programs will resolve the issues. Such work is now in progress. And, of course, the question of excessively high concentrations of heavy metals in certain effluents might be another example of a local situation requiring a local response that would be inappropriate elsewhere.

J.P. Bruce, Canada

In Canada national policy was directed to reducing the phosphorus that might enter lakes from waste water of cottages to delay the need to remove nutrients from sewage effluents; to reduce the cost of nutrient removal from effluents entering eutrophic lakes and reduce the phosphorus content of sewer overflows. This has resulted in a 30% overall reduction and a 15% overall reduction in P with beneficial results. N.T.A. has been used in Canada. It is estimated from observed data that with full replacement of phosphate by N.T.A. the levels of N.T.A. will be at least 2 or 3 orders of magnitude less than the concentration at which any carcinogenic effects may occur. No increased concentrations of metals in water have so far been observed as a result of N.T.A. usage. Finally sodium citrates which are biodegradable have been shown in both Canada and USA to be acceptable builder for detergents. Canada is encouraging their use as a phosphate substitute.

Reply

This additional information is useful. The Canadian situation is yet another example of a specific response to a local situation. I would only correct the implication that N.T.A. is carcinogenic at some level; there is no evidence whatsoever that N.T.A. is carcinogenic at any level.

T. Helfgott, USA

If N.T.A. does degrade completely it yields nitrates, which can be a problem in eutrophication and in ground waters. I believe overall treatment will eventually be needed to remove more contaminants from discharges to lakes and rivers.

Reply

Calculations have been made for the U.S. showing that if N.T.A. were used as a substitute for phosphate in detergents it would increase the amount of nitrogen going into our waters by about 1%. This increment seems negligible under most circumstances. The problem of excess nitrate in ground waters occurs only in certain localities. I would emphasize that any action that is acceptable or even desirable in one part of the world might be undesirable somewhere else, depending on local environmental conditions.

D.H.A. Price, UK

In the matter of risk benefit/analysis who should be responsible for decisions on acceptability? Both risk and benefit varied regionally. In the UK phosphates did not result in any significant problems — if they did the preferred solution might be removal of all phosphate from sewage effluents. Voluntary control was realised in the UK; this permitted continuous pressure for increased biodegradability. Did legislation tend to freeze the position and hold up progress? Was the control of detergent biodegradability in other countries confined to domestic products or did it include detergents used industrially?

Reply

In reply to the question of whose judgment is to be used in evaluating the benefit/risk ratio, I think that no one individual or group can be recognized as both adequately informed and free of bias. Therefore the decisions must be made jointly through wide discourse. We make the first decision, of course, since we have the data, and it often happens, even very early in the evaluation process, that we decide not to use a material. We also have the final responsibility, because we will have to answer to the consequences of any mis-judgement that we might make. If we do decide to use a material, that decision will be subject to review by government regulatory agencies, consumer spokesmen, the scientific community, and, finally, by each consumer. I believe that we need continuing discourse among all knowledgeable and interested parties to ensure progress toward optimum decisions.

I am not aware of any detergent-related situations where legislation has impeded desirable technological changes. To the best of my knowledge, different countries have taken different attidues toward legislative requirements for detergent biodegradability, and have differed in applying these requirements to both industrial and domestic products. However, we are now seeing in the U.S. some undesirable consequences of legislative action that was over-hasty and inadequately based on technical expertise. Some localities have banned phosphates from detergents. Since the Federal Government has not permitted the use of N.T.A., these localities are now being supplied with detergents containing high levels of sodium carbonate or sodium metasilicate.

We consider these highly alkaline products too dangerous to be handled casually in the home. More effective discourse among consumers, legislators, and industry might have led to a better result than this.

P.M. Higgins, Canada

Canada has embarked on a national regulatory program to restrict the phosphate content of detergents in order to reduce the total phosphorus loading to lakes and streams. Construction of treatment facilities to remove phosphorus in domestic sewage is also being accelerated. The P_2O_5 in laundering products was reduced to 20% in August 1971. A further reduction to 5% P_2O_5 is scheduled for December 1972. It is considered that phosphate reduction in laundering products has more benefits than risks. In this policy N.T.A. is regarded as one likely substitute for phosphate.

R. Cabridenc, France

Before a product is made currently available on the market, it is necessary to know its biological properties (biodegradability and toxicity) especially when it is a new product which is intended to be used on a large scale, as trinitrilo acetic acid. There is a great need for the study of the toxicities of such products, but it is often difficult to measure them in laboratory scale experiments because there may be interferences between the product itself, its metabolites and other substances present in the water. Toxicity may be — acute, — long term toxicity — or it may appear after concentration along a food chain. Many organisms are concerned : among the most important we find : bacteria responsible for biological purification processes, algae responsible for photosynthesis, fish. Such toxicity tests are indispensable, but it seems first necessary to come to a national and international agreement on the method.

Meanwhile we may be satisfied when we see the efforts done by the manufacturers, who, fully aware of the severity of the problem, subject their products to many tests to verify their properties before putting them on the market.

Study of N.T.A. biodegradation is a problem which is difficult to solve because biodegradation is defined as the rate of destruction of certain structural elements of the molecule and this rate is affected by the variables interfering with metabolisation.

Since different biodegradation test methods are used it is not surprising that results vary widely from author to author. Surely nitrilo-triacetic acid has some good qualities and meets the requirements needed by the substitutes for detergent phosphates.

But it seems necessary to avoid the use of new products when their effects towards environment are not completely known.

Although they are in some measure responsible for eutrophication of some waters, polyphosphates show some advantages. As their nuisance value is already known, they are easy to control, they are non-toxic towards living organisms, and a partial removal of phosphorus from sewage is possible.

Attitude towards Nonionic Detergents as a factor in water Pollution, relative to Anionic Detergents

by R. Cabridenc

Chef de Service à l'Institut National de Recherche Chimique Appliquée

91 – VERT-le-PETIT France

Studies on nonionic surfactants is proceeding as their production and use increases.

The behaviour in nature of anionic and nonionic surfactants may be compared and consideration given to rules concerning their production, commercialization and discharge.

Anionic surfactants still represent the larger part of products entering into the composition of detergents. The presently most used products are alkylsulphates, alkylsulphonates and alkylbenzene sulphonates.

The most commonly used nonionic surfactants are those formed by reaction of a molecule with a labile hydrogen bond (alcohol, acid, phenol, amine, sugar, etc . . .) in a polyether hydrophobic group derived from ethylene oxide.

There is general agreement that in some cases anionic surfactants are responsible for the foaming in streams and treatment plants, for decreased efficiency of purification, and for the complete or partial inhibition of the growth of water flora and fauna, generally because of lowering of the surface tension of the water. Their toxicity towards humans and mammals or the toxicity of their biodegradation products does not appear to be of importantce so far as their normal use is concerned.

Nonionic surfactants may also cause foaming, or stabilize or increase foaming caused by other products. They are also toxic towards the flora and fauna of receiving waters.

With the object of limiting nuisance it is essential to give very close attention to the waste surfactant problem, and to avoid the manufacture of all products which are not non-toxic or not rapidly biodegraded in sewage works or receiving waters.

"Biodegradability" has no absolute meaning; it is measured in laboratory experiments by a metabolisation rate under conditions which do not always closely simulate natural conditions.

Biodegradability may be considered from three different points of view:

1. Chemists regard biodegradation as occurring when the molecule undergoes a small modification, which may bring no significant alteration to its physical, chemical and biological properties.

2. Hygienists regard biodegradation as the loss by the molecule of those properties responsible for causing nuisance in the environment, i.e. when surface active properties disappear, whatever the reasons.

3. Biochemists regard a molecule as biodegraded when it is completely mineralised into carbon dioxide and water.

There are many biodegradation test methods, built on similar principles but showing significant differences in their procedures. In every case one must first extract the surfactants from the laundering product, then separate anionic from nonionic compounds with an ion exchange resin. Then a known quantity of the surfactant is introduced into the system, and its fate is observed by one of the three following criteria: 1. Oxygen uptake, 2. Physical properties of the medium or 3. Specific chemical methods.

The biodegradation rate is modified by numerous factors: nature and concentration of bacteria, their degree of acclimation, surfactant concentration, pH, temperature retention time; medium used (especially its organic matter content).

The two kinds of procedures are – static methods and continuous flow systems.

Among the static methods are the closed bottle technique with no oxygen added where biodegradation is measured by the weight of oxygen taken per unit weight of the compound; and the open bottle techniques, derived from the river water die away test, where oxygen is added by aeration or shaking, and biodegradation usually measured by analytical methods.

Continuous flow techniques are used to simulate field treatment processes. They need laboratory scale activated sludge or trickling filter units, fed with natural or synthetic sewage containing a known concentration of the surfactant. Biodegradability is usually assessed by a analytical method.

The most commonly used analytical technique for anionic surfactants is the methylene blue method of Longwell and Maniece[1].

In many countries legislation or agreement with industrialists exists to prevent the use of anionic products with a biodegradability of less than 80%.

For nonionic compounds, particularly for polyethoxylate surfactants the problem is much more complex. We can use the different biodegradability test methods and show the disappearance of the products according to the principles already described. Some authors consider the possibility of using the oxygen uptake rate, the decrease of the foaming capacity, or the lowering of the surface tension. But in the interests of uniformity we think that it would be better to use biological conditions (medium, outfit, concentration, inoculum) close to those used for anionic surfactants, and to utilize a simple chemical analytical method. Analytical techniques exist but are not quite satisfactory because of their limitations and their complexity. They depend upon the disruption of the polyoxyethylene chain or its shortening, but give no indication concerning the evolution of the lipophilic part of the molecule. The surface active properties of the product are modified but it is not completely metabolized. Chemical determinations do not always agree with the results of measurement of the foaming potential. So, two oxyethylated alkylphenols, one with a linear chain, the other with a branched chain will give relatively similar results when the concentrations are below the bacteriostatic threshold; but the branched chain will remain in the medium much longer than the other.

On the other hand we may paradoxically consider the same branched ethylbenzene group as biodegradable if it is transformed into a polyethoxylate and as resistant if it is transformed into an alkylbenzensulphonate. The chemical analytical methods convenient for nonionic surfactants may be divide into two classes:

1. methods in which the nonionic surfactants are transformed into anionics.

2. A complex with a heteropolyacid is formed and the concentration of the complex or one of the reactants in the system is determined.

Nonionic surfactants can be transformed into anionic surfactants by reaction of their terminal hydroxy group with some anhydrides giving an acid which reacts with methylene blue and is measurable by the Longwell and Maniece technique. If the hydrophobic part of the molecule is shortened the value obtained by the Longwell and Maniece technique is low so this reveals biodegradation of part of the molecule. But it has the disadvantage that many normal bacterial metabolites (phenols, alcohols, amines, etc . . .) give positive interferences.

In spite of its imperfections the most commonly used method is based on the formation of a precipitated complex of the polyethoxylate hydrophilic group with an heteropolyacid. A preliminary purification step by extraction with a more or less specific solvent (methylisobutylketone, ethylacetate) is necessary; it eliminates the non-foaming biodegradation intermediates such as ethylene glycol. Gravimetric, colorimetric, turbidimetric, potentiometric or chromatographic techniques may be used in the final stage of analysis.

In Pitter's method[3] polyethoxylates are complexed with phosphotungstic acid, and tungsten is measured by colorimetry after treatment with hydroquinone.

The methods of Crabb, Persinger, Greff, Stezkorn & Leslie[4,5] are based on the formation of a complex with ammonium thiocyanate, dissolution into benzene followed by a colorimetric finish. The technique is simple but lacking in sensitivity. The results are not closely related to those given by the foaming potential or surface tension measurement. The method does not work well with nonionic products of high molecular weight.

Wickbold's method[6,7,8] is often used in Germany, Switzerland and France. The surfactant is extracted in special conditions with ethylacetate, then a complex with potassium tetraiodobismuthate is precipitated, and bismuth is measured with sodium pyrrolidone dithiocarbamate by a colorimetric or better still by a potentiometric method. This method can be used with polyethoxylate nonionics ranging from 6 to 30 molecules of ethylene oxide containing more than 7 carbon atoms in the hydrophobic chain.

The results are reasonably near to those given by the foaming potential but not identical with them. This method is not very sensitive (300 μg per sample). Large volumes of solution are therefore required. It needs much care in the preparation and use of the reagents. Repeatability is not satisfactory with all products, and the results are often high.

The method of Patterson, Scott, Hunt & Tucker[9,10,11] is commonly used in the United Kingdom. It begins with extraction of the surfactant from water into chloroform followed by thin layer chromatography on a silica substrate. The spray reagent contains bismuth. Quantitative results are given by comparison with dilutions of a standard surfactant. This technique gives interesting scientific indications, because it shows the biodegradation intermediates. It is reasonably easy to use, but it seems difficult to base legislation on it because it is lacking in precision.

In conclusion it seems necessary to examine the nonionic surfactants problem with the aim of obtaining a general agreement on a biodegradation test method. The technical problem is not completely solved but there is much discussion at national and international levels, especially in the OECD, to get a satisfactory analytical method. It seems feasible that in the near future, it will be possible to base legislations or agreements between industrials and governments on a biodegradation test method for polyoxyethylene nonionic surfactants.

REFERENCES

1   LONGWELL, J., and MANIECE W.D., – Analyst. 80 – 167-71 (1955).

2   HAN, K.W., – Tenside 4 – 43-5 (1967).

3   PITTER, P., – Chem. and Ind. (1962) 1832-3.

4   CRABB, N.T., and PERSINGER, H.E., – J. Am. Oil Chemists' Soc. 41 – 752-3 (1964).

5 CREFF, R.A., SETZKORN, E.A., and LESLIE, W.D. – J. Am. Oil Chemists' Soc. 42 – 180-5 (1965).

6 WICKBOLD, R., – Communication a Essen 18.2.62.

7 WICKBOLD, R. – Vom Wasser 33 – 229-41 (1966).

8 WICKBOLD, R. – Tenside 8 – 61 (1971).

9 PATTERSON, S.T., HUNT, E.C., TUCKER, K.B.E., J. and Proc. of inst. Sew. Purif. 190 (1966).

10 PATTERSON, S.T., SCOTT, C.C., TUCKER, K.B.E. – J. Am. Oil Chemists' Soc. 44 - 407 (1967).

11 PATTERSON, S.T., SCOTT, C.C., TUCKER, K.B.E. – J. Am. Oil Chemists' Soc. 45 – 528 (1968).

SYNTHETIC DETERGENTS AND WATER. THE SOLUTION OF THE PROBLEM IN GERMANY

F. MALZ
43 Essen-Steele Eschenburg 74
Federal Republic Germany

Detergents Defined

By definition a detergent is anything that cleanses, but the term 'detergent' is now generally applied to packaged cleaning products used with water in household laundry — products which exhibit 'soapiness' without the disadvantages of ordinary soap when used in hard water. In ordinary soap some 90% of the total composition of the product may be the surfactant — the sodium salt of fatty acids. In synthetic detergent products, however, the surfactant accounts for a much smaller percentage of the entire product, i.e. 20% surface substance, 30–40% phosphates and 40–50% builders.

The efficiency of surfactants for specific cleaning tasks is upgraded by the presence of builders which add to their soil suspending power.

Today, the surfactant material most widely used are anionic substances, typified by alkyl benzene sulfonate.

'Cationics' used for sanitizing and softening, fall outside the field of household synthetic detergents with which we are concerned. In total we have about 75% Anionics, 20% Nonionics and 5% Cationics on the market.

Detergents and their Effect on the Water Supply System

Between 1950 and 1960, in many fields of application in household and industry 'natural' soap products were replaced by synthetic detergent raw materials. In the Federal Republic of Germany, the development on the detergent market began in 1954. By 1960, the market share of soap had declined to an insignificant percentage. Synthetic detergents still show an upward trend on the market. Developments have been similar in all industrialized countries of the globe.

Although the technical advantages offered by the detergents were considerable, the water supply system was seriously affected. The tetrapropylene benzene sulphonate in common use resisted biological degradation and it was very soon found in increasing concentrations in sewage works effluents, river and surface waters, in the ground water, and in some cases it became a hazard to safe and controlled water supply.

As ascertained in numerous tests on the basis of the decrease of 'methylene blue activity', the rate of biological degradation of TBS — which accounts for more than 80% of the anion-active detergents used — reached a maximum of 25% under German sewage treatment plant conditions. If drinking water and water for industrial use was obtained from river or surface waters, it was likely to contain MBAS (methylene blue active substance) in considerable concentrations. For example, the drinking water for the city of Essen is for the most part procured by artificial soil filtration of the Ruhr river water. Due to the long spell of dry weather in the summer of 1959, there was no additional supply from the reservoirs. Because of the need to cycle the water at times through treatment plants and receiving streams, mineral substances as well as MBAS, as shown in Fig. 1, accumulated in the drinking water within a few weeks. The presence of detergents

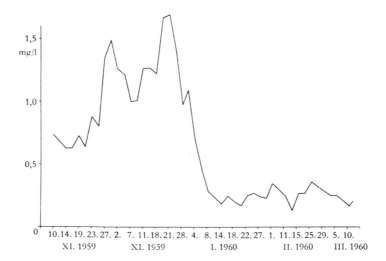

Fig. 1. Contents of anion-active detergents in Tap water

was apparent from considerable foaming when the taps were turned on. Difficulties were encountered by many industries, particularly by the beverage industry which depends on water of high purity.

The problem of foaming was not only the most apparent but in many sections of water supply the most difficult to cope with. In rivers and canals the increasing consumption of detergents showed in the foamy tracks behind the ships and excessive foaming at weirs and locks. In fine-bubble aeration plants the input of O_2 was diminished by over 30%. As a result of this excessive foaming, the activated sludge was floated in the activated sludge plant and remained ineffective. Anaerobic sludge treatment was biologically inhibited or indirectly impeded by the foam.

At first, anti-foaming agents were used, however, they did not solve the problem.

In the Federal Republic of Germany, it would only have been possible by considerably enlarging existing and projected treatment plants to remove 80% the biologically 'hard' and slowly degradable TBS. From the beginning, government and industry agreed that the only way to tackle the problem was to seek a compromise under the aspects of national economy between realizable sewage treatment technique and changeover on the production side to biologically 'soft' detergents.

Changeover to Soft Detergents and relevant Legislation

The German Detergent Legislation of September 1961 was enacted on Oct. 1st, 1964. An amendment specified that anionic surfactants in washing and cleaning products must be at least 80% degradable in a standardized laboratory test which simulates German sewage purification practice. Nonionic surfactants have so far not been touched by the law because of their much lesser importance and the difficulty in assessing them analytically. The German industry had three years between promulgation and actual enforcement of the law to develop suitable biologically 'soft' detergents.

The surfactant most widely used in Germany and everywhere at that time in detergent products was the sodium tetrapropylene benzene sulfonate. It has a relatively high biological stability and is the branched structure of the alkyl chain of the molecule. It was found that the biodegradability is greatly improved by replacing the branched by the corresponding linear alkyl chain. The linear C_{10}–C_{13} alkylbenzenesulfonate used by the detergent industry consists of a great number of isomers and homologues. The sulfophenyl group can be attached at the second, third, etc. C-atom of the alkyl chain. The degradation rate of these individuals varies somewhat. Work carried out at Chemische Werke Hüls proved that the compounds with longer alkyl chains are more rapidly degraded than those with shorter chains. The softer a LAS constituent is the more harmful it may be to fish and aquatic life so that a compromise between biodegradability and fish toxicity is probably best for practical purposes.

Years before the change, extensive laboratory and field work was carried out by the detergent industry, municipal and river board authorities etc., to assess the biodegradability and the environmental behavior of the new 'soft' detergents. On the other hand, the state of the surfactant contamination of the German rivers before the conversion had to be established in order to have a reliable base of comparison for the time after.

In all this work a modification of the internationally recognized method of Longwell and Maniece was used to determine very small quantities of anionic surfactants.

The standardized test to measure the biodegradability of a detergent is given in the governmental regulation of Dec. 1st 1962. This test is intended to reproduce conditions of sewage treatment.

The method of measurement employs a small activated sludge plant in laboratory scale (Fig. 2). The equipment consists of a storage vessel A for the synthetic sewage,

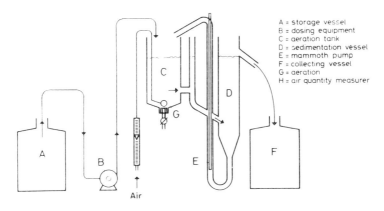

A = storage vessel
B = dosing equipment
C = aeration tank
D = sedimentation vessel
E = mammoth pump
F = collecting vessel
G = aeration
H = air quantity measurer

Fig. 2. Testing device

dosing pump B, aeration vessel C, separator D, airlift pump E to recycle the activated sludge, and vessel F for collecting the treated effluent. Vessels A and F hold at least 24 liters. Pump B must provide a constant flow of synthetic sewage to the aeration vessel at a flow rate of 1 liter per hour. The aeration vessel during normal operation contains 3 liters of mixed liquor. An aeration tube is inserted in the aeration vessel and the quantity of air blown through the aerator is controlled by means of a flow meter.

For the test a synthetic sewage is employed, consisting of 160 mg peptone, 110 mg meat extract, 30 mg urea, 7 mg NaCl, 4 mg $CaCl_2$ 2 H_2O, 2 mg $MgSO_4$, 7 H_2O, and 20 ± 2 mg MBAS per liter.

The synthetic sewage is freshly prepared daily.

Uncompounded surface active agents may be examined in the original state. Formulated products are analyzed for MBAS and soap content. The MBAS is extracted. If the sample contains less soap than MBAS the isopropanol extract is used. Further removal of the soap is necessary when the sample contains more soap than MBAS.

To operate the equipment the synthetic sewage must pass through the aeration vessel at the rate of one liter per hour; this gives a mean retention time of 3 hours. The rate of aeration should be regulated so that the contents of the aeration vessel are completely mixed while the dissolved oxygen content is at least 2 mg/l. Foaming must be prevented by appropriate means. Antifoaming agents which inhibit the activated sludge or contain MBAS must not be used. Airlift pump E must be set so that the activated sludge from the separator is continually and regularly recycled to aeration vessel C. Sludge which has accumulated around the top of aeration vessel C, in the base of settling vessel D, or in the circulation circuit must be brought into circulation at least once each day by brushing or some other appropriate means. When sludge fails to settle, its density may be increased by addition of 2 ml potions of a 5% solution of ferric chloride, and repeated if necessary.

The effluent from separator D is accumulated in vessel F for 24 hours, following which a sample is taken after thorough mixing. Vessel F must be carefully cleaned.

The MBAS content (in mg/l) of the synthetic sewage is determined immediately before use.

The MBAS content (in mg/l) of the effluent collected over 24 hours in vessel F should be determined analytically by the same method, as soon as possible after collection. The content ratio must be determined to the nearest 0.1 mg MBAS/l.

As a check on the efficiency of the process the C.O.D. of the filtered synthetic sewage in vessel A is measured at least twice weekly, as well as that of the filtrate of the effluent accumulated in vessel F. The reduction in C.O.D. is expressed in percent.

The reduction in C.O.D. should level off when a roughly regular daily MBAS degradation is obtained, i.e. at the end of the running-in period shown in Fig. 3.

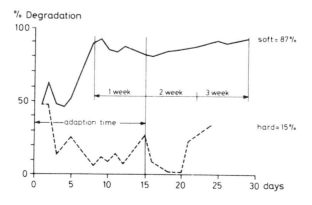

Fig. 3. Biol. Degradation

The loss on ignition of the dry matter in the activated sludge in the aeration tank should be determined twice a week (in g/l). If it is more than 2.5 g/l, the excess activated sludge must be discarded.

The test is performed at room temperature; this should be steady and should never fall below 18°C nor exceed 30°C.

The percentage degradation of MBAS must be calculated every day on the basis of the MBAS content in mg/l of the synthetic sewage and the corresponding effluent accumulated in vessel F.

The degradability figures thus obtained should be presented graphically as in Fig. 3.

Degradability of the MBAS should be calculated as the arithmetic mean of the figures obtained over the 21 days which follow the running-in period, and during which degradation has been regular and operation of the plant trouble-free.

The German regulation takes account of the soap content when determining the biodegradability.

$$A = a + \frac{b \cdot c}{100} \cdot \frac{100}{a + b}$$

A = biodegradability of the anion-active material
  (MBAS + soap) in percent
a = soap content in percent
b = MBAS in percent
c = biodegradability of MBAS in percent

By the German regulation of Dec. 1st 1962 a product which contains anion-active material (MBAS + soap) will pass the test when the biodegradability is at least 80%. The biodegradability of soap is taken as 100%.

Effect of the change from ABS to LAS

It was natural not to be expected that the mountains of foam on sewage plants, at weirs and locks would disappear immediately after the Detergents Law had come into force in Oct. 1964, since considerable quantities of washing and cleansing agents with hard detergents, were kept in households and at retailers, and had first of all to be used

TABLE 1

Sewage Plant	Degrading of Detergents in %	
	until October 1964	from the end of October 1964, to the beginning of 1965
Welver	23.8%	41.7%
Nottuln	16.2%	40.8%
Selm	19.6%	61.2%
Rhynern	24.2%	53.3%
Bad Sassendorf	18.4%	40.3%
Marl-East	15.1%	53.0%
Average	19.5%	48.4%

up. Table 1, which contains the results of investigations from some biological sewage plants of the Lippe Association, shows, however, that the changeover from hard to soft, i.e. biodegradable detergents, had an effect already at the end of 1964 and the beginning of 1965. Before the Detergents Law became effective, the average degradation was 19.5%. This increased to 48.5% within a few months of the Law's coming into force.

By the end of 1965 and the beginning of 1966 the situation had, however, substantially improved. Where faultless and adequate biological purification of municipal sewage is carried out, the good and satisfactory effects of the Detergents Law are clearly and unmistakeably recognizable. In Table 2, the absolute values (mg/l) of anion-active detergents (MBAS) in the discharges of a number of biological sewage plants of the Lippe Association have been compiled.

TABLE 2

Content of Detergents in the Discharges of Biological Sewage Plants

Sewage Plants	1 before 1955	2 1962 – 1964	3 1965 – 1966
Selm	a biological	4.9	1.2
Nottuln	sewage plant	8.2	2.1
Rhynern	did not yet	6.4	0.8
Lüdinghausen	exist	5.2	1.6
Marl-East	"	7.0	2.3
Bad Sassendorf	"	4.4	0.8
Welver	"	7.0	1.6
Kemminghausen	1.4	7.3	0.8
Lünen	2.5	5.6	1.6
Picksmühlenbach	1.9	6.5	1.5
Soest	0.7	3.9	1.1
Holzwickede	0.5	4.8	1.5
Lohner-Klei	0.3	1.7	0.5
Brandholz	0.3	2.4	0.7
Average	1.1 mg/l	5.4 mg/l	1.2 mg/l

Column 1 contains the results of investigations of the years before 1955, i.e. before the use of detergents in washing and cleansing agents in the Federal Republic increased so rapidly. Column 2 shows the figures ascertained in the effluents from biological sewage plants during 1962 to 1964, when hard detergents were used. Column 3 contains the analytical values in the biologically treated sewage in 1965/66, i.e. after the Detergents Law had come into force.

These analysis statistics clearly show that by 1965 the absolute content of detergents in the effluents from biological plants had returned to the level of the years prior to 1955.

In the sewage plants of the Lippe Association and the Emschergenossenschaft mentioned an average degradation of 22% was found during 1962 to 1964. In 1965, the detergent content of the effluents averaged 1.2 mg/l representing 73% degradation.

In connection with this figure proportional of 73% it might be objected that the figure of at least 80% required in the regulation of December 1st, 1962, supplementing the Law concerning Detergents in Washing and Cleansing Agents, has not been achieved. In this connection our investigations, confirmed by other laboratories, have shown that in many cases the full extent of degradation of detergents in a biological sewage plant itself can no longer be ascertained, since up to 25% occurs in sewers before the sewage reaches the sewage plants, depending upon the type and length of the sewer (whether closed or open).

LAS has also cut German river pollution. A good example for the improvement of the detergent situation is the Neckar, the river which passes Heidelberg before reaching the Rhine. It carries a substantial proportion of sewage pollution in its upper part. The foaming problem was highly aggravated by the many weirs and locks for navigation, which generated copious foam even at relatively low surfactant concentrations.

TABLE 3

River Neckar, Percent Decrease in Surfactant Load (g MBAS/s)
from 1964 to 1970 (1964 = 100)

River Section	% Decrease in Surfactant Load (1964 = 100)			
	1965	1966	1967	1970
1. Wernau-Hofen	55	60	75	65
2. Aldingen-Hessigheim	50	61	78	67
3. Lauffen-Zwingenberg	41	58	82	74
4. Rockenau-Heidelberg/Mannheim	26	28	85	72

The surfactant contents of the Neckar were recorded from 1960–1967 at corresponding river flows on a 112-mile stretch at 30 sampling points.

Between 1960–1964 the concentrations varied between 0.40 and 0.73 ppm MBAS (extremes 0.18 and 1.27). With the introduction of soft detergents the concentrations diminished in 1965 and 1966 to 0.32 and 0.34 ppm MBAS respectively. Probably due to the operation of new treatment works, only 0.21 ppm MBAS was found in 1967 in spite of a substantial increase in detergent consumption. This represents a 67% median decrease in detergent concentration of the Neckar river as a whole since 1964. The investigation showed that the proportional decrease is especially high in river sections with the greatest sewage pollution.

In the meantime the results of a series of interesting investigations have come about in the degradation of detergents in rivers and lakes before and after the coming into force of the Detergents Law.

According to Bucksteeg's investigations, the detergent load in the Ruhr from Hagen to its confluence with the Rhine has decreased considerably. Comparative observations of the Ruhr with an equal flow showed a detergent load of 1390 kg/day before the changeover and 435 kg/day after the changeover from hard to soft detergents. It can be seen from these figures that the detergent load of the Ruhr had decreased by 70%.

It is interesting to note that before the changeover from hard to soft detergents there was an increase in the load from Hagen to the confluence of the Ruhr with the Rhine downstream. After the changeover from hard to soft detergents not only a smaller load was present, but there was also a clear decrease in load in the Ruhr towards its confluence. This fact shows that a further degrading of detergents also takes place in the river.

The same observation could also be made in the case of the Emscher. Before the coming into force of the Detergents Law, the water of the Emscher at its confluence with the Rhine at an average flow of 18 m^3/s contained approximately 5–6 mg of detergents per liter. From the middle of 1965 onwards the values were only 2–3 mg/l.

Very extensive investigations of the waters of a series of German rivers were carried out by Fischer.

Investigations of some tributaries of the Rhine, random samples of which were taken at the same points, mostly before the confluence (in alphabetical order), gave the following results.

TABLE 4

	mg MBAS/l 1964	Average 1965	Decrease % in 1965	% in 1970
Agger	0.18	0.12	33	61
Ahr	0.30	0.13	57	67
Lahn	0.51	0.21	59	76
Lippe	0.88	0.39	56	67
Sieg	0.33	0.17	48	70
Wupper	2.25	1.11	51	76

In summarizing these investigations of the various rivers in the Federal Republic one can say that the decrease in the detergent content between 1964, before the coming into force of the Detergents Law, and 1965, ranges from a minimum of 13% to a maximum of 59% and improved in 1970 to some 76%.

On the basis of the results of the investigations obtained in the sewage plants as well as in the inland waters the road taken by the Federal Republic to solve the detergents problem has led to the expected success. It must, however, be strongly emphasized that the final solution of the detergents problem in our rivers can only be reached when an increasing number of biological purification plants for all domestic as well as industrial sewage containing detergents have been built and function efficiently.

Present Situation

In 1965, approximately 360,000 tons of household detergents were sold in the Federal Republic of Germany. The changeover from ABS to LAS showed the expected results within one year, as already mentioned. The effect of the measures taken by the FRG becomes very obvious if the present state of MBAS concentrations in rivers and biologically purified sewage waters is viewed against developments on the detergent market.

In 1970, 510,000 tons of detergents were sold in the FRG. Despite this considerable increase in sales, the MBAS concentrations in the effluents from treatment plants did not rise. Furthermore, the MBAS concentrations in the rivers of the FRG are much below the values found in 1964, i.e. prior to changeover from ABS to LAS.

In 1970, as compared with 1964, the MBAS concentration, for instance, of the Rhine between Wesel and Emmerich was lower by 40%. The river Ruhr, the important drinking water supply source in the Ruhr area, in 1970 showed an MBAS concentration in the lower course of the river which, as compared with 1964, was lower by 75%. The MBAS concentration in the river Neckar before its confluence with the Rhine, was 72% below the 1964 values. The river Moselle which to some extent takes sewage effluent from outside the FRG only showed a 47% lower concentration in 1970 than in 1964. Otherwise, the MBAS concentrations in the rivers Lippe, Wupper, Lahn, Sieg, Ahr and Agger were lower by 60–76% as against 1964.

Cooperation with Foreign Countries

In all these years, the FRG has been in close contact with relevant authorities and experts abroad and within the OECD has for years been actively engaged in establishing appropriate test methods. The latter project was brought to a close at the end of 1970 when 13 participating countries, assisted by 29 laboratories after many years of discussions and tests agreed on a recommendation which has now been released by OECD to all member countries. This recommendation proposes that the testing of the biodegradability of anion-active detergents be carried out according to two methods, the first method of which is a screening test while the second method as a practical test method allows a definite decision.

The proposed system to be used is as follows:

(a)   the use of the *screening test* and the acceptance of products whose biodegradability determined by this test reaches the requisite percentage; and

(b)   the use of the *confirmatory test* for any products which may not have passed the screening test in order to confirm or disprove the first results obtained.

The results of the confirmatory test are to be the only ones to be taken into consideration in the refusal or acceptance of products not accepted by the screening test.

The screening test is a static test of the 'open flask' type and the confirmatory test is based on the simulation of conditions existing in biological sewage treatment works.

The German Committee 'Detergents and Water' has taken an active part in working out these two test methods. Now that the final report of the OECD has been released, the Committee has informed the Federal Government that it would have no objection to the OECD recommendation being incorporated in an amendment of the German legislation on the testing of the biodegradability of detergents.

The EEC Commission for the Elimination of Trade Obstructions has also begun discussions regarding the testing of the biodegradability of detergents. The initial discussions have been discouraging in so far as the problem is being re-discussed, with several methods being put forward, although the OECD paper which has just been released proposes sound and simplified test methods for use on an international scale.

The Committee 'Detergents and Water' has also commented on this and stated that it considers the OECD recommendations appropriate and useful for general use.

The introduction of additional test methods, or even biological tests with fish to determine toxicological facts as routine tests, in my opinion bypass the actual task of testing the biodegradability. However, these tests are useful for basic research and should have been used prior to the introduction of new products on the market.

According to observations in the FRG to date, it appears that LAS and its decomposition products have not yet had any lethal effect on fish; the actual MBAS concentrations are considerably below toxic concentrations.

Future Work

In the FRG, the detergents problem may be considered as being solved. LAS products available on the market are biodegradable by about 95%, as is evident from current comparative control tests.

Non-ionic detergents which have about a 20% share of the market are under constant control. At present, the Committee 'Detergents and Water' is working on an analytical method for the determination of non-ionics in water and sewage. As soon as such an analytical method is available, biodegradability tests will be worked out. In view of the relatively low market share of non-ionics, the introduction of relevant legislation in the FRG is not considered necessary; on the other hand, in developing new products, the industry – in line with present requirements – is striving for highest possible biodegradability. Tests conducted so far in the field of non-ionics have shown that the concentration in rivers lies at about 0.02 mg/l. In biological treatment plants, biodegradation rates reach about 60%. The concentration in sewage is approximately 1 mg/l. In comparison, the MBAS concentration lies at about 5–10 mg/l.

Another problem in connection with 'Detergents in Water' are the phosphates contained in the detergents. Replacement of these phosphates by other substances with a similar effect is being discussed worldwide. In the FRG, this question is also being given attention. The Committee 'Phosphates and Water' which is doing research on phosphates from other sources, is also looking into the detergent phosphate problem.

The opinion prevails that, for instance, NTA would not solve the problem. Legislation regarding the phosphate content of detergents is not being considered at the present time.

Summary

The great problem of detergents and water has been solved by the changeover from biologically hard to biologically soft detergents. Legislation based on the practical test method has proved successful. Nevertheless, the FRG will follow the OECD recommendations regarding the testing of biodegradability. The market share of non-ionics is still rather on the low side so that legislation controlling their biodegradability does not yet seem to be required. Problems involving detergent phosphates will be discussed in connection with eutrophic problems.

The course pursued by the FRG in order to find a solution to the problem has proved correct and successful.

Anionic Surfactants and the Environment
by R.D. Swisher

Introduction

About ten years ago, in the early 1960s, the detergent industry of the world began its change to linear alkylbenzene sulfonate (LAS) as a major surfactant in its laundry detergent formulations. This action was undertaken solely in the interest of pollution abatement, and was unprecedented not only in that intent but also in its magnitude. LAS has provided excellent performance in laundry detergents and, more important, has been highly successful in meeting the environmental requirements. This success results from favorable biodegradation characteristics, and it is appropriate that information, both old and new, on this subject be brought together at this time.

From Soap to Synthetics

Some ten years prior to the introduction of LAS the detergent industry was passing another equally important milestone. Soap, one of the first products of chemical industry, had been a principal cleansing agent for several thousand years, despite its serious shortcomings in hard water. Then ABS was invented — the alkylbenzene sulfonates. The decade of the 1950s saw a rapid, dramatic replacement of soap because ABS gave a better detergent at a lower cost and because, once facilities were built, it was readily available in the amounts needed whereas the natural fats fluctuated widely in supply and price.

From TBS to LAS

At about that same time we also began noticing an increase in foaming wherever sewage entered the environment. Since household detergents are generally discarded to the sewer after use, along with the other household wastewaters, the new surfactants contributed to this problem to a significant extent. As a result the decade of the 1960s saw the second great change by the detergent industry. Its major surfactant had been TBS, a variety of ABS wherein the alkyl groups were drived from the olefin tetrapropylene. TBS was replaced by a new surfactant, LAS, or linear alkylbenzene sulfonate, in which the alkyl groups were straight chains instead of the highly branched ones found in TBS (Fig. 1).

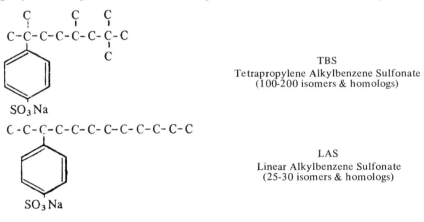

TBS
Tetrapropylene Alkylbenzene Sulfonate
(100-200 isomers & homologs)

LAS
Linear Alkylbenzene Sulfonate
(25-30 isomers & homologs)

Figure 1. Alkylbenzene sulfonates: typical structures of major types

Because of its branched structure the TBS was rather resistant to bacterial action. In contrast the LAS, having linear side chains, was more susceptible. Specifically, it was readily attacked by the bacteria in sewage and in the general environment.

This replacement of TBS required altogether some ten years to accomplish. First, laboratory studies were needed to find a material that would not only be degradable by bacteria, but that would also meet other necessary requirements. It had to be non-toxic. It had to be inexpensive. Its raw materials had to be available in quantities of around 500 million kilograms per year. And it had to wash clothes at least as effectively as TBS.

Finding and proving out such a material took somewhat over half of the ten years. The remaining time was necessary for process development and design and construction of the new manufacturing facilities. As it turned out, the industry had to develop not only the new product, but before that could be produced it had to develop the raw material for the new product. And before that, the raw material for the raw material for the new product.

It goes without saying that changes of this sort cannot be made at a moment's notice.

303

These two alkylbenzene sulfonates, TBS and now LAS, have been the leading surfactants used by the modern detergent industry. Other surfactants are used in rather smaller volumes, whenever their special properties for special uses offset their generally higher cost. This group includes such products as alkyl sulfates, aliphatic sulfonates, ethoxylates and ethyoxylate sulfates, among others.

Surfactant Structure and Biodegradability

During the past twenty years of study on this problem in academic, governmental and industrial laboratories throughout the world, an immense volume of information has been accumulated regarding the susceptibility of surfactants to biodegradation by bacteria and other organisms.[1] One generalization which has emerged is that surfactants with linear carbon chains are usually readily biodegradable. Presence of a single branch in the chain, for example a methyl group, does not appreciably diminish biodegradability. But an accumulation of branches as in TBS usually results in greater resistance.

The linear primary alkyl sulfates are the most rapid of all surfactants to degrade; they fall into a special category by themselves in this regard. All the other linear surfactants, including LAS, are measurably slower in comparison. However, this difference is of little significance under practical conditions, since they are all satisfactorily degraded in sewage treatment plants and in the receiving environment. In the specific case of LAS this becomes evident upon detailed consideration of its biodegradation in the laboratory and in the outside world.

LAS Biodegradation Pathways

The LAS molecules lose their surfactancy properties very early in the biodegradation process, and hence become non-toxic to aquatic life, non-foaming, and unresponsive to the methylene blue analysis [2,3]. (They are virtually non-toxic to mammals even before degradation [4].) It has been well established that biodegradation proceeds on beyond this early stage and goes to completion with conversion to carbon dioxide, water, sodium sulfate and bacterial protoplasm [5].

One important biodegradation pathway begins with oxidation at the end of the chain, followed by beta-oxidation of the chain, removing two carbons at a time down to the vicinity of the benzene ring. The metabolic pathways for the ring biodegradation are those commonly found in the degradation of other natural or man-made benzene derivatives. Cain and his group have shown that the usual catechol derivatives are formed, with subsequent cleavage of the ring either between or adjacent to the two hydroxyl groups.[6]

Although it requires a set of enzymes different from those involved in the chain degradation, nevertheless LAS ring degradation is 80-90% complete in a 6-hour continuous flow activated sludge treatment, and 85-95% in 24-hour semicontinuous.[7] Ring degradation occurs in each of the five isomers of C_{12} LAS when tested individually, and in each of the several homologs studied.[8] It appears that the sulfonate group may be split off either before or after ring degradation, and that either inorganic sulfite or inorganic sulfate may be formed. As often happens, different species of bacteria may prefer to use different pathways.

Ultimate Biodegradation of LAS

Beyond the disappearance of the ring, the degradation of the LAS proceeds still further, as determined by the carbon analyzer method developed by Janicke.[9] We have found that 94-97% of the soluble organic carbon content of the LAS is removed in the 24-hour semicontinuous activated sludge treatment (Table 1). This almost equals the glucose-nutrient broth growth medium used in this test, which averages about 95-97% removal. These figures are consistent with several earlier studies using other parameters, all indicating that bacterial growth always yields a small amount of nondegradable metabolic products even when pure natural foods are used.[10] Apparently nothing is 100% biodegradable, at least not in these short term operations.

In the Official German biodegradation test, as in other laboratory continuous flow activated sludge systems, difficulties often occur in establishing and maintaining steady state operation. Reflecting this, Janicke[11] has reported considerably lower and more erratic values for LAS carbon removal, anywhere from zero to around 80%. The lower maximum, 80% instead of 96%, is attributable to the milder conditions of the Official German procedure compared to the semicontinuous. Likewise the carbon removal for the natural food (peptone-beef extract) was lower also, 85-90% in the Official German compared to the 95-97 found in the semicontinuous.

TABLE 1. SOLUBLE ORGANIC CARBON IN SEMICONTINUOUS
ACTIVATED SLUDGE EFFLUENTS

		Effluent SOC, mg/liter		
Period	Samples	Control	LAS	LAS Net
A Apr 1970	15	9.2 ± 0.4	12.1 ± 0.7	2.8 ± 0.5
B May 1970	14	10.1 ± 0.7	10.5 ± 0.5	0.4 ± 0.7
C Sep 1971	20	5.8 ± 0.6	6.3 ± 0.4	0.5 ± 0.4
D Oct 1971	20	6.3 ± 0.5	6.8 ± 0.5	0.4 ± 0.2
E Nov 1971	18	5.9 ± 0.6	6.5 ± 0.6	0.6 ± 0.4
B–E, ave.	72			0.5 ± 0.2

Influent LAS (20 mg/liter C_{12} LAS)	11.6 ± 0.6 mg/liter SOC
SOC Removed (Period B–E)	$95.7 \pm 1.7\%$
Influent glucose-nutrient broth	196 ± 9 mg/liter SOC
SOC Removed (Period B)	$94.8 \pm 0.4\%$
SOC Removed (Period C)	$97.0 \pm 0.3\%$

± figures show 95% confidence limits

LAS and the Environment

We can expect that incompletely degraded LAS and natural foods may escape from full scale continuous flow sewage treatment systems in similar manner. But it is also reasonable to expect that their degradation will continue in the receiving environment, eventually reaching and probably exceeding the 95% and 97% observed in the semicontinuous tests. Proof of this in the outside world in terms of the non-specific carbon analysis or benzene ring analysis does not seem possible, because the contributions from LAS are so overwhelmed by the much larger amounts of carbon and benzene rings of natural and non-LAS origin. Even analysis for the original intact LAS by the methylene blue method is subject to such interference; gas chromatography must be used if environmental LAS is to be identified unequivocally.

As a specific example we can look at the Illinois River, which receives much of the sewage of Chicago and downstream cities. It has been monitored for many years, and in particular at Peoria, about 250 km downstream.[12,13] During the days of TBS detergents the anionic surfactant content, measured as methylene blue active substances (MBAS), averaged around 0.5 mg/liter. Conversion to LAS in the US was completed in 1965. By the end of 1966 the MBAS level had dropped to less than half of that, and by early 1968 to about one tenth. Still more important, most of the MBAS still detectable was not LAS but other materials, man-made or natural. The actual LAS content was below 0.01 mg/liter (Table 2).

TABLE 2. ANIONIC SURFACTANTS IN THE ILLINOIS RIVER

	MBAS, mg/liter		LAS mg/liter
Period	Extremes	Averages	
1953-1965	0.4-0.9	0.55	–
1966	0.1-0.3	0.22	–
1968	0.04-0.11	0.06	<0.01

Data from Sullivan (12,13)

Similar successes have been reported from Germany[14,15] and the UK.[16] In Japan the change from TBS has been more recent and more gradual, and as yet the results have been considerably less than spectacular.[17] This is traceable to the large volumes of untreated sewage entering the rivers at the sampling points. In numerous laboratory studies the biodegradation of LAS has been found to parallel the degradation of the major organic components of the sewage. Thus, wherever we find LAS in the environment we can be sure that untreated or partially treated sewage is present at that point also, carrying it. If that point is in the US and if we find 10-20 mg/liter LAS, we know we are dealing with something like 100% raw sewage. Or if we find 1-2 mg/liter, the water at that point contains some 10% sewage at least.

Past, Present and Future

In summary, the detergent industry's change from TBS to LAS has corrected detergent-related environmental foam problems just as anticipated. Furthermore, the biodegradability of LAS goes far beyond the degree necessary for that goal.

Looking into the future, we can be confident that LAS will eventually be replaced by some more economical, more efficient product or process. This will occur in the normal course of events as detergent research progresses. What the next improvement will be and when it will occur is not all evident at this time.

REFERENCES

1. SWISHER, R.D., Surfactant Biodegradation, Dekker, New York, 1970.
2. BORSTLAP, C., Intermediate Biodegradation Products of Anionic Detergents; their Toxicity and Foaming Properties, in Chemistry, Physics and Application of Surface Active Substances, Gordon and Breach, London, 1967. Vol. 3, pp 891-901.
3. SWISHER, R.D., O'ROURKE, J.T. and TOMLINSON, H.D., Fish Bioassays of LAS and Intermediate Degradation Products, J. Am. Oil Chemists' Soc., 41, 746-752 (1964).
4. SWISHER, R.D., Exposure Levels and Oral Toxicity of Surfactants, Arch. Environ. Health, 17, 232-246 (1968).
5. Reference 1, pp 274-294.
6. CAIN, R.B., WILLETTS, A.J. and BIRD, J.A., Surfactant Biodegradation: Metabolism and Enzymology, Presented at 2nd International Biodeterioration Symposium, Lunteren, September, 1971.
7. SWISHER, R.D., Biodegradation of LAS Benzene Rings in Activated Sludge, J. Am. Oil Chemists' Soc., 44, 717-724 (1967).
8. SWISHER, R.D., LAS Benzene Rings: Biodegradation and Acclimation Studies, Yukagaku, 21, in press (1972).
9. JANICKE, W., Die indirekte Ermittlung des biologischen Abbaugrades nicht-ionogener Detergentien durch Bestimmungen des organisch gebundenen Kohlenstoffs, Gas-Wasserfach, 109, 246-249 (1968).
10. Reference 1, pp 83-84.
11. JANICKE, W., On the Behaviour of Synthetic Organic Substances During the Treatment of Waste Liquors. IV. Progress of the Biodegradation of "Soft" Anionic Synthetic Detergents, Water Research, 5, 917-931 (1971).
12. SULLIVAN, W.T. and EVANS, R.L., Major US River Reflects Surfactant Changes, Environ. Sci. Technol., 2, 194-200 (1968).
13. SULLIVAN, W.T. and SWISHER, R.D., MBAS and LAS Surfactants in the Illinois River, 1968, Environ. Sci. Technol., 3, 481-483 (1969).
14. HUSMANN, W., Solving the Detergent Problem in Germany, Water Pollution Control, 67, 80-90 (1968).
15. HEINZ, H.J., and FISCHER, W.K., LAS Cuts German Water Pollution, Hydrocarbon Process., 47 (3), 96-102 (1968).
16. WALDMEYER, T., Analytical Records of Synthetic Detergent Concentrations, 1956-1966, Water Pollution Control, 67, 66-79 (1968).
17. TOMIYAMA, S. and OBA, K., Detergent Situation in Environmental Hygiene, Yukagaku, 20, 46-56, 890 (1971).

THE CURRENT STATUS OF TECHNOLOGICAL DEVELOPMENTS IN WATER RECLAMATION

M. R. HENZEN, G. J. STANDER, L. R. J. VAN VUUREN

National Institute for Water Research of the C.S.I.R.,
Republic of South Africa, P.O. Box 395, Pretoria

INTRODUCTION

Wastewater reclamation whether employed for augmentation of water supplies or for the abatement of pollution has received extensive attention since 1960.

In the USA, almost one hundred different advanced waste treatment processes and process variations for treatment and disposal of waterborne wastes are under study[1] notably Lake Tahoe, Pomona, Blue Plains, Lebanon and others. In South Africa, the Windhoek water reclamation plant commissioned in January, 1969, marked the first deliberate step towards the reclamation of purified sewage effluents for unrestricted reuse.[2] In July, 1970, a 5.5 mgd (2495 m^3/day) reclamation plant treating humus tank effluent for the production of high quality fine paper was commissioned at the South African Pulp and Paper Industries near the town of Springs,[3] whilst the 1 mgd (454 m^3/day) Stander Experimental and Demonstration Reclamation Plant at Daspoort, Pretoria, was officially inaugurated in November, 1970.

The reclamation of wastewaters comprise the removal of a very wide range of pollutants to restore the original quality of the water. Although great advances have been made in the technological developments of water reclamation, there are certain limitations in the performance of the processes developed which to some extent restrict their applicability on large scale and consequently, require refinement of the specific techniques and/or the development of alternative processes.

This paper deals with the current status of the art of water reclamation as applied in the USA, the Republic of South Africa and elsewhere.

CLASSIFICATION OF WASTEWATER CONTAMINANTS AND SELECTION OF PROCESS UNITS

Selection of advanced treatment processes is to a large extent dependent on the character of raw waters and the quality requirements of the specific beneficial uses envisaged for the reclaimed water. From a technological point of view, the predominance of specific pollutants play an important role in process selection and combinations of processes. An indication of the available unit processes and their potential application is given in Figure 1.

REMOVAL OF NITROGENOUS COMPOUNDS

The nitrogen concentration in raw sewage effluents is normally of the order of 20 mg/l. This nitrogen is present as soluble organic and ammoniacal nitrogen. Most of the

307

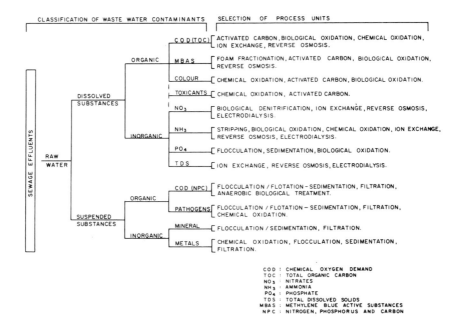

Fig. 1. Classification of wastewater contaminants and selection of process units.

nitrogenous compounds are converted to ammonia and nitrates by conventional biological purification processes. The presence of ammoniacal nitrogen in effluents is regarded as undesirable, particularly because of its high ratio chlorine demand which effects the economics of water reclamation.

Nitrate nitrogen above 10 mg/l is objectionable in drinking water because it can result in methaemoglobinemia in babies.

Removal of ammoniacal nitrogen by physical-chemical means

Ammoniacal nitrogen is converted to the gaseous form by the addition of lime to a pH of about 11.0. By counter- or cross-current air stripping in a suitably designed tower using a water to air ratio of the order of 1 : 2500 volumes, the ammonia may be stripped out of solution.[4] Removal efficiencies of 90 per cent or more can readily be achieved under favourable conditions. Ammonia stripping towers are, however, subject to hazards such as freezing up during winter seasons, and the build-up of calcium carbonate scale on the packing material which is provided for mass transfer.

The presence of scale is aggravated by the presence of supersaturated calcium carbonate after the lime treatment stage. At South Lake Tahoe it was found that the scale could be flushed off using a water jet, except in the more inaccessible parts of the tower. Other centres, however, found the scale to be hard, adherent and difficult to remove.

Ammonia stripping has been estimated at South Lake Tahoe to cost 2.9 cents (U.S.) per 1000 gallons of wastewater treated. This figure does not include the cost of lime treatment. In cold weather, this figure could be increased by fifty per cent, owing to the increased air requirements at low temperatures. Under South African conditions where winters are less severe, the average costs are more favourable.[5]

In general, ammonia stripping may be regarded as feasible where climatic conditions are favourable. The scaling problem can be counteracted by suitable design which permits cleaning by flushing. Furthermore, it is important that the effluent from the lime treatment stage should contain the minimum of supersaturated calcium carbonate. This could be achieved by operating this unit on the solids contacting principle.

Removal of nitrogen by biological denitrification processes

High efficiency biological denitrification demands efficient process control to ensure complete nitrification of ammonia as a primary step since unoxidized nitrogen compounds will pass through the denitrification process unaltered and impair overall nitrogen removal efficiency. Complete nitrification in conventional biological treatment systems presents complex problems. To overcome these problems several biological systems for the removal of nitrogen have been reported in the literature of which the three-stage biological denitrification system seems to hold the most promise.[6] It is claimed that this system can produce an effluent containing 2 mg/l, or less, of total nitrogen using methanol in the denitrification stage. This technique has been investigated at the Robert A. Taft Water Research Center and in large pilot plant operations at the University of Notre Dame and Manassas, Virginia. The process has also been evaluated on a 1 mgd scale at Hazel Crest, Illinois.[6]

The capital cost of such a system is estimated to be about 25 per cent more than a conventionally designed activated sludge process of equivalent capacity. It is further claimed that the use of high purity oxygen instead of air in the first and second stages of the three-stage system could bring the capital cost almost exactly in line with that for conventional activated sludge systems.

The operating cost for the methyl alcohol as a carbon source in the denitrification stage is estimated at 1.6 cents per thousand gallons for a wastewater containing 20 mg/l of nitrogenous material.

McCarty has also shown that column reactors may be used successfully for nitrogen removal using methanol as a carbon source.[7] These claims are based on laboratory scale investigations and although the technique appears to hold much promise, it still remains to be investigated on full-scale to prove its practical applicability.

Nitrogen removal by selective ion exchange

Ion exchange resins used for water softening are not satisfactory for the removal of ammonium ions from dilute solution.[8] Certain zeolites have unusual selectivity for the ammonium ion, notably the naturally occurring mineral clinoptilolite. Pilot studies at Battelle-Northwest indicated a cost approaching 10c/1000 gallons (U.S.) while at Lake Tahoe costs were estimated at 15c/1000 gallons for a 7.5 mgd plant.

Selective nitrate removal by ion exchange will only be feasible when new resins are synthesized with a high selectivity for nitrate over other anions present in the water. In addition, a suitable process for treating the nitrate-laden regenerants must be developed.[9]

Chlorination of Ammonia

Ammonia can be oxidized to nitrogen gas by chlorinating to breakpoint with either chlorine gas or sodium hypochlorite. The removal of ammonia by chlorination would, however, not be economically feasible for the concentration levels of ammonia encountered in wastewaters in general. In addition, the hydrochloric acid, which is produced from chlorine gas, requires subsequent treatment with lime or sodium hydroxide in order to stabilize the product. Hypochlorite is more expensive than chlorine gas, but is much safer to handle. The cost for the removal of 10 mg/l ammonia using chlorine amounts to 3c/1000 gallons (U.S.),[10] while the corresponding figure under South African conditions is about 8 cents.

REMOVAL OF PHOSPHORUS

Three basic processes are available for the removal of phosphorus, viz. biological treatment, chemical precipitation and a combination of both these.

Biological removal is effected by phosphorus uptake by cell synthesis. This technique is somewhat unpredictable so that the chemical or combined biological-chemical routes are preferred.[11]

The chemicals which effectively precipitate phosphorus from solution include aluminium and iron salts and lime. With lime the removal is independent of phosphorus concentration but dependent on pH and bicarbonate alkalinity. The very low solubility of hydroxyapatite which forms with excess lime treatment favours the use of lime for phosphorus removal.

Addition of chemicals to activated sludge plants to remove phosphorus is referred to as mineral addition. The minerals can be added with equal efficiency to the primary effluent, a flash mix tank just prior to final sedimentation, the recycle of sludge, or at any point in the aeration tank. Inorganic phosphorus in sewage is about equally divided between polyphosphate and orthophosphate. Polyphosphate is rapidly hydrolysed to orthophosphate by the enzymatic action of activated sludge. Precipitation of phosphorus is much more efficient in the ortho form than in the poly form Therefore, less mineral supplement is required in the aerator than for an equivalent removal in the primary.

The advantages of mineral addition in activated sludge plants include: minimum capital expenditure; reduced chemical requirements; improved mixed liquor settling; and a large reduction in phosphorus recycle back to the main flow stream via the digester supernatant.

The mineral addition method of phosphorus removal can produce effluents containing as little as $1 - 2$ mg/l of phosphorus as P. One important aspect of the process, which may well prove to have widespread application in the future, is the fact that it can be superimposed on the three-stage biological denitrification process. This means that removal of nitrogen, phosphorus, carbon and suspended solids can all be achieved in one integrated process.

Dorr-Oliver Phosphate Extraction Process (PEP) is the only commercially advertised lime precipitation process for use in the primary stage. A solids contact type reactor-clarifier is used instead of a conventional primary settling tank. Clarifier underflow solids are recycled to the raw sewage to maintain $500 - 2000$ mg/l of suspended solids in the reactor. The PEP process is designed to effect a reduction of 80 per cent in the primary stage and, if followed by activated sludge treatment, an overall reduction of 90 per cent or more.[12]

A recent approach to phosphate removal is the so-called independent physical-chemical technique where excess lime is added directly to the raw sewage in the primary settling tank. Subsequent units in this process include lime recovery, filtration, carbon adsorption and ammonia stripping.[13]

REMOVAL OF HEAVY METALS

Sewage purification works situated in industrialized areas often handle substantial amounts of industrial wastes in admixture with domestic sewage. These industrial wastes may contain varying concentrations of heavy metals originating from electroplating and other metal finishing works. Experiments to establish the efficiency of the reclamation process for the removal of heavy metals were conducted at the National Institute for Water Research,[14] Pretoria. Various solutions of heavy metals were added to purified sewage effluent and the polluted effluent was passed through a laboratory-scale plant. The results of these tests are shown in Table 1.

TABLE 1.

The removal of heavy metals by the Stander Water Reclamation Process
(All concentrations in mg/l)

Metal added	Feed concentration	Reclaimed water	W H O Standard	
			Maximum desirable	Maximum permissible
Cd	10	<0.01	Absent	0.01
	20	<0.01		
Cu	50	<0.15	1.0	1.5
	100	<0.15		
Fe++	50	<0.2	0.3	1.0
	100	<0.2		
Mn	50	<0.1	0.1	0.5
	100	<0.1		
Ni	50	0.3	N.S.	N.S.
	100	0.3		
Pb	50	<0.02	Absent	0.05
	100	<0.02		
Zn	50	<0.5	5	15
	100	<0.5		

N.S. = not specified

Chromium in the hexavalent state (Cr^{6+}) was not removed by the plant. The addition of a sufficient excess of ferrous iron, however, precipitated the chromium as the hydroxide and reduced it to well below the specifications of the standards.

Heavy metals present in the form of complex cyanides proved more intractable and remained in solution after the excess lime treatment stage. The use of alkaline chlorination was found to be effective for the removal of cyanides. Activated carbon has a very limited adsorbtive capacity for cyanides. These studies showed that where Cr^{6+} and cyanides are present provision should be made for the incorporation of suitable unit processes to cope with these constituents in the above reclamation plant.

REMOVAL OF SUSPENDED AND DISSOLVED ORGANICS

Biological oxidation is the classical approach to the removal of organic material from domestic and industrial wastewater. There are, however, inherent limitations in regard to the processes employed in that some organics are not readily degradable, toxic materials could be present and retarded biological activity at low temperatures result in low efficiencies. Other physical-chemical processes to deal with these components are, therefore, resorted to in water reclamation, viz. flocculation and adsorption.

Flocculation

In water reclamation processes the removal of residual organic matter is generally regarded as one of the main objectives. Of the various methods available, one which has been applied with particular success in South Africa and elsewhere is flocculation. The two flocculants most commonly used are aluminium sulphate and hydrated lime. This process is employed in conjunction with either flotation or sedimentation. At Windhoek, South-West Africa, aluminium sulphate is used as a flocculant and the flocculated material separated by flotation. The reclamation plant at the South African Pulp and Paper Industries, Springs, makes use of aluminium sulphate and polyelectrolytes as flocculants.[15]

At the Stander Water Reclamation Plant lime is used as a flocculant and the sludge separated by sedimentation or flotation. Due to the efficiency of these methods in the removal of residual organic material serious consideration is being given to their possible use as a substitute for biological oxidation. Both lime and alum achieve good phosphate removals and lime is also used to raise the pH to above 11 prior to ammonia stripping. Lime has also been found most effective in the removal of bacteria and viruses.[16]

Adsorption

One of the most promising developments in recent years to overcome the limitations of biological processes has been the introduction of activated carbon. This can be employed in either the granular or powdered form. Granular carbon is generally used in a fixed or fluidized bed. Originally, carbon adsorption was considered only as a tertiary treatment for sewage effluents. More recently, research has been directed towards the treatment of settled raw sewage.[13]

One of the first large-scale applications of granular carbon to wastewater treatment was the South Tahoe Wastewater Reclamation Plant. This 7.5 mgd granular activated carbon plant treats secondary effluent after clarification by lime and mixed media filters. The carbon effectively reduces an influent BOD from 5–20 mg/l to 2–5 mg/l; COD from 20–30 mg/l to 2–20 mg/l; and colour from 20–50 to less than 5 units. The average dosage of carbon to accomplish this treatment has been 300 lb/million gallons of treated wastewater.[17]

Large-scale studies at Pomona have substantially confirmed the results obtained at Tahoe. Carbon dosage, however, was found to average about 350 lbs/million gallons. Here too, effluent quality has been good.

At Windhoek, S.W.A., the full-scale application of granular carbon beds for the removal of residual organics have also been demonstrated.[2]

These large-scale studies plus bench investigations firmly established that activated carbon can produce effluents with low organic contents and at a cost that is reasonable. To make the process economical it was recognized very early that multiple use of the carbon, in contrast to the single use practised in water treatment, was necessary. Current regeneration techniques using temperatures of 1600–1700ºF plus steam have been able to recover 92–95 per cent of the carbon. Some losses, both physical and chemical, do occur during regeneration. Attempts to regenerate carbon *in situ* with chemical oxidants or caustic washes have been unsuccessful.

The use of powdered carbon has, however, developed into a rival of granular carbon. Its finer grain size increases the kinetics of adsorption such that 90 per cent of its adsorption equilibrium is attained in less than 10 minutes.[18] Powdered carbon is dosed in slurry form, after which it is separated by sedimentation following polymer flocculation. Other methods of separation are being investigated. Powdered carbon has the advantages that: it costs about one-third that of granular carbon; unit costs are less and the dosage applied can be controlled efficiently. Powdered activated carbon has been used with considerable success in water reclamation studies in South Africa, particularly where high quality secondary effluents are treated.[19] The ability of activated carbon to remove refractory organics from wastewater has been clearly established, and it is now seen as an integral part of any water reclamation plant where a high quality product is required.

REMOVAL OF DISSOLVED INORGANIC SUBSTANCES

In both domestic and industrial usage there is invariably an increase in the concentration of dissolved salts which has to be reduced if the water quality requirements are to be met. A number of processes have been investigated for reducing the concentration of dissolved inorganic substances, viz. ion exchange, reverse osmosis, distillation, electrodialysis, freezing and electrochemical treatment. Only the first four of these processes are currently being given serious consideration as offering a practical solution to the problem.[20]

All demineralization processes produce a brine solution. The disposal of this brine represents a major technical problem in the development of demineralization technology. In coastal areas it may be feasible to discharge brines to the ocean. Solar evaporation in lined lagoons can be employed where climatic conditions are favourable. However, inland areas with limited potential for solar evaporation will require the development of more sophisticated techniques for brine disposal.[21]

Ion exchange

Ion exchange will almost certainly be an economic process for demineralization of wastewater, if the mineral solids do not exceed 1000–1500 mg/l. This development derives from the commercial availability of new anion resins which have a high selectivity for the chloride ion and requires less regenerant and rinse water thus yielding a more favourable ratio of product to feed. It is claimed that these anion resins are not subject to fouling by organics. Up to 50–60 per cent of the COD is removed from secondary effluent with no detectable loss of exchange capacity.

Studies at the Pomona Pilot Plant facility demonstrated that an effluent containing about 50 mg/l of TDS can be produced from a feed of about 800 mg/l TDS. The bulk of the residual TDS was silica which is not removed by a weak anion resin. Total costs for the process were estimated to be 24c/1000 gallons, excluding the cost for disposal of the brines. Another cost estimate cited for the Desal Process is 18c/1000 gallons for a 10 mgd plant.[21] The fraction of the total flow to be desalinated and blended to attain the desired level of TDS will obviously reduce unit costs.

Reverse Osmosis

Reverse osmosis is a membrane process in which water is forced to flow from a solution of high salts concentration to one of lower concentration.

Since the early beginnings of reverse osmosis much work has been directed towards the improvement of these membranes. Serious problems still to be overcome include: membrane fouling, membrane cost, greater fluxes and reduction of operating costs. Despite these problems, reverse osmosis is regarded as a process with tremendous potential in wastewater treatment. In theory most components in wastewaters can be largely removed in a single unit process.

Distillation

Distillation is today an accepted technique for the desalination of sea-water. Its application is increasing in the world's more arid regions, and it is estimated that some 90 mgd are being treated by this process. Distillation of wastewater, however, presents problems quite different from those associated with sea-water. Some treatment of the distillate to remove volatile substances would probably be required. Solids and organic substances in the wastewater can present further problems. These and other difficulties are currently under investigation.[21]

Electrodialysis

The technical feasibility of electrodialysis has been demonstrated both for brackish water desalination and wastewater demineralization. But, as with reverse osmosis, membrane fouling by wastewater solids and organics has deterred practical application. The process is being investigated at both the Lebanon and Pomona pilot plants of Advanced Waste Treatment Research Laboratories (AWTRL). Emphasis of the research is on controlling the membrane fouling by intensive treatment of the feed and by enzyme flushing of the membrane surfaces. The process could be economically attractive once the fouling problems can be solved since cost, exclusive of brine disposal, has been estimated to be 15–20c/1000 gallons.[21]

Currently, no single process of the several mentioned above is ready for full-scale application and no single process has a clear and obvious advantage over the others. However, ion-exchange, because of its highly developed technology in other fields, appears to be the process which will find earliest application. A modest breakthrough in reverse osmosis could find this process applied, particularly to certain industrial waste streams. It bears repetition that a suitable method has to be found for disposal of the brine concentrates from any demineralization process.[21]

REMOVAL OF PATHOGENIC ORGANISMS

The viruses of human origin are small in number in comparison with the numbers of bacteria that are excreted.

Water-borne diseases can be combatted by removal of pathogenic micro-organisms. The two major methods employed are physical removal and disinfection. During conventional coagulation, sedimentation and filtration, the bulk of the micro-organisms present in a raw water is removed efficiently. It has been shown conclusively that in rendering water aesthetically attractive, an additional benefit is attained by the simultaneous removal of micro-organisms.[22]

As a final control procedure, disinfection is practised extensively throughout the world. This can be achieved by means of chemicals, heat or irradiation. Chlorination is at present still by far the most common method of chemical killing of the larger portion (not necessarily all) of the harmful and objectionable organisms. Virus infection, to which public opinion has become increasingly aware during recent years, is also effectively destroyed by breakpoint chlorination.[23, 24]

At an investigation at a waterworks at the river Ruhr, it was established that the following viruses were found in the untreated river surface water: Coxsackie B5; Poliomyelitis Type I, II and III; Echo Type II. After flocculation in an 'Accelator' with 10–30 mg/l $Al_2(SO_4)_3$ + 2–3 mg/l activated silica + 5–15 mg/l $Ca(OH)_2$ and chlorination at a rate of 5–30 mg/l Cl_2 with a disinfecting period of 1 to 2 hours, no viruses could be found, neither on HELA-cells nor on monkey kidney cells.[25]

Tests for the complete removal of viruses had also been carried out at Gelsenkirchen, using mobile water purification units as used by the army, or in cases of emergency, such as floods, earth tremours etc. It has been found that the filtration or flocculation/filtration stage did not guarantee persistent virus elimination; viruses were completely inactivated in the second, the chlorination stage, provided a retention period of 30 minutes and a free residual chlorine content of 1 mg/l were maintained.[26]

As grossly polluted waters and sewage effluents in particular may be expected to carry a correspondingly high pathogenic load, a very strict monitoring programme such as for Windhoek is warranted. Reclamation plants for such waters should provide for multiple safety barriers where bacteria and viruses are not only physically separated by flocculation, adsorption and filtration, but also chemically destroyed by hydrolysis and oxidation. All these mechanisms are well catered for in advanced physical/chemical treatment.

A breakthrough of floc during sand filtration sufficient to cause a turbidity of even less than 0.5 Jackson units can be accompanied by a virus breakthrough. Polyelectrolyte doses as low as 0.05 ppm will increase the floc strength and prevent this breakthrough.[27]

The removal of virus by activated carbon can cause difficulties in that the adsorbed virus can be released as the carbon becomes spent with a consequent breakthrough of virus at relatively high levels. Final reliance for the inactivation of virus must always be disinfection. Experience at Windhoek showed that carbon filters must not be looked upon as a unit process to control virus but must retain its designed function for removal of residual dissolved organics. To implement this it is necessary to ensure that breakpoint chlorination is applied before and after carbon filtration with an 0.5 mg/l free available chlorine before distribution.

Shuval[28] speculates that a probable concentration of one $TCID_{50}$ virus per 1000 litres may be present in conventionally treated contaminated river waters.[29] He suggested, on the assumption that one in 1000 persons will ingest this one $TCID_{50}$ virus, and that this level of ingestion could be responsible for a proportion of the endemic viral diseases in communities consuming highly treated waters. Shuval, however, does not specify what is meant by 'highly treated waters'. Investigations in Windhoek have shown that virus remained in the water from Goreangab Dam after it had received conventional water purification treatment. This has shown that water even from a protected catchment can contain virus. This clearly demonstrates the need for more advanced techniques to be incorporated, especially disinfection procedures, in conventional water purification practice.

To date no change has been found in the epidemic disease pattern in Windhoek, S.W.A., and whereas virus has been found in as little as one litre of conventionally treated dam water, no virus has been isolated in 10 litres of reclaimed wastewater. If it is, therefore, deemed necessary to test more than 10 litres of reclaimed water for the absence of virus, how much more so that such a standard be applied to all drinking waters. This is indeed a formidable task!

SLUDGE TREATMENT AND DISPOSAL

Sludge disposal has always constituted a problem and can amount to as much as 50 per cent of the total treatment costs. One of the main difficulties is that of sludge dewatering, since water can account for 95 to 99.5 per cent of the weight. Present techniques include sedimentation, flotation, vacuum filtration, pressure filtration, centrifuging and land disposal. The latter can lead to contamination of underground waters and give rise to foul odours.[30]

When a sludge has been dewatered, the residue can often be incinerated. In the case of lime sludges, it is possible to recover the lime for reuse. At Lake Tahoe organic sludge and lime sludge are incinerated separately in such a way that no plume or smoke can be detected. Lime recovered in this way has been reported to have better phosphate removal properties than the new product. The recovered lime produces a sludge which is easier to filter and centrifuge because of the presence of inerts in the ash.

DISCUSSION AND CONCLUSIONS

In attempting to form a true appraisal of the concept of water reclamation, it is essential to think in terms of basic considerations.

Wastewater, of whatever origin, plays a dual role. It is both a potential source of pollution and at the same time a potential source of water which can be reclaimed for useful purposes.

Where water is in plentiful supply, wastewater will be seen as a pollution hazard and reclamation will afford a means of minimizing this risk. On the other hand, in the more arid regions reclamation will provide greatly improved utilization of the existing water resources.

If water is to be reclaimed for reuse, the particular purpose for which it is to be used is an important consideration in the selection of unit processes.

From the available technological evidence it is clear that there is a wide range of pollutants present in wastewaters which can be dealt with by means of diverse biological, physical and chemical processes.

The removal of ammonia by air stripping at pH values of 11.0 is subject to certain limitations as far as its large-scale application is concerned, but may find useful application where climatic conditions are favourable.

As far as the removal of nitrogenous compounds in wastewaters are concerned, the conventional activated sludge system presents problems in terms of efficient control. The three-sludge system, however, appears to offer a solution to this problem if its applicability on a large scale proves practical and economical.

Phosphorus removal is best achieved using physical-chemical means such as excess lime treatment. In addition, the removal of viruses, bacteria, suspended organics and heavy metals are enhanced by lime treatment.

For the removal of residual organics, including synthetic detergents, the use of activated carbon in granular or powdered form is an indispensible prerequisite in the production of high quality reclaimed water.

For sterilization purposes, the complementary function of excess lime treatment and breakpoint chlorination may be regarded as the most suitable line of approach in securing a pathogen-free reclaimed water.

Sludge handling will remain a costly and troublesome part of the treatment in spite of all the new developments. The sludge derived from reclamation processes appears to have a reuse potential in terms of the lime and carbon dioxide produced upon incineration. The feasibility of this, however, depends on the size of the reclamation plant and local factors. There appears to be a need for a critical evaluation of the parameters involved in the treatment, disposal and reuse of sludges.

From the various techniques available for the removal of dissolved solids, the ion-exchange method, currently appears to be economically the most feasible in the water reclamation field.

The psychological factor of the domestic use of reclaimed wastewater with resulting public prejudice cannot be ignored. It can, however, be overcome by the enlightenment of the public not only on all aspects of wastewater reclamation but also of the fact that all conventionally treated water is in essence wastewater reuse.

REFERENCES

1. Stephan, D. G. and Schaffer, R. G. Wastewater treatment and renovation status of process development. *Journal W.P.C.F.,* March 1970, Part I, pp. 399–410.
2. Van Vuuren, L. R. J., Henzen, M. R., Stander, G. J. and Clayton, A. J. The full-scale reclamation of purified sewage effluent for the augmentation of the domestic supplies of the City of Windhoek. Paper presented at the *5th Conference of IAWPR.* San Francisco and Hawaii, 26 July – 1 August, 1970. Published in *Advances in Water Pollution Research,* 1971.
3. Myburgh, C. J. South African Pulp and Paper Industries Limited's approach to the water shortage and effluent disposal problem. Paper presented at the *Convention : Water for the Future,* 16 – 20 November, 1970, Pretoria.
4. Farrel, J. B. Ammonia nitrogen removal by stripping with air. *Advanced Waste Treatment and Water Reuse Symposium,* Dallas, Texas, January 12 – 14, 1971, Session Two.
5. THE STANDER WATER RECLAMATION PLANT, Pretoria. (Studies in progress).
6. Barth, E. F. Nitrogen removal by biological suspended growth reactors. *Advanced Waste Treatment and Water Reuse Symposium,* Dallas Texas, January 12 – 14, 1971, Session Two.
7. St. Amant, P. P. and McCarty, P. L. Treatment of high nitrate waters. *Journal, AWWA, 61* (12), pp. 659–662, 1969.
8. Dean, R. B. Removal of ammonia nitrogen by selective ion exchange. *Advanced Waste Treatment and Water Reuse Symposium,* Dallas, Texas, January 12 – 14, 1971, Session Two.
9. Dobbs, R. A. Ion exchange for nitrate removal. *Advanced Waste Treatment and Water Reuse Symposium,* Dallas, Texas, January, 12 – 14, 1971, Session Two.
10. Dean, R. B. Other methods for removing nitrogen. *Advanced Waste Treatment and Water Reuse Symposium,* Dallas, Texas, January 12 – 14, 1971, Session Two.
11. Brenner, R. C. Combined biological-chemical treatment for control of phosphorus. *Advanced Waste Treatment and Water Reuse Symposium,* Dallas, Texas, January 12 – 14, 1971, Session Three.
12. Dissolved Nutrient removal from wastewater. *Advanced Waste Treatment and Water Reuse Symposium,* Dallas Texas, January 12 – 14, 1971, Session Four, p. 32.
13. Cohen, Jesse M. Physical Chemical treatment. *Advanced Waste Treatment and Water Reuse Symposium,* Dallas, Texas, January 12 – 14, 1971, Session Four.
14. Funke, J. W. and Coombs, P. Efficiency of the unit processes in a multiple unit reclamation scheme to cope with shock loads and toxic substances. *NIWR internal report.*
15. Van Vuuren, L. R. J., Funke, J. W. and Smith, L. Full-scale refinement of purified sewage for unrestricted industrial use in the manufacture of fully bleached kraft-pulp and fine papers. (*In press*).

16. Stander, G. J. and Van Vuuren, L. R. J. The reclamation of potable water from wastewater. *Journal W.P.C.F.* Vol. 41, No. 3, Part 1, March, 1969, pp. 355–367. (Paper presented at the Conference of the Institute of Water Pollution Control, East London and King William's Town, 13 – 17 May, 1968).
17. Smith, C. E. Recovery of coagulant, nitrogen removal and carbon regeneration in wastewater reclamation. *Final Report, F.W.P.C.A. Grant WPD-85* (June 1967).
18. Dissolved refractory organics. *Advanced Waste Treatment and Water Reuse Symposium,* Dallas, Texas, January 12 – 14, 1971, Session Four.
19. Stander, G. J., Van Vuuren, L. R. J. and Dalton, G. L. Current status of research on wastewater reclamation in South Africa. *Journal I.W.P.C.* Vol. 70, 1971, No. 2. (Paper presented at the Conference of the Institute of Water Pollution Control, Cape Town, March, 1970.)
20. Cohen, Jesse M. Demineralization of wastewaters. *Advanced Waste Treatment and Water Reuse Symposium,* Dallas, Texas, January 12 – 14, 1971, Session Five.
21. Dissolved inorganic removal. *Advanced Waste Treatment and Water Reuse Symposium,* Dallas, Texas, January 12 – 14, 1971, Session Four.
22. Kabler, P. W. Microbial considerations in drinking water. *Journal AWWA, 60,* pp. 1173–1180. (1968).
23. Chang, S. L. Waterborne viral infections and their prevention. *Bull. Wld. Hlth. Org., 38,* pp. 401–414. (1968).
24. Neefe, J. R., Stokes, J. and Baty, J. B. Inactivation of the virus of infectious hepatitis in drinking water. *Am. J. Pub. Hlth., 37,* pp. 365–372 (1947).
25. Primavesi, Carl A. Die viruseliminierung im Oberflächenwasser durch Flockung und Chlorung. *Gesundheits-Ingenieur,* 91 Jahrgang, p. 266, Heft 9, 1970.
26. Althaus, Von Helmuth. Eliminierung von Viren Versuche mit fahrbaren Wasseraufbereitungsanlagen. *GWF* 109 Jahrgang, Heft 22, 1968.
27. Robeck, G. G., Clarke, N. A. and Dostal, K. A. (1962). Effectiveness of water treatment processes in virus removal. *Journal AWWA, 54,* pp. 1275–1292. (1962).
28. Berger, B. B. et al. Engineering evaluation of virus hazard in water. *Journal San. Eng. Div. Proc. Am. Soc. of Civil Engrs.* SA 1, No. 7112, pp. 111–150.
29. Nupen, Ethel M. and Stander, G. J. The virus problem in the Windhoek wastewater reclamation project (in preparation).
30. Dean, R. B. Sludge handling. *Advanced Waste Treatment and Water Reuse Symposium,* Dallas, Texas, January 12 – 14, 1971, Session Four.

Discussion *by* F.M. Middleton and C.A. Brunner
U.S. Environmental Protection Agency,
Cincinnati, Ohio, U.S.A.

One objective of improving wastewater treatment methods has been to produce water that can be reused. The reuse application requiring the most careful treatment and evoking the most interest is direct potable reuse. This type of reuse, as mentioned by the authors of the paper, has been practiced intentionally at Windhoek, South West Africa. Although an ultimate goal of potable reuse may be a desirable one, there are a number of other reuse goals that should also be given consideration. The authors cite industrial reuse, but they do not mention agricultural, recreational, and non-potable domestic reuse. The volumes of water required for industrial and agricultural use can greatly exceed the volume for potable use. All non-potable reuses result in making alternate water sources available for potable use. Local circumstances, especially the availability and cost of multiple distribution systems, determine which reuse applications are feasible.

The authors discuss the types of materials for which removal methods have been studied. These include soluble organics, suspended solids, the nutrients (phosphorus and nitrogen), minerals, heavy metals, and pathogens. The authors further point out the need to consider disposal of sludges that result from some of the treatment methods, and brines that result from demineralization processes. Disposal of brines may present more serious problems than disposal of sludges.

The authors present considerable discussion of many advanced treatment methods that might be considered for reuse systems. Additional comments on certain of the processes may be of value. Breakpoint chlorination for ammonia removal is cited as not being economically feasible for the ammonia concentrations generally found in wastewaters. Although chlorination is more expensive than air stripping, its cost is not out of line with the cost of the more generally applicable alternative combination of nitrification and denitrification. The greatest objection to using breakpoint chlorination for ammonia removal in a reuse system is the resulting change in mineral content. From 100 to 200 mg/l of chloride is added to the water. Since the chlorination reaction produces large amounts of hydrogen ion, a base must be added to maintain the pH at a reasonable level. This means there will also be a sizable increase in the concentration of cation associated with the base, probably calcium. The total dissolved solids concentration may increase 300 mg/l or more. On the topic of heavy metals, the authors point out several difficult problems that might arise. When considering reuse, heavy metal exclusion from the water to be renovated may be a more practical solution than treatment for removal. Exclusion can be done either by prohibiting discharge of toxic metals into the collection system, or by segregating metal-containing waste-water from that to be used for renovation. The latter approach was used in the Los Angeles, California area for drawing up a long term reuse plan.

Emphasis in this paper is on treatment process developments rather than on treatment systems or water qualities for specific reuse applications. It might be useful to mention briefly some of the water quality requirements for the various reuse applications to give perspective on the use of the treatment methods. For agricultural reuse, water quality requirements vary considerably with the crop being irrigated. In many cases in the United States, however, conventional secondary treatment and chlorination is all that is necessary. There is no need to consider removal of phosphorus and nitrogen since these nutrients can substitute for a significant fraction of the fertilizer requirements. Certain heavy metals and excessive salinity must be avoided. With due consideration for these limitations, irrigation provides a reuse application that can utilize large volumes of wastewater. In the State of California alone, over 200 treatment plants supply water for irrigation.

Use of renovated wastewater in recreational lakes is rapidly gaining in interest in the United States. For unrestricted use of such lakes it is necessary to provide treatment that includes soluble organic removal equivalent to secondary treatment, a high degree of suspended solids removal, phosphorus removal to about the 0.1 mg/l level, and disinfection. Conversion of ammonia to nitrate is desirable because of the toxicity of ammonia to fish at high pH and interference with disinfection by chlorination. Experience at Indian Creek Reservoir, which receives the effluent from the South Lake Tahoe advanced waste treatment plant, indicates, however, that a high degree of ammonia removal may not be necessary. The treatment system for recreational reuse may be quite complex such as that at South Tahoe. At Santee, California, on the other hand, percolation through gravel is used to carry out much of the treatment. The renovated water in a recreational lake can serve as a source for other reuses. Higher cost for treatment can then be justified compared to recreational reuse alone.

Industrial reuse of municipal wastewater can cover a wide range of applications including cooling water, boiler feed water, and process water. The water quality requirements will also vary widely. Secondary effluent has been used for cooling although a higher degree of solids removal and phosphorus removal is desirable. Nitrogen removal is not necessary, but conversion of ammonia to nitrate is desirable to prevent corrosion of copper equipment. There is a very large potential market for renovated water to be used for cooling purposes. The other industrial reuses would generally require a higher quality of water. In addition to secondary treatment or the equivalent, solids removal by chemical clarification and filtration is likely to be necessary. Additional soluble organic removal and nitrogen removal may also be required. For boiler water feed various degrees of demineralization would be required depending on boiler pressure.

Non-potable domestic reuse is practiced in a few locations. Systems are likely to be small. The application is greatly restricted by the need for a complete dual distribution system.

Potable reuse obviously requires a high degree of removal of soluble and suspended organic pollutants, and pathogens. To accomplish the removals necessitates a complex treatment system such as that at Windhoek. Some degree of demineralization is also likely to be required. Nutrient removal is not particularly important, except for ammonia. In the total potable supply, which is likely to contain 50 percent or more of an alternate water source, the ammonia concentration must be reduced to the 0.5 to 1.0 mg/l range to prevent taste and odor. Because of the likelihood of a significant amount of an alternate water source, conversion to nitrate would probably result in an acceptably low nitrate concentration. Total nitrogen removal should not be necessary.

Reply

The authors have placed more emphasis on the technological developments in wastewater renovation rather than various re-use goals. These would obviously be dependent on local conditions regarding cost structure, availability of land, water etc. The authors agree with the views of the discussor as regards the various quality requirements for various re-use applications. In South Africa reclaimed water is extensively re-used for agricultural and cooling water purposes which require limited tertiary treatment without the need for sophisticated processes. Direct reclamation for recreational purposes such as at Lake Tahoe has not as yet found useful application in South Africa and reuse applications are thus more directed towards agricultural, industrial and potable requirements.

D.G. Stephan, USA

In a dual water supply system, both supplies should be disinfected to avoid infection from accidental use or cross-connection with the "non-potable" supply, but the "non-potable" supply need not be carefully controlled for removal of trace amounts of impurities which might be suspected of producing only long-term or chronic physiological effects.

When distilling waste waters, volatile organics which are taste and odor-causing will be carried over in the distillate and must be removed by adsorption or oxidation, etc. Also, ammonia will be almost quantitatively transferred into the distillate and this is of major concern because of ammonia corrosion of cupro-alloys.

J. McKendrick, Rhodesia

Can the zeolite mentioned be recovered so as to reduce costs? Clinoptilolite is naturally occurring. Are there any synthetic resins which show a similar selectivity for the ammonium ion?

Reply

Ion exchanges whether naturally occurring or synthetic may be regenerated chemically. It is understood that synthetic resins are being investigated but in general the costs of ion exchange are still prohibitive.

A. Osdor, Israel.

A new approach to the wastewater treatment and reuse is to separate by distillation practically all the water contained in the wastewater by flash distillation, with salt production, as described in my paper presented at the "Eighth National Symposium on Desalination En Boqeq 14 – 15 March 1971."

In the case of wastewater distillation, scaling is practically impossible to prevent if the heat transfer is effected through metallic walls. By using an intermediary liquid such as an oil, which is insoluble in water and has a lower vapor pressure than water, heat is transferred from the hot pure water produced to the incoming cold wastewater.

Pollution of the pure water produced by direct contact with the oil vaporizing the wastewater is avoided because the intermediary liquid (the oil) is insoluble in and easily separated from water and because the solubility of water in oil increases with the temperature increase. This dissolved pure water is given up to the wastewater on cooling the oil. A small quantity of pure water is lost, but contamination of the pure water is avoided.

Odors are removed from the wastewater by reason of the fact that at the boiling point of the solvent the solubility of gaseous constituents is zero. The system includes a small oil flash evaporation device in which a portion of the circulating intermediary oil is vaporized without vaporizing fats and other impurities which become concentrated in the oil. The vaporized pure oil is condensed and returned to the main system.

The cost of the drinking water produced will be approximately the same as that produced by sea water flash distillation. This is higher than that of drinking water supplied by conventional methods, but it includes the cost of wastewater disposal.

Moshe Bar-Levan, Israel.

What information was obtained from the continuous bio-assay tests of waters treated for drinking water supply?

Reply

The National Institute for Water Research in South Africa has conducted extensive tests on the long termed effect of reclaimed water using rats for bio-assaying. To date no adverse effects have been confirmed.

E. Shamash, Israel.

What are the authors' views on the use of magnesium oxide to remove phosphate? It would also remove some of N as Mg ammonium phosphate, colour, and act as a flocculating agent.

Reply

Low burnt magnesium oxide would have useful application particularly for effluents with a relatively low hardness in terms of magnesium. In the Witwatersrand area in South Africa the effluents originate from dolomitic areas and, therefore, require no magnesium supplementation if treated for phosphate removal using lime. In cases where a deficiency of magnesium hardness would occur the use of ferric salts rather than magnesium may be a better proposition.

The kinetics for the formation of magnesium ammonium phosphate would seem to require favourable chemical conditions which are hampered by the presence of several other constituents.

A. Feinmesser, Israel.

The concentration of boron in sewage is relatively high in Israel reaching 2 ppm or more. This is too high for irrigation of crops. Washing powders are the source of the boron. Restrictions have therefore been imposed on the addition of boron to washing powder limiting the concentration to 5% of perborate.

Reply

The average concentrations of Boron in purified effluents in South Africa is of the order 0.6 ppm. This is not regarded as critical for the moment, but a close watch is kept should this problem show any tendencies to cause alarm.

STEPS TO UNDERSTANDING
ECOLOGICAL ASSESSMENT OF POLLUTION

NEAL E. ARMSTRONG

Environmental Health Engineering Labs, The University of Texas at Austin

INTRODUCTION

Few people dealing with environmental pollution problems today would argue with the statement that engineers and biologists have had difficulty working together to protect the environment. The charges of 'lack of specific quantitative guidelines suitable for design criteria' on the one hand and 'suitable criteria cannot be developed because ecological systems are too complex' on the other hand have overshadowed the large amount of cooperation and understanding achieved between the two groups in recent years. However, it is imperative that this understanding and cooperation continue so that both groups may continue to develop the techniques found to be so useful in 'ecological assessment of pollution'.

It appears to the author that at least three steps are needed to foster greater communication between the engineers and biologists for development of more meaningful and useful techniques for ecological assessment of pollution. The first of these is for the engineer to achieve a better understanding of the nature and behavior of ecological systems (or ecosystems) and for biologists to become more familiar with treatment processes and the social, political, economic, and technical realities of waste management. The engineer has long been faced with the ecologists who use what Paulik (1967) calls the verbal model which '. . . owes much of its popularity and power to its ambiguity and unknown complexity' and which has produced few design criteria usable by the engineer. Consequently, many waste management decisions have had to be made by the engineer without input from the ecologist. But the ecologist has been faced too often with an engineer impatient for design criteria which carry with them the ecologists' guarantee of no environmental impact. The ecologists, born of a different problem approach philosophy, cannot provide the guarantee nor can they provide that kind of criteria.

The second step needed is the use of some unit of measure, which both the engineer and the biologist understand, to describe the functions of ecosystems whether they be on land, in receiving waters, in an activated sludge plant, a trickling filter bed, or an oxidation pond. One of the first difficulties encountered in interdisciplinary approaches is the jargon of each discipline and that of the sanitary engineers and the ecologists is no exception. A unit of measure common to both disciplines which could be used to describe treatment plant operation, waste characteristics, and ecosystem response is needed.

The third step involves the tools used for assessing the effects of wastes on ecosystems. The tools used to date provide information on how pieces of the ecosystem have responded or might respond to various types of wastes. None of these which have been studied very extensively allow prediction of the effects of wastes and that is exactly the information needed to provide rational design criteria for waste management systems.

Some little-used tools developed by the ecologists will be discussed here, and their potential for providing such information discussed.

These are not the only steps to be sure, but they appear to be important ones, at least to the author. In discussing these steps, the author has been more concerned with providing overall concepts in simplistic language than with great detail and thus has taken some liberties with ideas and hypotheses. Hopefully, this will not detract from the paper's usefulness at the workshop for which it is intended. More detail and information is available in the literature cited.

REVIEW OF ECOSYSTEM BASICS

Components of and Material Flow in Ecosystems

An ecological system may be considered in a systems analysis approach; that is, it may be broken down into components, and the behavior of those components may be considered as well as the interactions between the components. Consider for example the generalized ecosystem shown in Figure 1 which could be an aquatic ecosystem. There are two basic groups of components in this system, organic and inorganic material roughly corresponding to living and non-living material respectively (Odum 1959). In the organic category, the producers (plants and certain bacteria) convert inorganic minerals and gases into organic material with the aid of radiant energy in the process of photosynthesis. That organic material in the form of plant biomass is then either eaten by the consumers or passed to the detritus pool as dead plant material. Unassimilated portions of the plant biomass and dead consumer biomass become part of the detritus pool also. Some consumers feed on detritus organic material; thus, exchange between the consumer and detritus components proceeds in both directions. The decomposers (the bacteria and fungi) consume the detrital organics producing more decomposers or more detrital material, or mineralize the organic material into inorganic material. These inorganics may then be reassimilated by the producers and thus recycled.

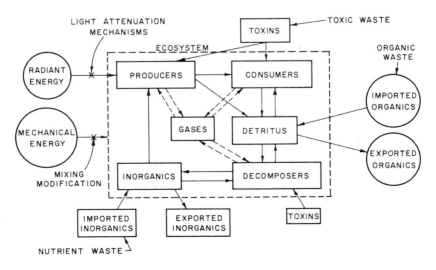

Fig. 1. Framework for Assessment of Ecosystem Response to Waste Inputs

Gases exchange with three components of this system principally. During photosynthesis, oxygen is liberated in the Hill reaction and carbon dioxide is consumed in the 'dark' reaction. At the same time in plant respiration, oxygen is consumed and carbon dioxide is produced. This is true also for consumer and decomposer respiration, although under anaerobic conditions other gases may be produced by the decomposers.

Oxygen may also be consumed by certain of the detrital or inorganic components by chemical oxidation but this is usually small. Because gases are exchanged in both photosynthesis and respiration, gaseous exchange is an important indicator of material exchange in the ecosystem, and the rate of gaseous exchange is used to calculate such activity as will be seen later.

Several types of imports and exports are included in Figure 1 representing the natural exchanges of material that occur with the external environment and also the points of impact of various types of waste material. Imported organics whether they be from natural sources or waste discharges become part of the detritus pool. Inorganic imports such as nitrogen, phosphorus, minerals, and heavy metals, become part of the inorganic pool. Exports of detritus and inorganics shown in Figure 1 would be to sediments since this ecosystem is 'closed'. If an 'open' ecosystem is considered, exports of all components may occur downstream complicating the picture considerably.

Two external physical energies act on an ecosystem — radiant energy and mechanical energy (Odum 1971). Radiant energy is, of course, essential for the process of photosynthesis, and without it the source of organic material is lost. Radiant energy is also essential for temperature control in natural systems. Mechanical energy provides the mixing which drives the exchange between components of the ecosystem. Too much mechanical energy may result in physical damage to biota, and too little may inhibit exchange between such components as the producers and inorganics as in a stratified lake or in the oceans.

Transfer Functions

The growth rates of each living component of an ecosystem change with concentrations of organic or inorganic material in a manner following the Monod relationship developed for bacteria and shown in Figure 2. This relationship is defined as:

$$\mu = \hat{\mu} \left(\frac{S}{K_S + S} \right)$$

where:

μ = specific growth rate, time^{-1}
$\hat{\mu}$ = maximum specific growth rate, time^{-1}
K_S = Michaelis constant, mg/l
S = substrate concentration, mg/l

Algae for example follow the Monod relationship for inorganic nutrients and radiant energy (DiToro *et al*, 1971) as do the consumers (for example the growth rate of sockeye salmon in response to zooplankton concentrations as shown in Figure 2) with prey (Warren 1971). Respiration rates, which are measures of an organism's activity follow similar relationships with gas concentrations (Warren 1971). Temperature may increase or decrease both the growth rate and Michaelis constant (such effects are fairly well defined for some organisms, DiToro *et al*, 1971), and toxins will usually decrease the growth rate

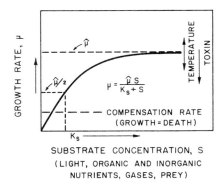

SUBSTRATE CONCENTRATION, S
(LIGHT, ORGANIC AND INORGANIC
NUTRIENTS, GASES, PREY)

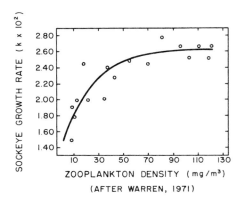

ZOOPLANKTON DENSITY (mg/m³)
(AFTER WARREN, 1971)

Fig. 2. Monod Representation of Growth Rate

(Figure 2). Thus, the Monod relationship may be considered a 'transfer function' between the inorganic, organic, or radiant energy input to an organism and its response.

It is well known, however, that the increase in some substrate beyond the point where the maximum growth rate has been achieved will result in a decrease in growth rate. The growth rate of algae, for example, is inhibited by high light intensities which may occur at mid-day in the upper layers of water (Ryther 1956). Heavy metals may become toxic at concentrations exceeding those permitting maximum growth rate.

Another phenomenon is the interaction of substrates at growth rate limiting concentrations and its effect on growth rates and Michaelis constants. Early evidence indicates that the interaction may be represented by (DiToro et al, 1971):

$$ \mu = \hat{\mu} \; \left(\frac{S_1}{K_{S1} + S_1} \right) \left(\frac{S_2}{K_{S2} + S_2} \right) $$

where:

S_1, S_2 = substrates 1 and 2 respectively, mg/l

K_{S1}, K_{S2} = Michaelis constants for substrates 1 and 2 respectively, mg/l

Energy Flow Through Ecosystems

Energy flow through ecosystems is shown in Figure 3 and is similar to material flow; the major and important difference is that no recycling occurs. The two basic laws of thermodynamics are obeyed in the process; that is, energy is conserved within the system and heat is lost at each transformation (Odum 1971).

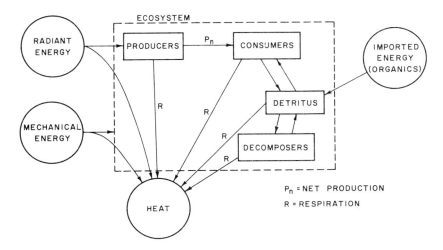

Fig. 3. Energy Flow in Ecosystems

Radiant energy is converted in photosynthesis to potential energy in the form of plant biomass. Only about half of the energy in the short wave portion of radiant energy may be used by plants in photosynthesis; only a few percent of this is actually converted to plant potential energy. The energy in plant biomass is used by the plants in respiration leaving some net amount (net potential energy production, P_n) which is used by consumers or becomes part of the detritus potential energy pool as does unassimilated or degraded energy from the consumers and decomposers. The detrital potential energy is used by consumers or decomposers. At each energy transfer step, energy is lost as heat during respiration, R, and the final product of energy degradation by the decomposers is heat. The energy transformations occur with less than 15 per cent efficiency (Odum 1959).

A significant asset for using energy flow is that the same unit of measurement is used throughout the system rather than organics and inorganics, metals, etc. Also, the amounts of energy entering the system from radiant energy, mechanical energy, energy in the form of organics or even inorganics may be estimated. Since the function of the entire system is dependent on energy inputs, the type of ecosystem resulting from combinations of these inputs may possibly be predicted. Such predictions will hopefully be a result of work now being conducted by the author which is an outgrowth of efforts by Odum, *et al.* (1970) to classify estuarine ecosystems based on energy flows through them.

Another very important reason for using energy as a basis for ecological assessment of waste discharge effects is that engineers are familiar with energy concepts. The energy equation (or Bernoulli's equation) provides the basis for flow calculations in pipes and open channels and is therefore a basic tool for the civil engineer. Energy concepts are also used in estimating wave forms (Neuman and Pierson 1966) and wave transitions near shore and around barriers (Ippen 1966). Servizi and Bogan (1963) have extended energy concepts to waste treatment practice. In short, the use of energy concepts in sanitary and civil engineering has had a long and successful history, and the extension of that framework to ecosystem analysis should not be difficult for the engineer.

Energy flow may be measured indirectly using the changes in oxygen and/or carbon dioxide that occur in photosynthesis production and respiration, since the amount of oxygen released in photosynthesis (or the amount of carbon dioxide consumed) is indicative of the amount of plant biomass or potential energy produced. Likewise, the utilization of organism biomass for metabolism (respiration, R) requires oxygen, and there is a simultaneous release of carbon dioxide. Various techniques are available for measuring these gaseous changes and estimating the energy flows occurring (Armstrong, *et al*, 1968). Another method is to measure the changes in population biomass and hence potential energy that occur over some period of time (Odum 1959, 1971).

PREDICTION OF ECOSYSTEM RESPONSE

Gross System Response

Ecosystems have been treated above as combinations of components with exchanges of energy and material occurring among the components in a systems analysis context. The 'behavior' or output of these systems has been found to be somewhat predictable given inputs to the system, some knowledge about the transfer functions for various components of the system and the various substrates to be considered. In essence, the 'black box' treatment has been used as pictured in Figure 4. If ecosystem function is divided into two processes, (1) the production, P, of organic material (or potential

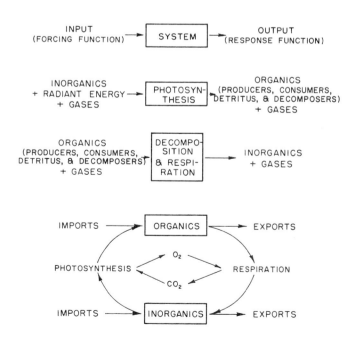

Fig. 4. Simplified System Analysis of Ecosystem

energy) through photosynthesis, and (2) its subsequent utilization by consumers and decomposers and the breakdown of this organic material (or potential energy) through respiration, R, these two processes can be readily measured by gaseous exchange techniques (see Figure 4). Production and respiration are then indicators of system activity and energy flow. If these two processes are coupled as shown in the lower part of Figure 4 and imports and exports of organics and inorganics are included, the behavior of this simple system can then be predicted for given initial states (Odum 1963). Two such initial states are shown in Figure 5, one with high organics (e.g. high organic waste discharge) and the other with high inorganics (e.g. high inorganic waste discharge). If these initial conditions are transitory (e.g. the waste inputs are instantaneous inputs) and there are no imports or exports, the final state of the ecosystem is likely to be at equilibrium. For the high organics initial state, respiration will be high until mineralization can occur and the inorganics produced be used in photosynthesis. The high respiration is simply the satisfaction of the biochemical oxygen demand. For the high inorganic initial state, photosythesis will be stimulated and plant biomass produced which, when used by the consumers and the decomposers, will result in an increasing respiration until the two processes reach equilibrium (Figure 5).

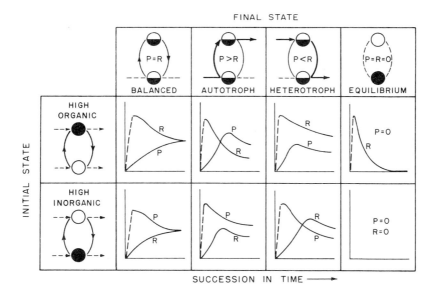

Fig. 5. Theorectical Ecosystem Succession Patterns (Odum 1963)

If the initial states are followed by continuous discharges (imports) of organics or inorganics, then non-equilibrium final states will result. For high initial organics and a continuous discharge of inorganics or a high initial inorganics and a continuous discharge of organics, the photosynthesis and respiration curves will cross as the ecosystem proceeds to a material (or energy) flow balance. For high initial and continuous organics discharge and for high initial and continuous inorganics discharge, the photosynthesis and respiration curves do not cross and photosynthesis dominates with inorganics discharge and respiration dominates for the organics discharge (Figure 5).

Combinations of these predictive curves or changes in the balance of inputs may be used to construct curves for photosynthesis and respiration below a waste discharge. For example, the pattern of photosynthesis and respiration below a domestic waste discharge in the White River, Indiana (shown in Figure 6), shows the transition from a high initial organics state to one of high continuous inorganics input (Odum 1956). This pattern would be expected downstream from a continuous domestic waste discharge as organic material is stabilized and inorganics are produced which are then used by the producers.

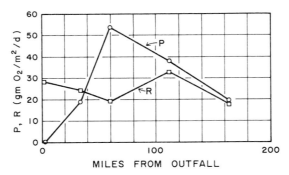

Fig. 6. Changes in Production, P, and Respiration, R, Downstream from a Waste Discharge

EFFECTS OF WASTES INTRODUCED INTO ECOSYSTEMS

Generalized Responses of Ecosystems to Wastes

The effect of any waste discharge on an ecosystem is usually stimulatory or inhibitory to growth of organisms. This is a result of increasing or decreasing the substrate concentrations of organic and inorganic materials, gases, ions, trace metals, radiant energy, prey, etc., or the level of temperature, such that the specific growth rates of the organisms in one of the ecosystem components are increased or decreased. The organisms in each ecosystem component may react differently to substrate concentrations depending on the shape of the Monod-type specific growth rate – substrate curve for those organisms. In those cases where the increase in substrate concentration is inhibitory or toxic, the net result is a decrease in growth rate and/or death. Decrease in growth rate at low or high substrate levels is a stress to an organism, and the physiological shunting of internal energy reserves to stress-resisting processes may over a long period of time decrease the organism's ability to perform necessary physiological or behavioral functions or to withstand further stress.

The increase or decrease of the specific growth rate for the organisms in one component of an ecosystem invariably affects the other components directly or indirectly. For example, the introduction of organic materials into the detritus pool as depicted in Figure 1 makes more detrital food available for the decomposers as well as those detritus feeding consumers, and there is a subsequent increase in their populations. There is a concurrent depletion of oxygen reserves in the gas phase. Mineralized organic matter increases the inorganic supply which then provides more substrate for the producers. With adequate radiant energy, the producers then increase replenishment of the oxygen in the gas phase and provide more food for the consumers. The net result of the organic material import is an increase of the populations in all components of the ecosystem, an increased utilization and subsequent replenishment of the oxygen supply, and a higher rate of material flow through the system Anyone familiar with the changes in water quality and biota downstream from a domestic waste discharge recognizes the above sequence of events as the oxygen sag just downstream from the waste discharge, the disappearance of sport fish and other organisms, the appearance of nuisance growths of algae further downstream, and the reappearance of sport fish or 'normal' biota even further downstream.

An increase in inorganic nutrients causes somewhat similar results in a different sequence. Plant growth is stimulated first and nuisance blooms may result. With plant death and decomposition by decomposers, oxygen depletions occur as well as recycling of the inorganics. Consumer populations are also stimulated.

The overall ecosystem responses to these impacts could have been predicted, at least qualitatively, from the patterns shown in Figure 5. By the use of Figures 1 and 2, material pathways and ecosystem component responses may also be qualitatively predicted, and a tool which provides a 'feel' for the consequences of waste discharges is available. This framework or portions of it (though perhaps not recognized as such) have been used to assess the effects of waste discharges on ecosystems in all studies to date.

Present Methods for Ecological Assessment

The present methods used to assess the ecological effects of pollution may be divided into four categories: (1) field investigations; (2) laboratory bioassays; (3) mathematical modelling; and (4) ecosystem modelling.

Field Investigations

The field investigations used to date have included: comparison of species lists from polluted and non-polluted areas: selection of a particular organism or group of organisms as an 'indicator' of certain types of pollution; comparison of populations of certain groups of organisms; quantitative comparisons of community structure using similarity (and dissimilarity) indices and diversity indices; and community activity using production and respiration rates (Armstrong and Storrs 1969). These approaches then have varied from the examination of species and/or their abundance in the various ecosystem components to the activities of ecosystems as a whole. Because of the difficulty of correlating community structure described by species lists and/or abundance of those species, quantitative indices which could 'sum' this structure into a single number which would have ecological significance have been employed (Pearson *et al.*, 1967). Recently correlations of diversity of benthic communities with naturally occurring environmental factors and the toxicity of wastes from various sources was achieved (Storrs, *et al.*, 1969, and Armstrong *et al.*, 1971). The indices least used by engineers but which hold great promise for understanding ecosystem function and the effect of waste on the ecosystem are production and respiration. Few engineering studies have been carried out in which these indices were used for developing design criteria. For the most part they have been used to estimate oxygen production and consumption in oxygen balance studies. But as noted above, they do indicate overall activity in ecosystems. Perhaps the most important reason that limits their use (and this reason applies to every biological index) is the question of what levels are desirable or undesirable. There is, of course, no easy answer to this question. The best answer available is to compare the polluted ecosystem with one not polluted and make some decision from the comparison. Some 'polluted' engineered ecological systems may be desirable, such as treatment plants and oxidation ponds.

Laboratory Bioassays

Laboratory bioassays used to assess the effects of pollution on ecosystems have been of two types: (1) biostimulation; and (2) toxicity. In the biostimulation test, the consequences of increasing the organic or inorganic materials in an ecosystem are simulated in some fashion. The most familiar of these tests is for biochemical oxygen demand (BOD) or the oxygen consumed during stabilization of organic material as described in *Standard Methods* (1971). Other types of biostimulation tests developed more recently are for assessing the impact of addition of inorganics to an ecosystem namely nitrogen and phosphorus (McGauhey *et al.*, 1968, 1969, 1970). The Environmental Protection Agency has an active program to develop a standard test for assaying the algal growth potential of inorganic additions in pure chemical form or with waste discharge (EPA 1971). These tests which include only two components of the ecosystem do not consider the total effects of inorganics additions.

Toxicity bioassays have been the subject of much investigation in recent years as knowledge of the toxic effects of heavy metals and pesticides became more commonly known. Most bioassays now being conducted in monitoring or testing programs are acute toxicity bioassays testing the lethality of compounds or combinations of compounds (*Standard Methods* 1971). Chronic toxicity bioassays, which require more elaborate equipment and much longer periods of time, produce more meaningful results because 'safe' concentrations of compounds are determined in such tests (Klock 1961). Like the biostimulatory tests, the toxicity bioassays have been conducted on organisms from one component (consumers) of the ecosystem principally, but other components (producers) have received attention (Thirmurthi and Gloyna 1965). Since that component relies on other components for continued sustenance, only a part of the overall effects of the toxic compounds being tested is determined. Summaries of toxic effects determined by toxicity bioassays are available.

Mathematical Modelling

Several workers (Thomann *et al.*, 1970; DiToro *et al.*, 1971; Chen, 1970; Davidson and Clymer, 1966) have recently developed mathematical models for describing changes in populations of organisms in the various components of the ecosystem These rely heavily on earlier work by investigators of population dynamics in oceanic waters and are basically mass balance equations for ecosystem components. The DiToro *et al.* (1971) model includes the producers, the consumers represented by zooplankton, inorganics, and radiant energy. The transfer functions for each of the components were developed from data in the literature and have the Monod relationship form described earlier. The Thomann *et al.* (1970) model was developed principally for examining the dynamics of the nitrogen cycle (including producers) in estuaries. A third model is that developed by Chen (1970). His model does include all the ecosystem components listed earlier as well as the organic and inorganic imports, but in a somewhat different framework. The model developed by Davidson and Clymer (1966) includes the same components as the DiToro *et al.* model, but is somewhat simpler.

Relatively few ecologists have developed models of energy flow in ecosystems despite the recognition of its importance. Lindeman (1942) precipitated the approach with his classic paper on ecosystem trophic dynamics. His ideas were further developed by Ivlev (1945) and more recently by Odum (1971), Slobodkin (1961), Warren (1971), and Macfadyen (1964). All of these workers have developed conceptual and in a few cases, mathematical models of energy flow based on Lindeman's original concepts. Odum

(1971) has developed some of the most useful concepts for use by engineers. These concepts include the classification of estuarine ecological systems based on energy flows through them (Odum *et al.*, 1970), the development of passive electrical analog circuits for energy flow simulation in ecological systems (Odum 1960), and the development of electrical-type symbols for components of ecological systems (Odum 1971). His work on the use of 'pilot ecosystems' for modelling prototype systems (described below) has shown the utility of using small scale ecosystem models of receiving waters to show the impact of wastes discharges to these systems.

Ecosystem Modelling

A 'pilot ecosystem' bioassay (analogous to 'pilot plant') has been developed by Odum *et al.* (1963) for testing the effects of manipulating ecosystems including determining the effects of wastes on ecosystems. The approach is to 'model' the ecosystem much as physical models of rivers and estuaries have been constructed. A portion of the receiving water ecosystem is placed in a small tank in the laboratory or in outdoor ponds such that the same physical conditions (light, mixing, etc.) are experienced in the model as in the prototype. Then waste discharges or other alterations to the system are imposed on it. The ecosystem in the model has been found to behave much like the prototype system with energy pathways and biogeochemical cycles duplicated (Odum *et al.*, 1963; Odum *et al.*, 1970; Copeland and Fruh, 1970; and Armstrong and Gloyna, 1968). With a number of such model systems or 'pilot ecosystems' available, a number of levels of waste concentrations and combinations of waste constituents may be tested for their ecosystem response. Such tests have the advantages of delineating the total ecosystem response instead of only part and of being carried out in confined testing equipment with no discharges to the environment occurring. The response of the ecosystems in the models are, like their counterparts in the prototype, difficult at times to understand or even to model. The bioassay is still in a developmental stage, but if as much attention is devoted to it as has been to single species bioassays, the bioassay can be developed into a quite useful technique for assessing the effects of wastes on ecosystems.

DISCUSSION

The discussion of ecosystem components and material flow may have left the impression that ecosystems are in actuality simple systems; that impression should be dispelled because they are not. However, the treatment of ecosystems in a systems analysis approach has certainly elucidated the overall functions of ecosystems and has opened new approaches of study not only of ecosystem function but of the effects of waste discharges on these systems.

Prediction of ecosystem response to waste discharges is desperately needed for proper waste management program development, for without that knowledge design criteria are by definition incomplete. More intensified study of ecosystems by engineers and ecologists alike is needed so that the one group has a better appreciation of the other's problems. After all, engineers and ecologists are both members of the same world ecosystem with special concerns about particular subsystems. The engineer is concerned about the unassimilated and degraded material coming from an urban ecosystem from one consumer, man, and how to treat that waste in a specialized engineered ecosystem a treatment plant. The ecologist on the other hand is concerned about the receiving water ecosystem receiving the effluent of the plant. What is needed is the further recognition that these three ecosystems are part of a larger ecosystem which includes man, treatment plants, and receiving water ecosystems and which may also be considered in a systems analysis context.

The concept of energy flow in ecosystems is still quite new; very few ecologists deal directly with research in this area. Yet energy flow is basic to ecosystems, and as man is discovering, especially to him. It appears to be inevitable that the energy concept will ultimately be a focal point for engineers and ecologists. The present and future world interest in exploiting energy resources in the form of oil and gas, hydroelectric, solar, tidal, nuclear, and geothermal energy (National Academy of Science, 1969), the existing use of energy concepts in many areas of engineering already, and the recognition of the importance of energy to ecological assessment of pollution will force the engineer and the ecologist to think more and more in energy terms.

The use of energy flow as an assessment tool should receive more attention. Consider for example the usefulness of having ecosystem function and response, waste inputs, and physical energy inputs all in the same unit of measurement. Energy flow through ecosystems can be measured indirectly by gas exchange and organic material flow, waste inputs can be treated as energy equivalents (either as potential energy in the organic matter or a substrate of a Monod-type relationship effecting energy flow), and physical energy inputs (sunlight and mechanical energy) can be measured directly or estimated from flows, wave heights, etc. and back calculating in the energy equations. Thus, areal energy densities would be available for ecosystems inputs which could be integrated with ecosystem energy flow for correlation with ecosystem response. It may be possible to predict, based on the energy inputs alone, both the kinds of ecosystems that may exist and their response to changes in any of the energy inputs. This capability would be an ideal ecological assessment tool.

The tools used for ecological assessment of pollution in the future must demonstrate that the total ecosystem is being considered in the assessment. Laboratory bioassays should recognize that test organisms cannot in reality be considered separate from the ecosystem of which they are a part, and laboratory conditions in these bioassays should simulate as nearly as possible field conditions. Field studies should include sampling programs for members of all the ecosystems components, for without information on other components, the data for one component are not very useful. Mathematical models should include the major material and energy flow pathways if they are to be realistic and useful. 'Pilot ecosystems' should be developed further as an assessment tool. The ability to observe receiving water ecosystem response to wastes in a 'pilot ecosystem' just as one observes the treatment plant ecosystem response to wastes in a pilot plant promises to be the most realistic and useful tool developed to date.

CONCLUSIONS

At least three steps are needed for a proper understanding of ecological assessment of pollution. First, the engineer and the ecologist need to have a better understanding of ecosystem components and function. Second, a unit of measurement common to both disciplines needs to be employed to facilitate understanding of ecosystems and to foster more productive communication between the two groups. The suggested unit is energy. Third, tools used to assess the impact of waste discharges on ecosystems should stress or at least recognize the interconnections of ecosystem components.

REFERENCES

'Algal Assay Procedure Bottle Test,' National Eutrophication Research Program Environmental Protection Agency (1971).

Armstrong, N. E., Gloyna, E. F., and Copeland, B. J., 'Ecological Aspects of Stream Pollution,' Gloyna, E. F. and Eckenfelder, W. W. Jr. (eds.), in *Advances in Water Quality Improvement*, Water Resources Symposium No. 1, University of Texas Press, Austin, Texas (1968).

Armstrong, N. E. and Storrs, P. N., 'Biological Effects of Waste Discharges on Coastal Receiving Waters,' in *Background Papers on Coastal Waste Management*, National Academy of Sciences – National Academy of Engineers, Vol. I, Washington, D. C. (1969).

Armstrong, N. E., Storrs, P. N., and Pearson, E. A., 'Development of a Gross Toxicity Criterion in San Francisco Bay,' Jenkins, S. H. (ed.), in *Advances in Water Pollution Research*, Proceedings of 5th International Conference, Pergamon Press, London (1972).

Chen, C. W., 'Concepts and Utilities of Ecologic Model,' *Journal Sanitary Engineering Division, American Society of Civil Engineers*, 36, 1063–1097 (1970).

Copeland, B. J., and Fruh, E. G., (eds.), *Ecological Studies of Galveston Bay – 1969*, Final Report to the Texas Water Quality Board, Galveston Bay Study Program (1970).

DiToro, D. M., O'Connor, D. J., and Thomann, R. V., 'A Dynamic Model of the Phytoplankton Population in the Sacrament – San Joaquin Delta,' in *Advances in Chemistry Series*, No. 106, 'Nonequilibrium Systems in Natural Water Chemistry,' American Chemical Society (1971).

Ippen, A. T. (ed.), *Estuary and Coastline Hydrodynamics*, McGraw-Hill Book Co., Inc., New York (1966).

Ivlev, V. S., 'The Biological Productivity of Waters,' Translation of 1945 document by Wicker, W. E., *J. Fish. Res. Bd. Can.*, 23, 1707–1759 (1966).

Klock, J. W., 'Bioassay Kinetics of Selected Marine Fauna,' Ph.D. Thesis in Sanitary Engineering, University of California, Berkeley (1960).

Lindeman, R. L., 'The Trophic Dynamic Aspect of Ecology,' *Ecol.* 23, 399–418 (1942).

Macfadyen, A., 'Energy Flow in Ecosystems and its Exploitation by Grazing,' in Grazing Terrestrial and Marine Environments, *British Ecol. Sco. Symposium 4*, 3–20 (1964).

McGauhey, P. H., Rohlich, G. A., and Pearson, E. A., 'First Progress Report – Eutrophication of Surface Waters – Lake Tahoe Bioassay of Nutrient Sources, 'Lake Tahoe Area Council, South Lake Tahoe (May 1968).

McGauhey, P. H., Rohlich, G. A., and Pearson, E. A., 'Second Progress Report – Eutrophication of Surface Waters – Lake Tahoe Bioassay of Nutrient Sources,' Lake Tahoe Area Council, South Lake Tahoe (1969).

McGauhey, P. H., Rohlich, G. A., and Pearson, E. A., 'Third Progress Report – Eutrophication of Surface Waters – Lake Tahoe Bioassay of Nutrient Sources,' Lake Tahoe Area Council, South Lake Tahoe (1970).

National Academy of Science Committee on Resources and Man, *Resources and Man*, W. H. Freeman and Co., San Francisco (1969).

Neumann, G. and Pierson, W. J., Jr., *Principles of Physical Oceanography*, Prentice-Hall, Inc., Englewood Cliffs, N. J. (1966).

Odum, E. P., *Fundamentals of Ecology*, W. B. Saunders Co., Philadephia (1959).

Odum, H. T., 'Primary Production in Flowing Waters,' *Limnol. Oceanogr.*, 1, 102–117 (1956).

Odum, H. T., 'Ecological Potential and Analogue Circuits for the Ecosystem ' *American Science*, 48, 1–8 (1960).

Odum, H. T., 'The Element Ratio Method for Predicting Biogeochemical Movements from Metabolic Measurements in Ecosystems,' Proceedings of the Conference on Transport of Radionuclides in Fresh Water Systems, University of Texas, Austin (1963).

Odum, H. T., *Environment, Power, and Society*, Wiley-Interscience, New York (1971).

Odum, H. T., and Chestnut, A. F. (eds.), 'Studies of Marine Estuarine Ecosystems Developing with Treated Sewage Wastes,' Annual Report for 1969–70 to National Science Foundation, Sea Grants Project Division by Institute of Marine Sciences, University of North Carolina (May 1970).

Odum, H. T., and Copeland, B. J., 'Functional Classification of the Coastal Systems of the United States,' Odum, H. T., Copeland, B. J., and McMahan, E. A. (eds.), *Coastal Ecological Systems of the United States*, U.S. Federal Water Pollution Control Administration, Washington, D. C. (mimeographed manuscript) (1970).

Odum, H. T., Siler, W. L., Beyers, R. J. and Armstrong, N. E., 'Experiments with Engineering of Marine Ecosystems,' *Publ. Inst. Mar. Sci., Texas*, 9, 373–403 (1963).

Paulik, G. J., 'Digital Simulation of Natural Animal Communities,' Olson, T. A. and Burgess, F. J. (eds.), in *Pollution and Marine Ecology*, Interscience Publishers, New York, pp. 67–85 (1967).

Pearson, E. A., Storrs, P. N., and Selleck, R. A., 'Some Physical Parameters and Their Significance in Marine Waste Disposal,' Olson, T. A. and Burgess, F. J. (eds.), *Pollution and Marine Ecology*, Interscience Publishers, New York, pp. 297–315 (1967).

Ryther, J. H., 'Photosynthesis in the Ocean as a Function of Light Intensity,' *Limnol. Oceanogr.*, 1, 61–70 (1956).

Servizi, J. A., and Bogan, R. H., 'Free Energy as a Parameter in Biological Treatment,' *Proc. Amer. Soc. of Civil Eng.*, 89, SA3, 17–40 (1963).

Slobodkin, L. B., 'Preliminary Ideas for a Predictive Theory of Ecology,' *Amer. Nat.*, 95, pp. 147–153 (1961).

Standard Methods for the Examination of Water and Wastewater – 13 Edition, Prepared by American Public Health Association, American Water Works Association, Water Pollution Control Federation, Publication Office: American Public Health Association, Washington, D. C. (1971).

Storrs, P. N., Pearson, E. A., Ludwig, H. F., Walsh, R., and Stann, E. J., 'Estuarine Water Quality and Biologic Population Indices, S. H. Jenkins (ed.), in *Advances in Water Pollution Research*, Proceedings of the 4th International Conference, Prague, Pergamon Press, London (1969).

Thirumurthi, D., and Gloyna, E. F., 'Relative Toxicity of Organics to *Chlorella pyrenoidosa*,' Report for U.S. Public Health Service, Water Supply and Pollution Control Division, by Center for Research in Water Resources, University of Texas, Austin (1965).

Thomann, R. V., O'Connor, D. J., and DiToro, D. M., 'Modeling of the Nitrogen and Algal Cycles in Estuaries,' S. H. Jenkins (ed.), in *Advances in Water Pollution Research*, 5th International Conference, San Francisco, Pergamon Press (July 1970).

Warren, C. E., *Biology and Water Pollution Control*, W. B. Saunders Co., Philadelphia (1971).

EVALUATION OF ECOLOGICAL
CONSEQUENCES OF MARINE POLLUTION

M. AUBERT and B. DONNIER
Institut National de la Santé et de la Recherche
Médicale, C.E.R.B.O.M., Nice, France

Discharges of chemical substances coming from urban areas and industrial activity into sea-water continue to increase. They lead to important modifications of the environment that one must evaluate as much in the field of nutritional capital destruction as in the field of public health. One must be concerned about the evolution of marine environment and the dystrophy threatening to disturb the biological balance.

Most chemical substances are metabolized by organisms living in the sea. By concentration processes, they can reach important amounts in the tissues of some species. This concentration process is increased by transmission through the different levels of marine trophic chains: thus, for some substances, it may lead to highly toxic phenomena for human beings who consume these marine products. In particular, this is the case with metallic salts of industrial origin such as those of mercury, cadmium, chromium, etc., and also of some pesticides.

These facts are pointed out at the level of marine species and also at the level of human consumers in the *Etude chimique des Côtes de France,* that we carried out. Industry is unprepared for removing highly toxic chemical substances from its waste waters and, besides, the industrial world, on economic pretexts, does not feel obliged to purify these waters.

In the field of chemical pollution, hydrocarbons deserve particular mention; the increasing energy needs, and the mineral-oil chemical industry demand increased output and transport of more and more important quantities of mineral-oil products. Either by drilling under the sea, or by accidents during transport, or by the breaking of pipes, wreckage of tankers etc., either by discharge of non-purified waste waters from refineries or by clearing tanks, a large part of sea is covered by a film of hydrocarbons, which has the action on marine life of destroying a part of the surface species. On the other hand, these hydrocarbons, metabolized by marine bacteria, supplement the organic matter that can have some beneficial influence upon productivity. If we take into consideration this metabolic activity, the biodegradation capacity of which is not yet entirely known, we could be less worried, at first sight, though we have little information on the release and the fixation of some of components such as benzpyrene which are undeniably carcinogenic. Among the products from the oil industry, detergents, by their massive use, are certainly a source of pollution. One may believe that the recent application of laws making a biodegradation rate for detergents compulsory would protect marine organisms from damage. Unfortunately, this policy is proving illusory because, in the first place, a biodegradation rate of 80% was accepted for detergents and the most toxic substances, such as alkyl-benzene, which form from 5% to 20% of the products are not biodegraded; secondly some products arising from the biodegradation of detergents have a higher toxicity than the detergent itself. We have no satisfactory solution in view except the return to the safe and classical product of the saponification of lipids. The United States make and promote new non-toxic detergents which could provide a solution to these problems.

What means can we use to study chemical pollutions?

Methods of analytical chemistry have been elaborated and adapted to the analysis of drinking, fresh, waste and marine waters. These methods are indispensable, because they allow a general view *a priori* of the quality and quantity of the main elements in waste waters.

Nevertheless, when we have to consider the toxicity of an effluent and its possible discharge into sea-water, many doubts arise. Systematic study does not take into account the evaluation of substances which do not form part of the product itself, such as disinfectants, used in some paper-factories to prevent the growth of moulds. In other cases, even after a survey by the manufacturers, it is not possible to know the nature of any additives, because they arise from secret processes. Besides, in complex effluents, some substances can be masked or they can occur in such small quantities that they cannot be determined. Moreoever, even if we know the exact composition of an effluent, it may be appreciably changed by the receiving medium. Thus, in the marine environment, effluents are submitted to different actions:

variation of pH, sea-water acting as a buffer,

variation of salinity,

precipitation by salts,

complexing action of natural organic substances in the marine environment,

influence of organisms living in the marine environment.

Chemical analyses can give information about the possible toxicity of waste waters, but the joint influence of different elements or the simultaneous effect of a mixture of effluents cannot be predicted.

Biological assays are in general use in many fields, either for checking drinking water quality (by breeding gudgeons in the water), by tests on fresh water with animals and plants and in sea-water. Valuable knowledge with regard to the direct ecological consequences of an industrial discharge into sea may be obtained. However, these assays give only a little information on the indirect consequences that such discharges can have upon the consumers of marine products.

If previously described methods allow some estimate to be made of the damage caused to the productivity of marine environment and consequently the degradation of fishing and of oceanic culture, they do not take into consideration all phenomena occurring during the discharge of chemically polluted effluents into the sea.

When these chemicals enter marine environments, their biological fate is dependent upon a double mechanism:

On the one hand, and it is the case, for instance, with hydrocarbons, they are attacked by marine bacteria which feed on them. There is, then, a progressive disappearance of the initial product which changes into various substances, the toxicity of which can be negligible, unchanged or increased. Some detergents are biodegradable in the marine environment, while others are non-biodegradable and persist indefinitely. Among biodegradable detergents, some recent investigations have pointed out that, in some cases, altered detergent compounds can be more toxic for the elements living in the sea than the initial compound.

There is another phenomenon in the opposite direction to biodegradation: this is the phenomenon of concentration which, because of the laws of nutrition in marine environment, can lead to very high levels such as 10,000 at the level of plankton. This phenomenon has been well studied in radiochemical pollution. Indeed, marine plants or animals concentrate some products in their tissues at very much greater rates than one can find during their diffusion into sea-water. These effects of concentration lead to serious diseases from some chemicals. They will be all the more important as the con-sumed organism is at a higher level in the biological chain; thus if we consider the dynamics of chemicals in sea-water in order to predict their effect we must take into consideration the phenomena of concentration.

For that purpose, we were led to re-create in the laboratory a system of trophic chains which begins with marine micro-organisms and ends at the higher mammals.

This methodology allows one to evaluate the destruction of type-organisms represent-ing different levels which are the most characteristic of marine trophic chains. It is thus possible to evaluate the destruction in the sea with regard to total biological resources. Furthermore, consideration can be given to the whole phenomena of concentration and biodegradation of the chemical substances submitted to these tests and to judge the biological effects induced by the initial or secondary toxicity after their introduction into the environment and their successive absorption by the different organisms. Finally, by the introduction of a mammalian consumer, one may judge the toxic phenomena at the end of the chain and evaluate the risk incurred by man fed on polluted marine products. This aspect opens the way to determine the impact of marine pollution on public health. incurred by man fed on polluted marine products. This aspect opens the way to determine the impact of marine pollution on public health.

M. Aubert (1970) defined four types of chains that can be used for this experimental purpose: (1) general pelagic chain, i.e.: phytoplankton − zooplankton − fishes − mam-mals; (2) benthic chain with molluscs, i.e.: phytoplankton − molluscs − mammals; (3) benthic chain with crustacea, i.e.: bacteria − invertebrates − benthic fishes − crustacea − mammals; (4) general benthic chain, including: bacteria − inverte-brates − benthic fishes − mammals. These four marine biological chains end at a terrestrial final consumer, the 'white mouse' (the chains are described in Table 1).

A biological chain can be used for different purposes.

1. Toxicity thresholds

Each pollutant (pure chemical or industrial effluent) is used at different concentra-tions; we can determine with sufficient accuracy the sensitivity of each species to this pollutant, and so determine the toxic threshold for each level.

The results therefore allow a short-term measure of the consequences of the pollutant discharged into sea-water on the survival of species and they allow one to determine the risks of loss with regard to nutritional capital.

2. The possible transmission of toxicity

After the determination of the direct toxicity thresholds of an effluent, we choose the maximum concentration allowing the survival of all levels of the marine chain: thus, the toxic substances will be able to be transmitted from one level to another and it may be concentrated throughout the different levels without breaking the biological chain by the death of one of the intermediates.

This initial amount established, the experiment consists in feeding the upper levels of the chain by the lower ones previously poisoned by chemically polluted waters. All the levels are, nevertheless, fed in a medium also polluted so as to re-create the natural conditions of the polluted environment.

The results permit evaluation of the risks of consumption by man of marine products and, also the possible dangers with respect to public health.

TABLE 1

Marine Trophic Chains

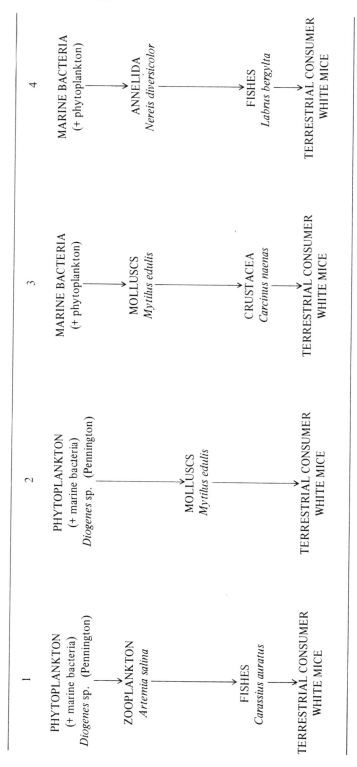

1

PHYTOPLANKTON
(+ marine bacteria)
Diogenes sp. (Pennington)

ZOOPLANKTON
Artemia salina

FISHES
Carassius auratus

TERRESTRIAL CONSUMER
WHITE MICE

2

PHYTOPLANKTON
(+ marine bacteria)
Diogenes sp. (Pennington)

MOLLUSCS
Mytilus edulis

TERRESTRIAL CONSUMER
WHITE MICE

3

MARINE BACTERIA
(+ phytoplankton)

MOLLUSCS
Mytilus edulis

CRUSTACEA
Carcinus naenas

TERRESTRIAL CONSUMER
WHITE MICE

4

MARINE BACTERIA
(+ phytoplankton)

ANNELIDA
Nereis diversicolor

FISHES
Labrus bergylta

TERRESTRIAL CONSUMER
WHITE MICE

3. The biodegradability of a pollutant

When discharged into the sea, a pollutant can be used as a substrate for marine bacteria and can therefore be biodegraded. Such an effect can lead to a variation of the toxicity of the pollutant towards the different marine levels of the biological chain.

To evaluate this biodegradability, the pollutant is brought together with sea-water enriched with marine bacteria during periods varying from 0 to 9 days. The pollutant is then introduced at three different concentrations: a non-toxic dilution and two toxic dilutions chosen near the threshold.

The results show the toxicity variation of the pollutant after contact with sea-water.

In the field of secondary pollutions, one must also refer to the action of pollutants upon the inter-species relations: the regulation of biological life can be done through chemical substances discharged into marine environment by the organisms living in it.

Aubert (1970) defined these substances thus: 'The chemical telemediators are substances synthesized by marine animal or plant species, liberated into the environment and acting at a distance upon the behaviour or the biological functions of the same species or of other species.'

By experiments described in previous publications, we have shown, isolated and determined the chemical structure of some telemediators. We are now interested in anticipating the possible fields in which these telemediators might act. If we take into consideration the main functions of marine species, we can imagine that nutrition, reproduction, defence or movement may proceed by chemotaxis. These various functions involve biological mechanisms which may be phenomena of nutritional attraction, of sexual attraction, of attraction creating commensalism and symbiosis, of migratory attraction, or the capacity of synthesis of substances allowing some kinds of metabolisms or allowing aggression or defence. It can easily be understood that these biological mechanisms can act through chemical substances, which will warn some organisms of the presence of other organisms necessary for reproduction for instance, for nutrition possibly, or which will regulate vast migrations leading some species through marine immensities.

In this paper we shall limit ourselves to the description of facts concerning microbiology or planktonology observed in our laboratory.

In this field the most-known fact concerns biological balance existing among some phytoplanktonic species, for instance, the antagonism which may be noted between the populations of Diatoms and those of Dinoflagellates. Aubert (1966) showed the evolution of the populations of these two genera as a function of time, when this antagonism appeared especially significant. More recently in our laboratory, J. M. Pincemin (1971) in experiments *in vitro* has brought out this interspecific opposition. He described the decrease of a Diatom *Asterionella japonica* population, by putting the latter together with *Glenodinium monotis*, and was also able to show that the presence of the culture medium where this Diatom was living had a favourable action upon the growth of the Dinoflagellate. Taking into account the slowness of the production of Dinoflagellates with regard to the production of Diatoms, this regulation by the substances coming from each species and discharged into the environment where the other species lives brings few changes to the possible invasion of the environment by the population of Diatoms, and so a biological balance between the two species is kept.

A more precise phenomenon is shown by observing the mechanism of dynamic balance between Diatom populations and terrestrial bacteria discharged into marine environment. Since 1961, our work has brought out the antagonism between two substances from some species of Diatoms and many terrestrial bacteria carried into marine environment by streams or waste water. (We have published numerous articles on this subject.) We have been able to isolate and analyse some chemical mediators responsible for this antagonism. Our experiments on this subject have brought out the presence of a *nucleoside* and of a *fatty acid* secreted by some species of Diatoms and acting as an antibiotic on a number of terrestrial bacteria. It is therefore a biologically and chemically demonstrated fact which therefore concerns two levels of the oceanic biomass.

Another example of this twofold mechanism concerning marine biology is the antagonism between marine bacteria and some terrestrial bacteria. This fact had been pointed out by ZoBell and Rosenfeld, and Krassil'nikova. Gauthier (1970) has recently demonstrated the interest of this interspecific antagonism, bringing out the existence of a chemical mediator which has been found to be a *lipopolyoside*. In that case also a twofold mechanism, biologically and chemically, has been proved.

If we analyse these phenomena more thoroughly, we reach much more delicate mechanisms entailing a combination of a greater number of species and acting on very specific metabolic functions.

We shall give two examples from our recent research, by summing up our latest publications on this subject. From a series of experiments carried out *in situ* and *in vitro*, we have been able to demonstrate that the proximity of a species of Dinoflagellate *Prorocentrum micans* and a species of antibiotic-producing Diatom *Asterionella japonica* stops the synthesis of the antibiotic substance secreted by this Diatom which blocks the antibacterial action usually due to this phytoplanktonic species. In collaboration with D. Pesando (1970, 1971, 1972) we have been able to specify the nature of the chemical mediator secreted within the environment by the Dinoflagellate and acting at a distance. It is a *protein* that we have found in the cells of *Prorocentrum micans* and in its culture medium. This specific action, whose various mechanisms have been demonstrated by us, is carried out by a threefold biological process. Each level has been specified and chemically identified. This threefold process with regard to the twofold process points to the existence of *secondary type* telemediators, whose action is indirect, independent of the *primary type* telemediators, whose action is direct.

In the above-mentioned experiment, we have been able to demonstrate the successive stages of the phenomenon by discovering the origin of the initial releaser; other similar or opposite actions may be brought out. Thus, instead of a blocking agent which stops the antibiotic secreted by the Diatom, we have shown the action of a mediator which, on the contrary, induces this synthesis. Indeed, so that the Diatom may synthesize the antibiotic, we have noted that, in the sea water where Diatoms live, some substances are needed. We have pointed out these substances, but we do not know their chemical structure and the source of release.

These phenomena concerning plankton and bacteria are rather similar to those demonstrated in our laboratory by Gauthier (1970) as regards the relations between marine and terrestrial bacteria. It appeared that marine bacteria (*Chromobacterium* and *Pseudomonas*) producing some substances endowed with antagonistic power to some terrestrial bacteria can stop this secretion by the presence of other marine bacteria belonging to different species (*Achromobacter, Flavobacterium*).

Moreover, as in the field of Diatom antibacterial action, the induction phenomena of this secretion may be realized through the action of telemediators originating from other marine bacteria belonging to different species as well.

Independently of these theoretical considerations, we should like to point out the fragility of these mechanisms, which, as mentioned, are the basis of the biological balance in the sea. It is quite obvious that the introduction of various industrial chemical substances into the marine environment entails structural modifications of the biological components of sea-water; owing to the destruction or the modification of these 'metabolites' or 'signals', an alteration of the oceanic balance may occur quickly as a consequence of those polluting phenomena.

Through researches concerning the toxicity of various chemical substances, we have observed the blocking of the growth of Diatom *Asterionella japonica* by

0.004 ml/l of domestic fuel,

4 mg/l of a detergent (alkylbenzenesulfonate),

0.004 ml/l of phosphene ⎫
0.05 ml/l of perfektion ⎬ pesticides

To far lower rates when the growth is normal, we have studied the action of those various pollutants upon antibiotic production:

As regards domestic fuel, up to 3×10^{-7} ml/l, the productivity of cultures is normal, but the antibiotic activity of the cells is reduced by half. Pollutant action occurs at the level of the secretion of active substances. Indeed, if we add the pollutant to the culture a quarter of an hour before testing the cell antibiotic activity we observe the same activity inhibition, whereas, if the pollutant is added to the extracted antibiotic, we do not notice any activity modification.

As regards pesticides, there is an increase of antibiotic activity, which seems to be bound to an alteration of the cell walls which release active substances, faster. At low concentrations and for a normal productivity there is no modification of the antibiotic power of *Asterionella japonica*.

The observation of such phenomena has led us, during systematical studies of pollution, to find out something about the secondary actions. It is quite obvious that if a chemical pollution disturbs these balances and reduces antibiotic activity within the marine environment, the risk of bacterial pollution will be increased.

Among secondary pollution, we must also quote eutrophication and distrophication. The almost unlimited technical means at man's disposal enable important trans-formations to be made in sea-shores, producing partitions within marine environment which, to a great extent, suppress the currents along wide coastal areas, and hence hydrological exchanges. The accumulation of wastes in these areas withdrawn from the normal phenomena of diffusion and metabolism leads to a hyper-productivity of some phytoplanktonic or algal species, followed by an inversion of the biological balance: this can result in the production of 'red waters' due to the proliferation of some species of Dinoflagellates; some of them are highly toxic, because when used as food by edible species, they induce human poisoning or, carried into the atmosphere by aerosols arising from wind over polluted stretches of sea, they induce serious respiratory toxi-allergic troubles among coastal populations.

Later on, this distrophication of species by the resorption of organic matters originating from the death of this too flourishing plant biomass leads to anaerobic phenomena with production of H_2S and destruction of living species. Some marine areas begin to reach that point because of excessive transformations of sea-shores and discharges of waste waters into relatively enclosed places. These areas are not too numerous yet, fortunately, but in view of some plans for huge building along the sea coast which is bound to induce important ecological consequences, we must be cautious of their consequences upon the biological balance of the sea.

From these reflections, what can we conclude? These various observations lead us to more and more elaborate studies of the marine environment and of the metabolic phenomena that occur in it. They are numerous and not easily accessible, and, in the conditions now prevailing, we must make rapid progress because sources of pollution do not stop spreading and they increase perhaps more quickly than we know.

In advance of the harmful influence of man upon the marine environment and its future consequences, which can easily be foreseen it is necessary to develop technical processes to measure and to study carefully the extension mechanisms of pollution. These studies involve many subjects: physiology, microbiology, chemistry, nuclear sciences, adapted to oceanography. Above all, very severe regulations must be adopted by governments to control not only the existing pollution but also the establishment of new industries.

SUMMARY

In order to study the biological effects of pollution we have re-created, in CERBOM, a system of trophic chains, which, from the marine micro-organisms, ends at the mammals.

This original methodology allows one:

To evaluate the destruction produced at the level of type-organisms most characteristic of marine trophic chains. In this way, it is possible to evaluate the destruction of the capital of biological resources of sea.

To take into consideration the whole phenomena of concentration and biodegradation of chemical substances submitted to these tests and to judge the biological effects induced by the initial toxicity or appearing as a secondary manner, these chemical substances going through the medium and being absorbed successively by different living elements.

Lastly by the introduction of a final mammal consumer, one succeeds in judging the toxic phenomena at the end of the chain and evaluating the danger incurred by man fed on polluted marine products. This aspect shows the impact of marine pollution on public health.

Four types of chains were used for this experimental purpose: (1) General pelagic chain, i.e.: phytoplankton – zooplankton – fish – mammals; (2) Benthic chain with molluscs, i.e.: phytoplankton – molluscs – mammals; (3) Benthic chain with crustacea, i.e.: bacteria – invertebrates – benthic fish – crustacea – mammals; (4) General benthic chain, including: bacteria – invertebrates – benthic fish – mammals.

In the field of secondary pollution, we must speak of the action of pollutants upon the inter-species relations: the regulation of biological life may be carried out through chemical substances, spread in marine environment by the beings living it it. These substances are thus defined:

The chemical telemediators are substances synthesized by animal or plant marine species, liberated into the environment and acting at distance upon the behaviour or the biological functions of the same species or of other species. The Authors show how biochemical factors maintain ecological equilibrium. These effects are governed by telemediators, some of them having been isolated and chemically characterized. These mechanisms are finely balanced but essential to the maintenance of oceanic life; they may be disturbed by pollutants.

REFERENCES

Aubert, M., 1966 – Le comportement des bactéries terrigènes en mer. Relations avec le phytoplancton. *Thèse*. Université Marseille, pp. 5–285.
Aubert, M., Aubert, J., Gambarotta, J. P. et Daniel, S., 1968 – Inventaire National de la Pollution Bactérienne des Eaux Littorales des Côtes de France. *Rev. Intern. Oceanogr. Med.* Tome I, II, III, IV.

Aubert, M., Aubert J., Daniel, S. et Gambarotta, J. P., 1969 — Etude des effets des pollutions chimiques sur le plancton. Dégradabilité du fuel par les micro-organismes telluriques et marins. *Rev. Intern. Oceanogr. Med.* Tome XIII–XIV, pp. 107–124.

Aubert, M. et Gambarotta, J. P., 1969 — Etude de la biodégradabilité des produits chimiques toxiques vis-à-vis de la chaîn biologique marine. *Rev. Intern. Oceanogr. Med.* Tome XIII–XIV, pp. 73–106.

Aubert, M., Aubert, J., Gambarotta, J. P., Donnier, B., Barelli, M. et Daniel, S., 1969 — Etude générale des pollutions chimiques rejetées en mer. Inventaire et études de toxicité. *Rev. Intern. Oceanogr. Med. Suppl. 69.* Tome I pp. 1–72, Tome II pp. 1–35.

Aubert, M., Aubert, J., Donnier, B., et Barelli, M., 1970 — Etude générale des pollutions chimiques rejetées en mer. Inventaire et études de toxicité. *Rev. Intern. Oceanogr. Med. Suppl. 70.* Tome III, pp. 1–225.

Aubert, M., 1970 — Pollution des océans. *Conférence de la Faculté des Sciences de Paris dans le cadre du cycle 'LES Océans'.* Paris. 23 avril 1970. Mises à jour Gauthier-Villars. Edit. 1970, vol. 4, pp. 247–268.

Aubert, J., Pesando, D. et Gauthier, M., 1970 — Danger des pollutions vis-à-vis des mécanismes biochimiques conditionnant les relations interespèces. *Ve Symposium Européen de Biologie Marine,* Venise, 4–13 oct. 1970, pp. 1–7.

Aubert, M., Aubert, J., Donnier, B. et Barelli, M., 1970 — Utilisation de la chaîne trophodynamique dans l'étude de la toxicité des rejets d'eaux chimiquement polluées. *Conference Technique de la F.A.O. sur la pollution des mers et sur ses effets sur les ressources biologiques et la pêche,* Rome (Italie), 9–18 décembre 1970, rapport MP/70/E–49, pp. 1–7.

Aubert, M., Pesando, D. et Pincemin, J. M., 1970 — Médiateurs chimiques et relations interespèces: Mise en évidence d'un inhibiteur de synthèse métabolique d'une Diatomée produit par un Péridinien (Etude 'in vitro'). *Rev. Intern. Oceanogr. Med.* Tome XVII, pp. 5–21.

Aubert, M., 1971 — Télémédiateurs chimiques et équilibre biologique océanique: Première Partie: Théorie générale. *Rev. Intern. Oceanogr. Med.* Tome XXI, pp. 5–15.

Aubert, M. et Pesando, D., 1971 — Télémédiateurs chimiques et équilibre biologique océanique: Deuxième Partie; Nature chimique de l'inhibiteur de la synthèse d'un antibiotique produit par une Diatomée. *Rev. Intern. Oceanogr. Med.* Tome XXI, pp. 17–22.

Aubert, M., Pesando, D. et Pincemin, J. M., 1972 — Télémédiateurs chimiques et équilibre biologique océanique. Quatrième Partie: Seuil d'activité de l'inhibiteur de la synthèse d'un antibiotique produit par une Diatomée. *Rev. Intern. Oceanogr. Med.* Tome XXV, pp. 17–22.

Donnier, B., 1972 — Etude des pollutions chimiques au moyen de chaînes trophodynamiques marines. Application à des effluents de papeterie. *Thèse.* Nice, pp. 1–80.

Gauthier, M., 1970 — Propriétés antibactériennes de micro-organismes marins. *These.* Nice, 116 p.

Pesando, D., 1971 — Etude chimique et structurale d'une substance lipidique antibiotique produite par une Diatomée marine: *Asterionella japonica. Thèse.* Nice, 95 p.

Pincemin, J. M., 1971 — Action de facteurs physiques, chimiques et biotiques sur quelques Dinoflagellés et Diatomées en culture. *Thèse.* Aix-Marseille, 119 p.

Pincemin, J. M., 1971 — Télémédiateurs chimiques et équilibre biologique océanique. Troisième Partie: Etude 'in vitro' des relations en populations phytoplanctoniques. *Rev. Intern. Oceanogr. Med.* Tome XXII–XXIII, pp. 165–196.

Discussion *by* N.E. Cooke, Canada

It is impractical to consider all the animals or species in a part of an ecosystem. If there are species in the area we are studying and want to establish the population density of half of these species over the period of a year, sampling the area once a month, we should collect about a minimum 30 individuals from each species each time to take statistical variation into account. We would have 300 species x 30 individuals x 12 sample times = 108,000 individuals as a minimum. Of course many species will have many more individuals in the collected samples, sometimes in excess of 1,000. Take as a rough average of 200 then the figure rises to 720,000. Merely to identify this number of organisms will require a large staff. We do not have to be so comprehensive as the authors have suggested. In our work we have been able to find test organisms living in the mud. Only a very small fraction of the new chemicals are manufactured in quantities large enough to affect the environment. The effects of many new chemicals can be predicted because they are biodegradable. Only a small residue of potential troublesome compounds remain.

Reply by N.E. Armstrong

I agree with Mr. Cooke that it is impractical to consider all the animals or species in a part of an ecosystem because it is extremely difficult to sample all of these organisms or species. However, for the part of the ecosystem being sampled, there are methods to insure that most of the species have been collected and estimates of the populations of the species are representative. I do not concede, however, that we can be less comprehensive in sampling if we are to obtain a realistic estimate of the species and their populations present or to compute a diversity index based on the assemblage of species. Based on work in two large bays in which phytoplankton, zooplankton and benthic animals were collected, we found that approximately four hours were required per sample for both phytoplankton and zooplankton and about eight hours per sample for benthic animals to identify the organisms and to count the number of organisms in each species or classification level. For some types of samples, of course, it is impractical to key organisms to species or sometimes even to genus. The zooplankton is an excellent example for many larval forms cannot be identified. One is able to make these identifications and quantifications with a relatively small staff, on the order of one or two biologists at most who are familiar with the biota in the area being sampled; we have found it extremely difficult to obtain competent biologists to make these identifications because the number of biologists in taxonomy are relatively few and those in the field are somewhat unwilling to make these routine identifications. This is a problem also pointed out in the NAS-NAE report, "Waste Management Concepts in the Coastal Zone."

J.B. Sprague, Canada

With the newer methods of biological monitoring being developed it is not necessary to identify large numbers of individuals to species and study population changes of each species. The "diversity index" is relatively quick and yields a single number for comparison of changes. In the further simplification proposed by Cairns and colleagues a "sequential comparison index" is obtained by noting whether one individual in a sample is the same as the previous individual or different. This method yields a measure of diversity index, which can apparently be learned in an hour.

Reply by N.E. Armstrong

If the only information one is trying to obtain is diversity index data, then the newer methods of biological monitoring may indeed facilitate obtaining such data. However, one of the features of the diversity index which has not been investigated in great detail is the variation in the absolute values obtained from the index with varying degrees of identification. That is, how will the diversity index change in numerical magnitude if one were to base the computation on identifications to genus only, as compared to species? Also, how would the diversity index change with addition or rarer species which would only appear after many samples? One obtains at best a relative measure of diversity and must keep the limitations of the diversity index being used in mind when comparing values within an ecosystem and certainly when comparing values among ecosystems.

J. McN. Sieburth, USA

Dialysis culture studies by Arne Jensen and colleagues of the Norwegian Institute of Seaweed Research Trondheim and studies by my students and myself appears to be one approach for obtaining information on toxicity, tolerances and stimulation of marine organisms and microorganisms in the absence of predation. The use of fritted glass cages would permit one to assay the effect of soluble substances (both monomeric and polymeric), on captive populations in the absence of predation.

S.E. Jorgensen, Denmark

A model leading to a prediction using a suitable safety factor chosen in accordance with the risk involved, is in my opinion the only answer to the problem. What alternatives are there?

J.B. Sprague, Canada

On the question of simultaneous action of two or more pollutants on an ecosystem, there is evidence that such effects can be predicted by assuming that they are strictly additive.

Reply by M. Aubert and B. Donnier

We believe that bioassays and the study of trophodynamic chains appear to be the most likely method to determine toxicity.

We also think that statistical studies carried out with the tested species will increase the precision of the method. However, we are not optimistic about chemical pollution of the sea because low rates of certain chemical products are able to exert very toxic effects due to their concentration in marine species. The examples of Minamata disease and Hitai-Hitai disease are convincing and tragic demonstrations of our fears.

AUTHOR INDEX

SUBJECT INDEX